Fibrinogen
Recent Biochemical and Medical Aspects

Fibrinogen

Recent Biochemical and Medical Aspects

Proceedings
International Symposium, February 17–18, 1981
Max-Planck-Institute for Biochemistry
Martinsried/Munich, Germany

Editors
A. Henschen · H. Graeff · F. Lottspeich

W DE G

Walter de Gruyter · Berlin · New York 1982

Editors

Agnes Henschen, Dr. med.
Friedrich Lottspeich, Dr. phil.
Max-Planck-Institut für Biochemie
D-8033 Martinsried bei München

Henner Graeff, Professor Dr. med.
1. Frauenklinik der Universität München
Maistraße 11
D-8000 München 2

CIP-Kurztitelaufnahme der Deutschen Bibliothek

Fibrinogen: recent biochem. and med. aspects; proceedings, internat. symposium, February 17–18, 1981 / Max Planck Inst. for Biochemistry, Martinsried/Munich, Germany. Ed. A. Henschen ... – Berlin; New York: de Gruyter, 1982.
ISBN 3-11-008543-7
NE: Henschen, Agnes [Hrsg.]; Max-Planck-Institut für Biochemie <Martinsried>

 Printing: Gerike, Berlin. – Binding: Lüderitz & Bauer, Buchgewerbe GmbH, Berlin. – Printed in Germany.

PREFACE

This volume is a record of the contributions to the symposium entitled "Fibrinogen-Recent Biochemical and Medical Aspects". The conference was held at the Max-Planck-Institut für Biochemie in Martinsried outside München on February 17. and 18., 1981. About 100 scientists with a special interest in fibrinogen research convened from ten different European countries.

The object of the conference, as with that of the previous three fibrinogen workshops which were held in Martinsried, three, four, and eight years earlier, was to review the progress in the biochemistry and pathobiochemistry of fibrinogen and to facilitate the exchange of ideas and experience between investigators from different fields. A summary of the main topics reflects the broad range from theoretically to clinically oriented research: structure and evolution of fibrinogen and fibrin, conversion of fibrinogen to fibrin, influence of carbohydrate components and calcium ions, molecular variants, inherited abnormal fibrinogens, biological effects and determinations of fibrinopeptides and other fibrinogen-derived peptides, affinity and complex formation between fibrinogen and fibrin, biochemistry and clinical relevance of fibrinogen degradation products and crosslinked fibrin derivatives, and finally the interaction of fibrinogen-fibrin with a plasminogen activator, fibronectin and platelets.

Informal discussion was an important part of the workshop and each contribution could be discussed in detail. It is the hope of the editors that some highlights of the enthusiastic participation in the exchange of information and ideas are reflected in these proceedings. It is also hoped that this volume, by correlating and combining scientific and clinical aspects of fibrinogen research, will enhance the spread of information from theoretical biochemistry to clinical medicine and vice versa. These efforts could help to reduce the lag phase between the instant when the implications of certain discoveries are recognised in the research laboratory and the time when these findings can be adapted to clinical analysis or treatment of patients.

The conference was sponsored by the Max-Planck-Gesellschaft and the Sonderforschungsbereich 51 of the Deutsche Forschungsgemeinschaft. It was also most generously supported financially by the companies Bayer, Behringwerke, Deutsche Kabi-Vitrum and Luitpoldwerke. For all this the organisers wish to express their sincere gratitude.

At this Martinsried meeting it was already felt that additional conferences on fibrinogen ought to be organised in the years to come. The editors therefore strongly wish and hope that further volumes from fibrinogen conferences will follow.

Martinsried and München, November 1981

A. Henschen
H. Graeff
F. Lottspeich

CONTENTS

I. PHYSICOCHEMICAL ASPECTS

II. BIOCHEMICAL ASPECTS

V. FIBRINOGEN - FIBRIN COMPLEXES

VI. FIBRIN(OGEN) DEGRADATION AND CROSSLINKED FIBRIN DERIVATIVES

I. PHYSICOCHEMICAL ASPECTS

PARACRYSTALLINE STRUCTURE OF NATIVE HUMAN FIBRINOGEN

Rotraut Gollwitzer, Wolfram Bode, Hans-Josef Schramm, Dieter Typke and Reinhard Guckenberger

Max-Planck-Institut für Biochemie, Abteilungen Bindegewebsforschung, Strukturforschung I und II
D-8033 Martinsried b. München, W.Germany

Introduction

The overall structure of fibrinogen is still unknown and a controversial subject. Several contradictory models have been discussed during the last decade. However, the fragments obtained by enzymatic and chemical cleavage and the amino acid sequence of human fibrinogen favour a three-domain structure as proposed by Hall and Slayter in 1959 (1) based on their electron microscopic experiments. This trinodular structure has been confirmed in recent studies by Krakow (2) and by Fowler and Erickson (3), using shadow casting and negative staining techniques.
A detailed three-dimensional structure can best be obtained by single-crystal diffraction methods. Cohen and coworkers (4,5) succeeded in producing crystals of partially digested bovine and rabbit fibrinogen, but an X-ray structure analysis of these crystals has not been published up to now. Electron microscopy has been used (4,6,7) to gain information about the shape and especially the alignment of individual molecules in these fibrinogen crystals and in fibrin fibers.
In this paper we present electron microscopic studies of paracrystalline arrays of native human fibrinogen and describe the interpretation of the electron micrographs by LASER-diffraction and by autocorrelation (8).

Fibrinogen - Recent Biochemical and Medical Aspects

Methods and results

Small, birefringent needles (Fig. 1) were obtained from native human fibrinogen by vapour diffusion at 20°C. The fibrinogen was freshly prepared from citrated blood according to Blombäck (9). These paracrystalline needles grew to a maximal length of 0.5 mm and a maximal diameter of less than 0.01 mm. They consisted of native fibrinogen as proven by their clottability and by SDS disc electrophoresis (Fig. 2) of reduced samples. The A-α-chain of our paracrystalline material is not detectably shortened, indicating that the paracrystals are in fact formed by native fibrinogen. In comparison, spherulites obtained from human fibrinogen digested by Pseudomonas proteases or by α-chymotrypsin under the conditions described by Cohen et al.(6), possess no band in α position and shortened β-chains.

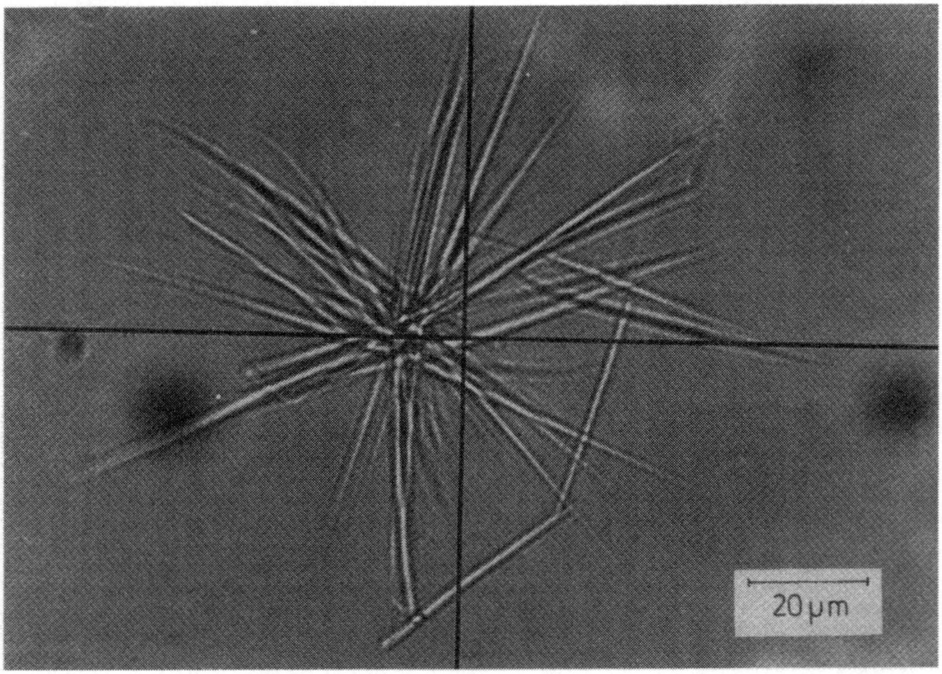

Figure 1: Paracrystalline needles of native human fibrinogen crosslinked with 0.1% glutaraldehyde.

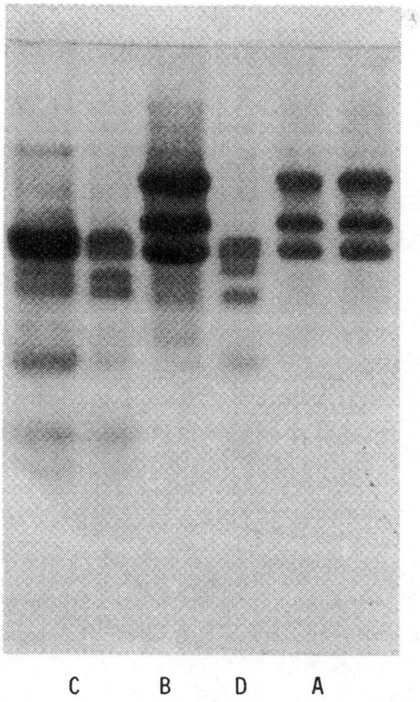

Figure 2: SDS-gel-electrophoresis (5.5% gel) of reduced samples: (A) native human fibrinogen; (B) paracrystals from human fibrinogen; (C) spherulites from human fibrinogen digested by α-chymotrypsin; (D) spherulites from human fibrinogen digested by pseudomonas proteases.

At an early stage of their growth the paracrystals are suitable for electron microscopy. When negatively stained with ammonium-molybdate at pH 7, they reveal a characteristic, periodic cross striation pattern, showing no polarity (Fig.3). One prominent continuous light band is adjacent to two less continuous and less distinct bands per period. The bands seem to consist of spherical, unstained globules of an approximate diameter of 3-5 nm. Each triad is separated from the next by a larger interval. The identity period along the horizontal axis of the micrographs (corresponding to the long axis of the needles) is 22.5 nm. There is only weak periodicity in the direction perpendicular to the main axis of the paracrystals indicating relatively poor order in this direction. To gain more information about the periodicity in the perpendicular direction, the micrographs were studied by LASER-diffraction and by correlation with an incoherent light optical correlation device (10). When diffraction was carried out with a plane wave of LASER light, it was necessary to reduce the diffracting area of the electron micrograph to ca. 5 mm in diameter in order to obtain clear diffraction spots in the perpendicular direction (insert in Fig. 3). In the horizontal direction, the spots corresponding to the 3rd and 4th orders of the 22.5 nm periodicity are easily observed. However, even with this small aperture, clear diffraction spots in the perpendicular direction were only obtained for a few regions of the image. Although the whole area of the paracrystal in the

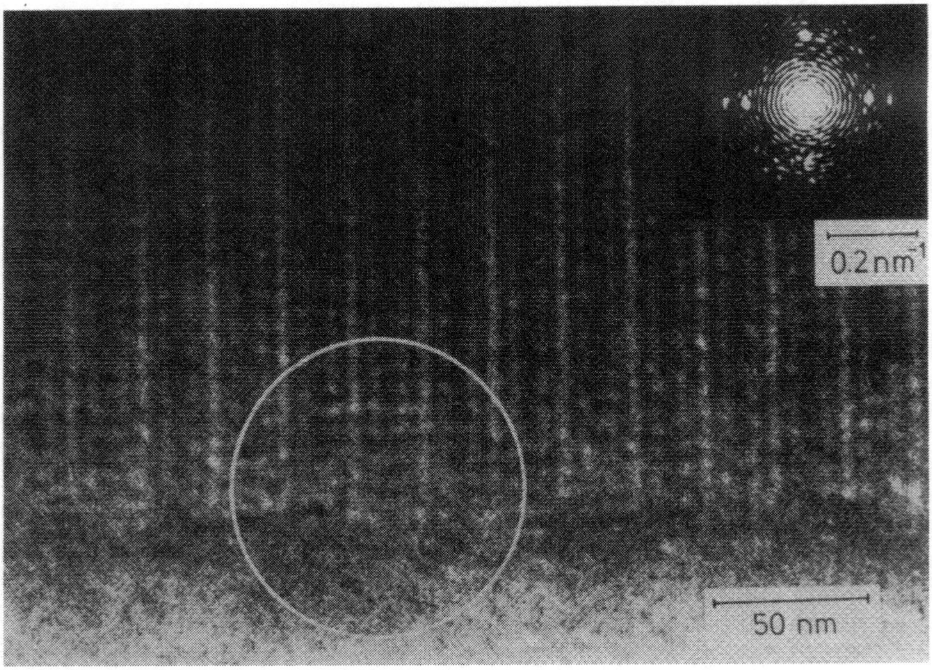

Figure 3: Electron micrograph of paracrystals from native, human fibrinogen, negatively stained with 2% ammonium molybdate, pH 7.0, containing 0.5% glucose. Electron-optical magnification: 53 000. Inset: light-optical diffraction pattern of the encircled region.

original electron micrograph (ca. 20x80 mm, corresponding to ca. 400x1600 nm of the paracrystal) was scanned. The observed optical diffraction maxima obtained from different regions of the electron micrograph, correspond to vertical periodicities in the micrograph in the range of 6-8 nm, with periodicities from 7 to 7.5 nm predominating.

With the correlation device, the 2-dimensional autocorrelation function of an image region or the cross-correlation function of different image regions can be determined. The autocorrelation function is in principle the Fourier transform of the diffraction pattern. It allows comparisons of different subregions within a studied region and can be used to determine periodicities or characteristic distances. An advantage of our

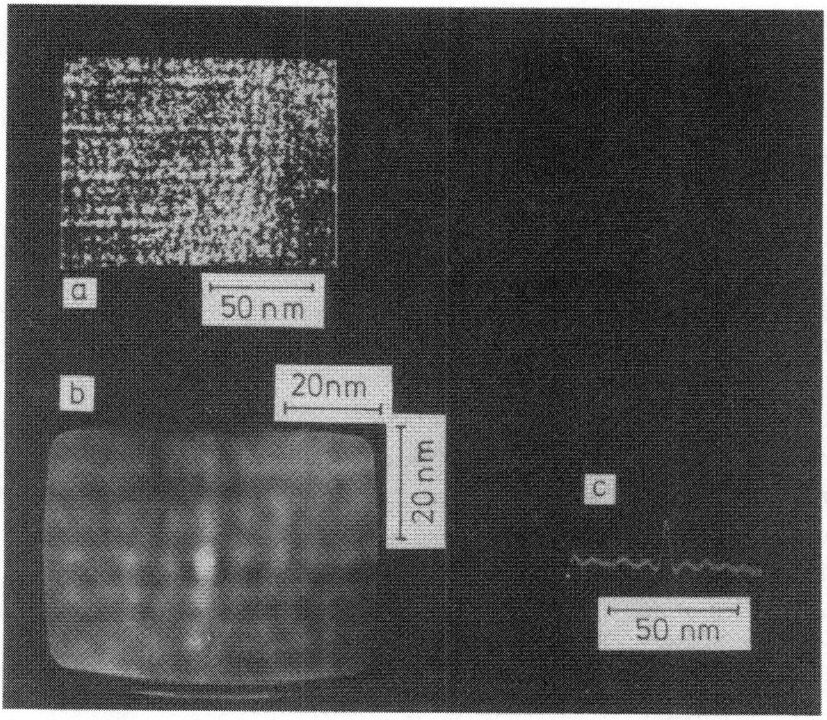

Figure 4: Autocorrelation of an image region of the electron micrograph not contained in Fig. 3. a) image region with enhanced contrast; b) autocorrelation function of a); c) horizontal section through b).

correlation device ist that pictures can be manipulated by electronic means for contrast enhancement.
In Figure 4 the contrast-enhanced image region and its autocorrelation function are displayed on TV-screens. A horizontal section through the autocorrelation function containing the highest peak is displayed on an oscilloscope. The Y direction of this trace corresponds to the light intensities, and the X direction to the distances along the line of intersection between the section and the autocorrelation function. A typical separation of 7.5 nm over at least four periods was observed in the direction perpendicular to the main axis of the paracrystal.

Discussion

A tentative alignment of the single fibrinogen molecules in the paracrystal was made from these data on the basis of a trinodular structure of the molecule (Fig. 5). Due to the twofold symmetry of fibrinogen, the molecular centers should be placed on the broad unstained bands which are apparent mirror lines in the electron micrograph. The most probable position of both outer domains of each molecule are the first faint bands of the two triads on either sides.
A model must allow for the feature that the apparent globular structures within thin and broad unstained bands seem to be arranged in lines perpendicular to the striation. Consequently the central globular structures can in principle belong to two terminal globuli of the same line or of an adjacent line, thus leading either to a straight or to a bent molecule (as shown in Fig. 5). Between the central and the terminal globuli there is a broad stained band of ca. 13 nm width, accomodating the invisible rodlike coiled-coil-structures proposed by Doolittle (11) to connect the three globuli. The C-terminal part of the α-chain, ca. 400 residues, could extend beyond the terminal globuli and be positioned between the central domains of ad-

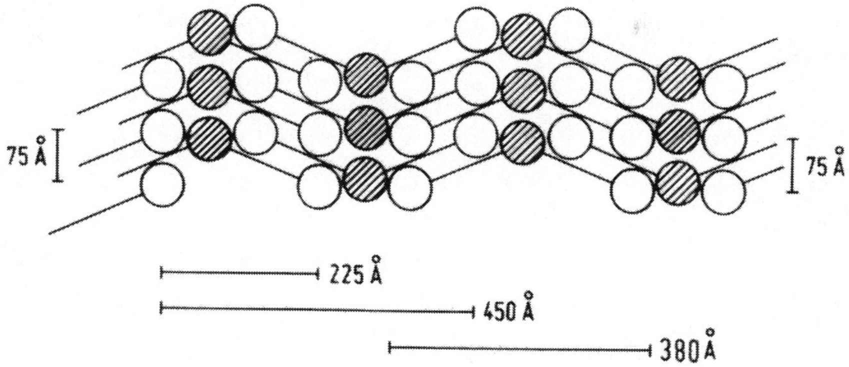

Figure 5: Tentative alignment of fibrinogen molecules in the paracrystal.

joining fibrinogen molecules.
Thus individual molecules can extend from about 380 Å to about 450 Å. Neighbouring molecules overlap for about 1/3 of their length.

Surprisingly, this proposed arrangement of fibrinogen molecules within the paracrystal is similar to that in fibrin, where due to the creation of new affinity centers, one E- and two D-domains of two different molecules are packed closely together.

However, we wish to point out that this tentative alignment of the molecules within the paracrystal does not necessarily need to resemble that of fibrin molecules within their respective fibers. In general there are only a few intermolecular contacts in protein crystals, stabilizing an almost undistorted protein conformation of individual molecules, in contrast to three-dimensional protein aggregates where protein molecules mostly are connected via large contact area(see e.g. the contact area in the trypsin-trypsin inhibitor complex (12)).

Our model corresponds roughly to the fibrinogen models proposed by Cohen et al. (13), based on their crystals and aggregates containing modified bovine and human fibrinogen. However, our electron micrographs do not display additional globular subunits. These different features could be due to the reduced length of the fibrinogen chains of their partially digested material.

Acknowledgements

We thank Miss I. Mayr, Mrs. E. Weller and Mrs. Z. Cejka for their expert technical assistance and the "Deutsche Forschungsgemeinschaft" SFB 51, project B9 and B22 for financial support.

References

1. Hall, C.E., Slayter, H.S.: J. Biophys.Biochem.Cytol.5, 11-16 (1959).
2. Krakow, W., Endres, G., Siegel, B., Scheraga, W.: J.Mol. Biol. 71, 95-103 (1972).
3. Fowler, W.E., Erickson, H.P.: J.Mol.Biol. 134, 241-249 (1979).
4. Cohen, C., Slayter, H., Goldstein, L., Kucera, J., Hall,C.: J. Mol. Biol. 22, 385-388 (1966).
5. Tooney, N.M., Cohen, C.: J.Mol.Biol. 110, 363-385 (1977).
6. Weisel, J.W., Warren, S.E., Cohen, C.: J.Mol.Biol. 126, 159-183 (1979).
7. Weisel, J.W., Cohen, C., Tooney, N.M, Kaplan, J., Armani, D.: J. Mol.Biol. 143, 329-334 (1980).
8. Gollwitzer, R., Bode, W., Schramm, H.J., Typke, D., Guckenberger, R.: Proteins and Related Subjects 28, 307-310; Protides of Biological Fluids 28th Colloquium 1980, H. Peeters Ed., Pergamon Press (1980).
9. Blombäck, B., Blombäck, M.: Ark.Kem. 10, 415-443 (1956).
10. Guckenberger, R., Hoppe, W.: Proc.7th Europ.Congr.Electr. Microsc. The Hague, Vol. 2, 696 (1980).
11. Doolittle, R.F.: Horizons Biochem.Biophys. 3, 164-191 (1977).
12. Huber, R., Bode, W.: Accounts Chem. Res. 11, 114-122 (1978).
13. Weisel, J.W., Phillips, G.N., Cohen, C.: Nature 289, 263-267 (1981).

DETERMINATION OF FIBRIN STRUCTURES BY LIGHT SCATTERING

Walther Burchard, Michael Müller
Institute of Macromolecular Chemistry
D-7800 Freiburg, Germany

Introduction

Light scattering is a traditional technique used in polymer science and has been used to investigate fibrinogen and fibrin (1-5) in as early as 1947 when this technique was utilized by Debye (6) in formulating the detailed principles. So the question may be posed concerning what can be learned, given the number of years this technique has been used, that is essentially new to this field. One important answer deals with the increased accuracy of measurement due to the utilization of lasers as extremely monochromatic light sources as well as the development of new and fast detection systems. Although proceeding at a rather frenetic pace, development will continue over the next few years, and this will most likely lead to exciting new insights in the mechanism of fibrin network formation. This, however, is only half of the answer. The other half concerns the development of theory which having made large advances, now offers considerable assistance particularly in the interpretation of branched structures.

In the following, an operational review of the principles of structure determination by static and dynamic light scattering will be offered. Theory will be avoided whenever possible. The intention of this contribution is to facilitate an understanding of essential points so that experiments can be judged without necessitating expertise in light scattering. Relevant points will be illustrated by some selected results from

Fibrinogen - Recent Biochemical and Medical Aspects

fibrin. The conclusions, however, will not be discussed in detail. Rather, such results will be presented by Dr. M. Müller and Professor I. Witt in their respective contributions.

The Static Structure Factor

Since the interest here is in structures, the first thing which can be measured is the molecular weight, M_w, of the sample. In the absence of thermodynamic interaction between particles, the molecular weight is directly proportional to the scattering intensity at zero angle of observation. Zero angle is defined as the direction of the primary beam of the light and is thus not directly accessible to measurement. The scattering intensity at finite angle is often lower than at zero angle and extrapolation to zero angle has to be accomplished before M_w can be determined. Usually this is done with a Zimm plot (7), i.e. a plot of the reciprocal scattering intensity against $\sin^2 \theta/2$. At small scattering angles this plot in linear. An angular dependence occurs when the structural dimensions are larger than $\lambda/20$, with λ being the wave length of the light. This is caused by internal interference of light that is scattered by the various scatter centres in the samples. According to Debye (8), the angular dependence is described by the structure factor, S(q), which is given by the following equation:

$$S(q) = \sum_{i}^{N}\sum_{j}^{N}\left\langle \frac{\sin qr_{ij}}{qr_{ij}} \right\rangle \qquad \{1\}$$

where r_{ij} is the distance between two scattering centres, i and j, N is the total number of scattering elements in the sample and q is connected to the scattering angle, θ and the wave length, λ by

$$q = (4\pi/\lambda)\sin\, \theta/2 \qquad \{2\}$$

Finally, the angle brackets ⟨ ⟩ denote the average over possible distance fluctuations since the distance, r_{ij}, may vary due to the collision with solvent molecules.

Because a double sum is difficult to read, it is therefore useful to examine a very simple structure which will immediately offer a clearer understanding. Such a structure is a rigid three centre array with two different bond lengths, r_{12} and r_{23}. The structure factor of this simple rod model can be expressed without difficulties as [1]

$$S(q) = 3 + 2\left[\frac{\sin qr_{12}}{qr_{12}} + \frac{\sin qr_{13}}{qr_{13}} + \frac{\sin qr_{23}}{qr_{23}}\right] \qquad \{3\}$$

This function is plotted in Figure 1, where the contribution

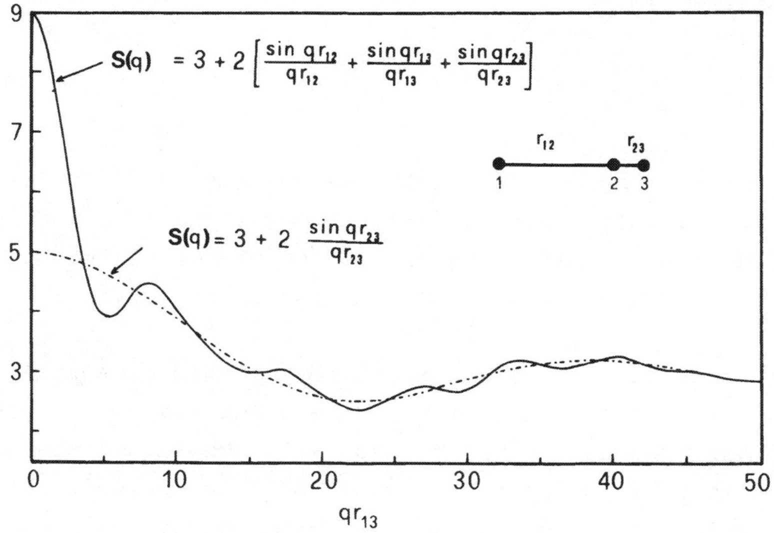

Figure 1: Angular dependence of the structure factor, S(q) for a rigid three scattering centre array, (solid line) and of the contribution arising from the shortest distance r_{23} (dotted line). $q = (4\pi/\lambda)\sin\theta/2$

[1] For N=3, the double sum contains nine elements. Three arise from the isolated scattering centres and the other six terms result from the different distances between the scattering elements, where the distance, e.g. from 1 to 2 and 2 to 1 have to be taken separately.

arising from the shortest distance, r_{23}, is also given. The three distances in this example are $r_{12} = (4/5)r_{13}$, $r_{23} = (1/5)r_{13}$ and $r_{13} = r_{12} + r_{23}$, with the parameter of the abscissa being $qr_{13} = 4\pi(r_{13}/\lambda)\sin\theta/2$. It can be recognized that at small scattering angles all distances contribute equally to the structure factor but at large angles, or more correctly at large values of the scattering vector q, the scattering behaviour is fully governed by the short distance in the structure. Thus, light scattering is analagous to viewing the structure through a window. When the window is widely opened at small angles, the whole structure is seen. When it is opened to a small slit at large scattering angles, however, only parts of the structure, or short sections, can be seen.

Rod-Structure of Fibrin

In the present fibrin study, full use was made of this "window" effect. In addition, special attention was given to the scattering behaviour at small angles, on the one hand, and to the asymptotic behaviour at large angles, on the other.

Since at small angles all substructures contribute equally to the structure factor, only little direct information on the shape of the molecule can be gained, although this global information is still of great value, as will be discussed below. Therefore, initial attention will be paid to the asymptotic behaviour. This region will be observed when the structures are sufficiently large. (The angular distribution of the scattered light depends on the ratio, r_{ij}/λ, not only on distance r_{ij}.) Since the fibrin monomer has dimensions larger than 47.5 nm, when for instance, the Hall and Slayter (9) or Bachmann-Lederer model (10) is used as a basis, it is ideally suited to the present purpose. Aggregation of only a small number of these monomers produces dimensions in the order of the light wave length so that the asymptotic region is accessible to measurement.

By selecting low fibrin concentration and low thrombin concentration, the rate of association at a pH 7.4 and 0.3 m KCl ionic strength was decreased drastically, and the formation of fibrin structure could be easily followed by light scattering (11). Initially, as other investigators have found, rod-formations were noted. A rod has a characteristic asymptote in its structure factor which can be expressed in terms of scattering intensities as (12)[2]

$$\frac{R(\theta)}{Kc} \quad \frac{M_w}{L_w} \quad \frac{\pi}{q} \qquad \{4\}$$

Here, $R(\theta)$ is the Rayleigh ratio ($= r^2 i(\theta)/I$) with I and $i(\theta)$ denoting the primary beam and scattering intensity at angle θ, respectively. r is the distance of the detector from the scattering cell centre. K is an optical constant and describes the contrast of the sample scattering over the solvent scattering. Thus, a rod is easily detected when $R(\theta)/(Kc)$ is plotted against q and a horizontal line is obtained. This graph has been termed as a Holtzer plot (11) in recognition of the significant work of Alfred Holtzer concerning polydisperse and heterogeneous rods (12,13). It should be emphasized that no other structure shows this typical asymptote. The value of this asymptote is the mass per unit length, (M_w/L_w). It is the molecular weight of a rod section of 1 nm in length. This quantity, experimentally obtained without prior assumptions is of great value because it provides information relating to the lateral aggregation of fibrin strands. An example will serve to illustrate this point. Assuming the fibrinogen model of Hall and Slayter (9) or Bachmann and Lederer

[2] By the term rod, it is always meant a rigid thread with a diameter that is small as compared to the wave length. Thus, such a rod may have a periodic or non-periodic nodular structure. This however, will not be detectable by light scattering.

(10),which has a length of 47.5 nm, the fibrinogen molecular weight is M_w = 340 000 g mol^{-1} and therefore, M_w/L_w = 7160 g $mol^{-1}nm^{-1}$. In discussing fibrin, the staggered overlapping three bead model (9,14) is usually assumed. This therefore, increases the value for a double strand twofold, i.e. 14 320 g $mol^{-1}nm^{-1}$. When these rods aggregate end-to-end this ratio of mass per unit length will not change, since M_w and L_w increase proportionally. A higher value thus indicates a stronger increase of the mass than of the length, or in other words, a side-by-side aggregation.

Figure 2 shows a typical Holtzer plot of human fibrin structure measured 10 and 25 minutes after initiation with thrombin (15).

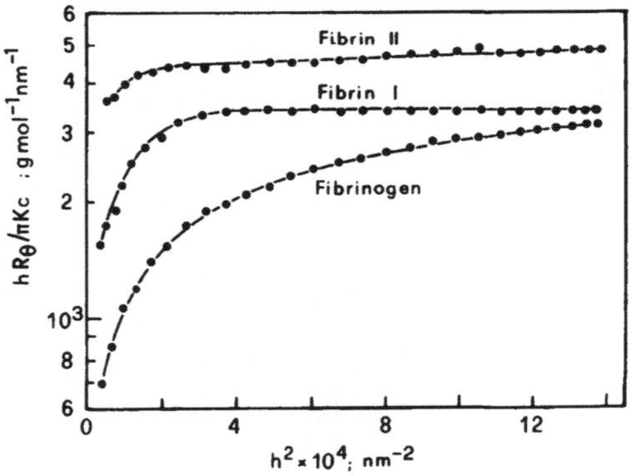

Figure 2: Holtzer plot of scattering intensities for measurements from fibrinogen and two fibrin intermediate structures. Fibrin I was recorded 10 minutes and Fibrin II 25 minutes after initiation with thrombin. (h has here the same meaning as q)

The third curve represents the scattering of fibrinogen. The rod-asymptote for the two fibrin intermediate structures is clearly developed. In addition, an increase in the mass per unit length with time is noticed. Fibrinogen does not exhibit the typical rod asymptote due to the fact that the molecule is somewhat too short as compared to the length of the light wave. The structure can be still described by the rod structure factor, however, and produces a value of $M_w/L_w = 3860$ (15). The mass per unit length value for the first fibrin intermediate structure is $M_w/L_w = 3500$ g $mol^{-1} nm^{-1}$. This is by nearly four times smaller than the value for a double strand staggered three sphere model discussed in the literature. It is even lower than the value for fibrinogen. Table I contains a list of the mass per unit length for human and bovine fibrinogen and some fibrin rods together with some other molecular data obtained by light scattering. Clearly the staggered overlapping sphere model must be discarded at least for the initial

Table I: Molecular weight M_w, rod length $(L_z^2)^{1/2}$ and mass per unit length (M_w/L_w) and $M_w/(L_z^2)^{1/2}$ of fibrin intermediates from Human fibrinogen (H1,H2) and Bovine fibrinogen (B1-B3) and of Human fibrinogen (H) and Bovine fibrinogen (B).

Sample	M_w 10^{-6} g $mole^{-1}$	$(L_z^2)^{1/2}$ nm	(M_w/L_w) g $mole^{-1}$ nm^{-1}	$M_w/(L_z^2)^{1/2}$ g $mole^{-1}$ nm^{-1}
H1	0.86	230	3500	3740
H2	1.69	380	4400	4380
B1	0.83	194	4700	4300
B2	1.10	220	5200	5000
B3	1.78	291	6300	6130
H	0.33	85.5	----	3860
B	0.33	60.6	----	5450

stage of the fibre formation. However, even a stronger assertion can be made: the length of a rod section of molecular weight 340 000 in the fibrin fibre can be calculated. This can be easily obtained by dividing this molecular weight by the mass per unit length. The result, L = 97 nm for the human fibrin monomer length, is approximately twice as large as the length of the Hall and Slayter model, and is larger than the length of fibrinogen. Thus it can be concluded that the fibrin monomer must have a more stretched conformation than fibrinogen.

Branching

The typical rod asymptote is obtained only at the onset of fibrin formation. At later times, when M_w exceeds $5 \cdot 10^6$, a slight deviation from the horizontal line at small angles is first observed. This becomes more and more pronounced with time, and eventually a marked decay over the whole angular region is found (11,15). The fibrillic structure evidently undergoes a change. Such change would be expected since the final clot is a polymer network and can only occur following branching.

Several model calculations were performed to ascertain whether the decaying asymptote was caused by branching (16,17). Such calculations were possible due to the recent progress in branching theory (16-19). (This theory has its origin in genealogy and evolution theory (20) and dates back to Reverend Watson, an Englishman who developed this theory in 1873 (21).) In the first model, crosslinking of fibrin strands was assumed, in the second, a random trifunctional branching of the chains at the end. In both cases, an asymptotic decay of the scattering intensity of q^{-2} is predicted by theory while the experiment shows a much stronger decay of q^{-4} (note: the rod has only a q^{-1} decay). However, when it was assumed that both ends of the fibrin monomer behave differently (i.e. the random trifuctional branching model was altered to a restricted branching, (the difference between cross-linking and terminal branching is depicted in Figure 3.), a q^{-4} dependence is in in fact obtained (17). It may, therefore be concluded that the fibrin clot cannot be produced by cross-linking.

This model of a trifuctional fibre terminal branching demonstrates many additional details which accord well with experi-

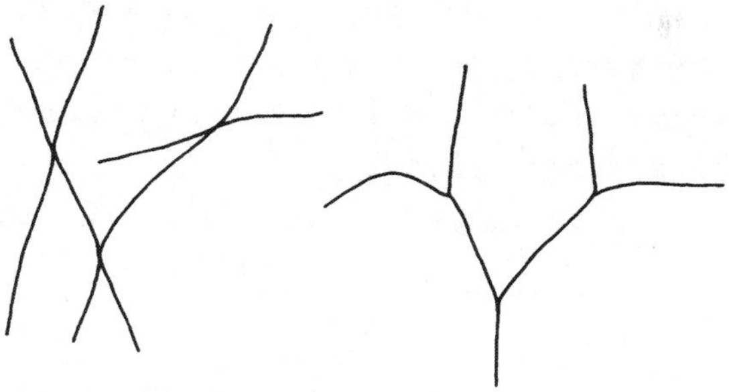

Figure 3: Sketch of cross-linked and terminally branched chains

mental results. Figure 4 exhibits the theoretical structure factor in the Holtzer plot. It may first be noted how the typical rod asymptote is approached with growing fibre length. Then, at approximately $M_w = 5 \cdot 10^6$, the previously mentioned upturn at small angles occurs, coinciding with the commencement of branching. Some experimental curves are exhibited for purposes of comparison in Figure 3 in the contribution by M.Müller.

Fibre-Length between Branching Points

In the present work, a discussion of the initial part of the scattering curves has heretofore been omitted. This portion of the curve gives the mean square radius of gyration $\langle S^2 \rangle_z$, which is an overall measure of the molecule dimensions. Its dependence on molecular weight, M_w, is however, sensitive to structure (22). Figure 5 gives an example. A steep increase of dimensions with M_w is found in the first stages of the reaction and at large M_w the curve becomes much flatter. The steep increase is typical for rods and the flat part results from branching where the mass (and thus the density) increases much stronger than the dimensions. The kink point, which is obtained by extrapolation of the two parts of the curve, may be intuitively interpreted as the point where branching commences. Upon further analyses, this tentative interpretation was proved to be correct. The full line in Figure 5

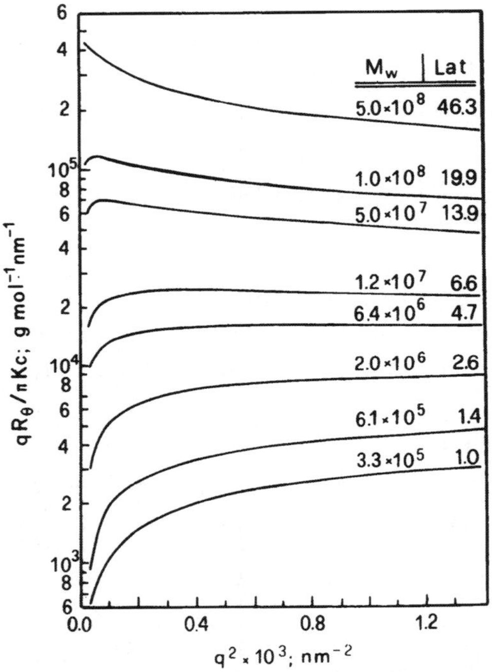

Figure 4: Calculated angular dependence of scattering intensities represented by a Holtzer plot according to the terminal branching model with restriction. The row Lat denotes the extent of lateral aggregation, i.e. the number of aggregated fibrin strands, and M_w denotes the corresponding molecular weights

corresponds to the theoretical prediction using the model outlined above and the quantitative fit produced a fibre length of 525 nm. This agrees well with the morphometric measurements on clots from EM pictures by Witt et al (24). With this theory then, a suitable tool for a quantitative comparison of branching in different fibrins is now available. The effect

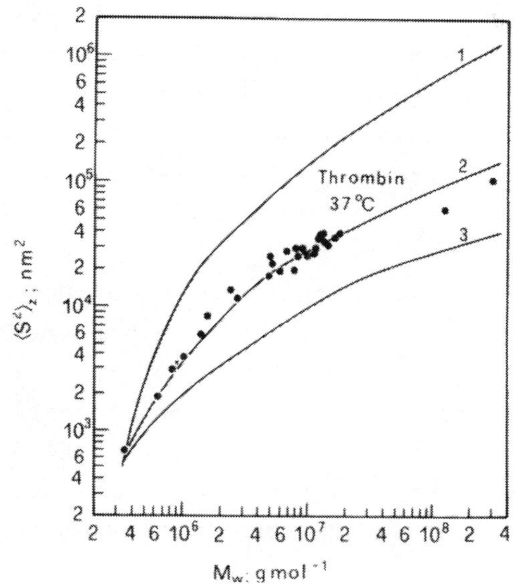

Figure 5: Molecular weight dependence of the mean square radius of gyration $\langle S^2 \rangle_z$ for soluble fibrin intermediate structures obtained at different times after initiation with thrombin. Curves 1,2, and 3 represent theoretical curves with no, weak and strong lateral aggregation, respectively, according to the restricted terminally branched fibre model.

of pH, temperature and different enzymes and the influence of fibrinogen from different sources has been previously studied and is discussed in another paper (15).

Pre-Gel versus Post-Gel Properties

At this point it is reasonable to ask whether studies on the soluble fibrin produce the same result as studies on the clot, or to what extent pre-gel properties allow prediction of post-gel properties. With regard to the branching density, or the fibre length in the network, light scattering measurements in the pre-gel state agree well with the morphometric measurements in the gel. Identical results were found for adult and foetal fibrin (23) as Witt et al (24) discovered from electron micrographs. In addition, the data obtained by Krause and Zimmermann (25) pertaining to networks produced by enzymes of snake venoms are confirmed by light scattering results of the pre-gel state in the present work. (15).

There is, however, no agreement concerning the extent of lateral aggregation. For instance, the EM pictures from adult and foetal fibrin networks exhibit much finer fibrils for the foetal fibrin clot (24). According the the present authors' light scattering results, however, there is no detectable difference in the lateral aggregation of the two types of fibrin. An examination of Figure 6 clarifies this point.

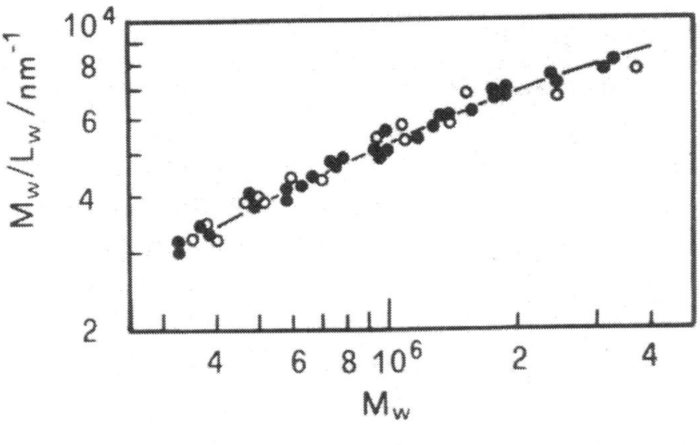

Figure 6: Increase of the mass per unit length M_w/L_w with molecular weight for foetal (o) and adult soluble fibrin

It can be noted from this picture that the lateral aggregation increases continuously with increases in molecular weight in the region of these light scattering measurements. Obviously, an equilibrium state has not yet been reached. Additionally from these measurements taken at the onset of fibrin formation no definite determination can be made regarding the final equilibrium fibre thickness. Unfortunately, at present an evaluation of the lateral aggregation only up to the point of incipient branching can be offered since the influence of large fibre diameters at later stages is not yet fully understood in theory.

Fibre Flexibility

Modern dynamic light scattering has allowed clarification of the question as to whether the fibres are in fact rigid or have some inherent flexibility. In dynamic light scattering the influence of solvent molecule collision on the motion and structure of macromolecules is observed. The collision with solvent molecules causes the well known Brownian motion, which is directly related to the translational diffusion coefficient of the macromolecules. When these Brownian particles are not rigid but possess a certain degree of flexibility random bending, torsion and twisting will be observed. For a flexible thread, this will eventually result in random coiling. All these random motions cause fluctuations in density, concentration and structure in a time scale of 10^{-9} to 10^{-5} seconds. These rapid fluctuations can be resolved by fast computers, usually termed as autocorrelators. The correlator traces the fluctuations in the form of a time correlation function,which decays exponentially with correlation time (26). Experimentally, the initial slope of the logarithmic

[3]The computer multiplies the scattering intensity at a time, t, with the scattering intensity at time $t+\tau$. This process is repeated 10^5 to 10^6 times to get a meaningful average. τ is the corrlation time. In the present experiments, the correlation function was measured simultaneously at 72 different correlation times τ).

correlation function is measured with the highest accuracy. This slope is termed the first cumulant of the correlation function. In the limit of small scattering angles this first cumulant, Γ, is simply related to the translational diffusion coefficient

$$\Gamma = Dq^2 \tag{5}$$

where $q = (4\pi/\lambda)\sin\theta/2$ is the value of the scattering vector, and also defines the static structure factor. At larger angles and with large structures, the "window" effect, outlined in the discussion of the asymptotic behaviour of the static structure factor, becomes effective: Only parts of the molecule are observed and consequently, the irregular <u>internal</u> modes of motion become evident. In measurement, this effect becomes noticeable by the observation that

$$\Gamma/q^2 \equiv D_{app}(q) \tag{6}$$

is now angular dependent and is thus considered only an apparent diffusion coefficient.

The apparent diffusion coefficient has been calculated for random coils (27) and rigid infinitely thin rigid rods (28) as well as for some branched structures (27). The present discussion will be limited to the random coil and rigid rod. Figure 7 exhibits the result of measurements from unbranched fibrin fibres which have been obtained by thrombin initiation at pH$\geqslant$9.5. At these pH values, stable products are obtained, with the fibres decreasing in length as the pH value is raised. The two solid curves represent the random coil and rigid rod, respectively. At small molecular weights (i.e. short fibres) the curves fall below the rod cumulant. This is theoretically expected when cylinders of a small axial ratio of length to diameter are present (28). At higher molecular weights, the experimental curve follows the rod behaviour at small angles,

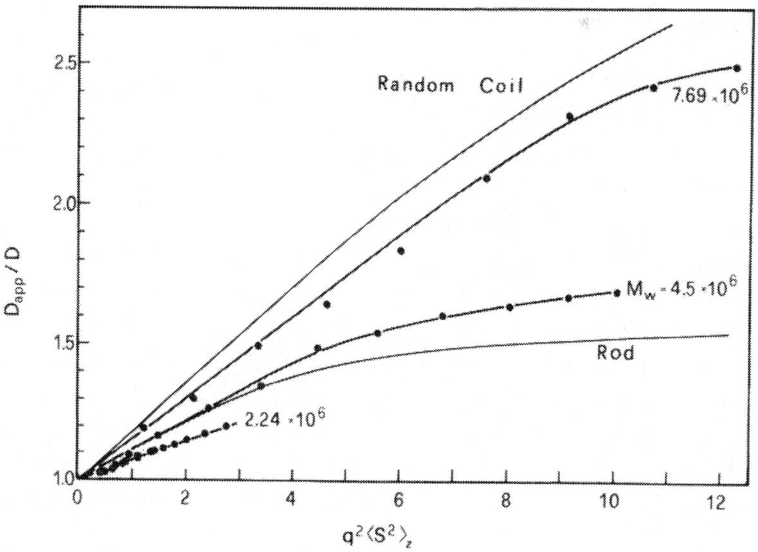

Figure 7: Angular dependence of rod-like fibrin fibres, obtained at pH$\geqslant$9.5, of the normalized apparent diffusion coefficient acording to dynamic light scattering. The solid lines are theoretical curves for random, flexible coils and for infinitely thin rods, respectively

while exhibiting deviations, however, at large scattering angles. The long fibres exhibit a behaviour between that of the flexible coil and the rigid rod. It must therefore be concluded that the fibrils must have some flexibility in spite of their appreciably large thickness. This fact must be kept in mind when suggesting models for fibrin, since evidently the majority of the present models cannot account for any sort of flexibility.

This work was kindly supported by the Deutsche Forschungsgemeinschaft.

References

1. Steiner, R.F., Laki, K.: Arch. Biochim. Biophys. 34, 24 (1951).
2. Katz, S., Gutfreund, K., Shulman, S., Ferry, J.D.: J. Amer. Chem. Soc. 74, 5706 (1952).
3. Casassa, E.F.: J. Chem. Phys, 23, 596 (1955).
4. Blombäck, B., Laurent, T.C.: Ark. Kemi 12, 137 (1957).
5. Geiduscheck, E.P., Holtzer, A.M.: Adv. Biol. Phys. Med. 6, 431 (1958).
6. Debye, P.: J. Phys. Colloid. Chem. 51, 18 (1947).
7. Zimm, B.H.: J. Chem Phys. 16, 1093 (1948).
8. Debye, P.: Ann. Phys. 46, 806 (1915); Phys. Z. 31, 419 (1930).
9. Hall, C.E., Slayter, H.S.: J. Biochem. Biophys. Cytol. 5, 11 (1959).
10. Bachmann, L., Schmitt-Fumian, W.W., Hammel, R., Lederer,K.: Makromol.Chem. 176, 203 (1975).
11. Müller, M., Burchard, W.: Biochim, Biophys. Acta 537, 208 (1978).
12. Holtzer, A.M.: J. Polymer Sci.: 17, 432 (1955).
13. Holtzer, A.M., Rice, S.A.: J. Amer. Chem. Soc. 79 4847 (1957).
14. Weisel, J.W., Warren, S.G., Cohen, C.: J. Mol. Biol. 126, 159 (1978), see also Weisel, J.W., Phillips, G.N., Cohen, C.: Nature 289, 263 (1981).
15. Müller, M., Lasarcyk, H., Burchard, W.: Int. J. Biol. Macromol. 3, 19 (1981).
16. Müller, M., Burchard, W.: Makromol. Chem. 179, 1821 (1978).
17. Burchard, W., Müller, M.: Int. J. Biol. Macromol. 2, 225 (1980).
18. Gordon, M.: Proc. Royal Soc. (London) A 268, 240 (1962).
19. Kajiwara, K., Burchard, W., Gordon, M.: Br. Polymer J. 2, 110 (1970).
20. See e.g. Harris, T.: The Theory of Branching, Springer-Verlag 1963.
21. Watson, H.W.: Educational Times, pp. 115 (1873).
22. See e.g. Burchard, W.: Light Scattering Techniques, in: Happey, F. (Ed.) Applied Fibre Science, Vol. 1, Academic Press (1978).

23. Müller, M., Burchard, W., Witt, I.: Thrombosis Research in press.

24. Tesch, R., Tolp, R., Witt, I.: Thrombosis Research 16, 239 (1979).

25. Krause, W., Zimmermann, P.: Klin Wschr. 50, 557 (1962).

26. For more details of the technique, see e.g. Pecora, R.: Dynamic Light Scattering, Wiley 1976.

27. Burchard, W., Schmidt, M., Stockmayer, W.H.: Macromolecules 13, 580 (1980).

28. Aragon, S.R., Pecora, R.: J. Chem. Phys. 66, 2506 (1977).

29. Müller, M., Burchard, W.: Int. J. Biol. Macromol. 3, (1981) in press.

PARTICULARITIES OF FIBRINOGEN, ROD-LIKE AND BRANCHED FIBRINS

Michael Müller and Walther Burchard
Institute of Macromolecular Chemistry
D-7800 Freiburg, Germany

Introduction

For more than thirty years fibrinogen has been a subject of intense biochemical and biophysical research (1,2). Still, a consensus on the shape of this important blood protein has not yet been reached. Since similar disagreements have not been observed with any other protein, fibrinogen holds a rather singular position with respect to blood proteins. Even more astonishing is the fact that the research has been concerned primarily with the unactivated fibrinogen molecule, although it is the fibrin network which has the relevant physiological function. Little is known of the structure of this network from direct measurements. Instead, conclusions were drawn mostly by simulation of the striking stratification of the fibrin fibres in electron microscopic pictures using specific assumptions for the fibrinogen molecule (3,4,5,6).

In the present contribution, new data are presented on the basis of light scattering (7,8) of fibrinogen as well as of fibrin intermediate structures. From these, a more detailed insight into the arrangement of the fibrin monomer units in the fibrin structures is gained and thus produces a better understanding of the fibrinogen fibrin transition.

Fibrinogen

Most of the physical investigations concerning polymers in the blood clotting process have dealt with the structure of fibrinogen, which is the nonactivated monomer for the fibrin polymer. The present interest in the special feature of fibrinogen was aroused, when the mean square radius of gyration (7) was measured and found to have a value $5^{1/2}$ times greater than would be expected for a rod conformation of 45 nm in length. In spite of concerted efforts to determine the structure, along with the application of various techniques, several contradicting models are still under discussion. Table I presents different fibrinogen structures, and for rod-like molecules, different length for the fibrinogen suggested by various authors.

Sedimentation	Shulman 1953 (9)	rod	L = 86 nm
Light Scattering	Katz et al 1952 (10)	rod	L = 66 nm
Flow Birefringence	Hocking et al 1952 (11)	rod	L = 67 nm
Viscosity	Edsall 1952 (12)	rod	L = 50 nm
EM	Hall and Slayter 1959 (13)	rod	L = 47.5 nm
EM	Bachmann et al 1975 (14)	rod	L = 45 nm
EM of Fibrins	Krakow et al 1972 (15)	rod	L = 43 nm
EM of Fibrins	Kay and Cuddigan 1967 (3)	rod	L = 69 nm
EM	Köppel 1966 (15,6)	spherical	R = 12 nm
Neutron Scattering	Marguerie et al 1976 (16)	spherical	R = 15nm
QELS	Serrallach 1977 (17)	rod	L = 90 nm
EM-Diffraction	Cohen et al 1981 (18)	rod	L = 45 nm
X-Ray Scattering	Lederer and Hammel 1975 (19)	rod	L=54.4 nm

Table I: Fibrinogen Structures Suggested by Various Authors

As can be noted from Table I, the suggested rod length for the fibrinogen molecule ranges from 45 nm to 90 nm or by a factor of two! It is important to note that nearly all measurements which have been performed in aqueous solution produce rod lengths greater than 45 nm. The experiments concerning X-ray scattering by Lederer and Hammel[19] will be discussed in detail since their conclusions, as well as the present authors', are both drawn from the evaluation of the M/L val-use[20,7]--the molecular weight per unit length of 1 nm. Lederer found a value of 6150 g/mol nm for M/L and with a molecular weight of the fibrinogen of 335 000 g/mol. This corresponds to a rod length of 54.5 nm. However, Lederer also pointed out that occasionally a value of M/L of only 4200 g/mol nm was obtained, which corresponds to a rod length of 80 nm. These findings cannot be interpreted by small amounts of aggregated material which might be present in solution, since an end-to-end aggregated molecule with a double length also has a doubled molecular weight. Thus, it can never explain a lower value for the molecular weight than for the fibrinogen itself. It should be stressed, however, that lower values of M/L for the intact fibrinogen with a molecular weight of 340 000 g/mol can only be expected for a more elongated molecule. Figure 1 exhibits the results of static light scattering measurements from human and bovine fibrinogen.

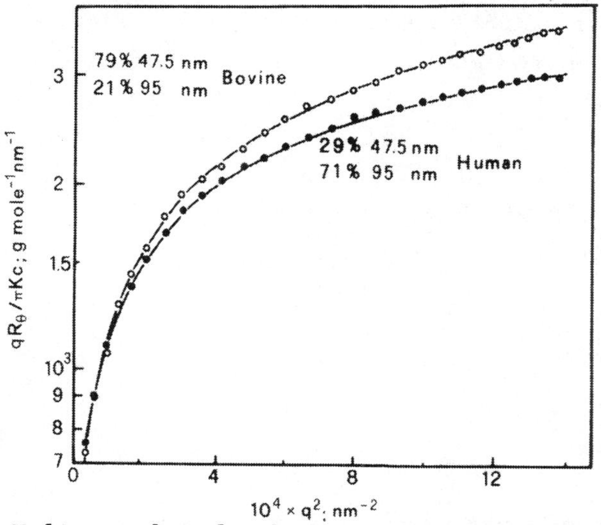

Figure 1: Holtzer plot for human and bovine fibrinogen. This exhibits the angular dependence of the scattering intensity as a function of q^2, $h=q=4\pi/\lambda \sin\theta/2$

A molecular weight of 340 000 g/mol was found for human and bovine fibrinogen. However, bovine fibrinogen exhibits a scattering behaviour which differs significantly from human fibrinogen. In particular, a much higher asymptote at large values of the scattering vector q is obtained, which represents the mass per unit length. Thus, this qualitative inspection already reveals an appreciable shorter rod for bovine than for human fibrinogen. In view of the strikingly contradicting models suggested for fibrinogen, a description of the scattering behaviour produced by two rod conformations for the fibrinogen present in different mass fractions in solution was attempted. The first was a short rod conformation of 47.5 nm and corresponded to the Hall-Slayter model (4) and the second was a long rod conformation with the doubled length of 95 nm, which was formed for the fibrin monomer in fibrillic fibrin (8,21). The full and open circles in Figure 1 are experimental data while the lines represent the least square fits with the theoretical formula for rods of Neugebauer (22).

These quantitative fits demonstrate that on average 71 percent of the long rod conformation and 29 per cent of the short rod conformation are present in the solution of human fibrinogen. For bovine fibrinogen under the same conditions, however only 21 percent of the long rod and 79 percent of the short are found. Evaluation of many different experiments led to the following mean values for the rod length L and the M/L values:

Bovine fibrinogen (see footnote)

$L_z = 64.3 \pm 1.6$nm $\quad M_w/L_z = 5450$ g/mol nm

Human fibrinogen

$L_z = 85.6 \pm 1.6$ nm $\quad M_w/L_z = 3860$ g/mol nm

Footnote: These values were obtained from the weight average molecular weight M_w, and the z-average mean square radius of gyration, $\langle S^2 \rangle_z = {}^wL^2_z/12$.

Parallel to these static light scattering experiments a great number of QELS (quasi-elastic light scattering) experiments were performed with human and bovine fibrinogen (23) This method allows the determination of diffusion coefficients and is sensitive to a change of the hydrodynamic radius of macromolecules (24).

Bovine fibrinogen

$$D_z = 2.17 \pm 0.015 \; 10^{-7} \; cm^2/sec$$

Human fibrinogen

$$D_z = 1.72 \pm 0.013 \; 10^{-7} \; cm^2/sec$$

These results are in complete agreement with the results of the geometric radius of the static light scattering experiments. Theory concerning diffusion coefficients predicts a lower value for the diffusion constants for a solution with a higher mass fraction of the long rod conformation.

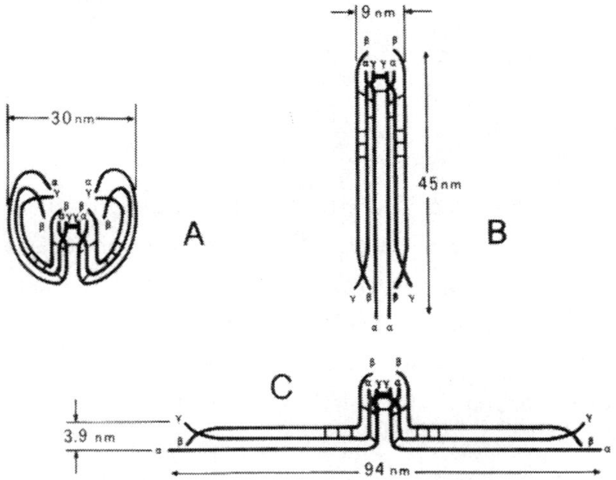

Figure 2: Three different isomeric structures of fibrinogen
A the banana-like model of Marguerie (16)
B the folded rod of Bachmann (13)
C the stretched rod of the authors (7,21,23)

Taking the various model suggestions for the fibrinogen molecule into account and following the data of light scattering experiments performed by the authors, a fibrinogen molecule which might have sections of flexibility so that it can undergo conformational changes is discussed. Thus, the banana-like structure with a circumference of approximately 90 nm, the short rod conformation of 45 nm and the long rod conformation of approximately 90 nm may be considered as isomeric structures of one and the same molecule.

Fibrin - Monomer

When the fibrinogen has sections of flexibility, then from the physiological standpoint the primary question concerns the determination of which conformation does the fibrin monomer have in the polymer. The data from fibrin intermediate structures can only be interpreted as consisting of a long rod conformation of ca. 90 nm of the fibrin monomer in the polymer (8,21).

Rod-Like and Branched Fibrins

The greatest advantage of the previously mentioned light scattering experiments are that these measurements can be performed during the aggregation process. Thus, the structural informations could be drawn by analysing the scattering behaviour at different times of the aggregation process (7). This aggregation process is usually understood to be as follows:
After the initiation process, the fibrin monomers form long fibrin threads by an end-to-end and parallel sid-by-side aggregation. After this, however, branching becomes observable and in the final stage network formation takes place (25). For a clearer understanding, the evaluation of a series of such

scattering experiments is demonstrated in more detail.

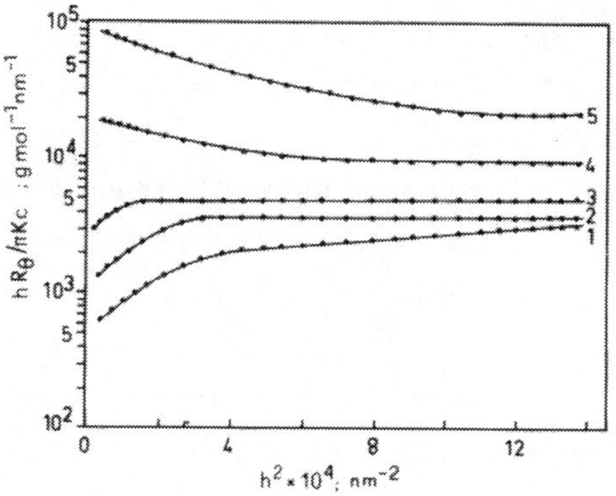

Figure 3: Holtzer plot of a series of scattering experiments $h=q=4\pi/\lambda \sin\theta/2$

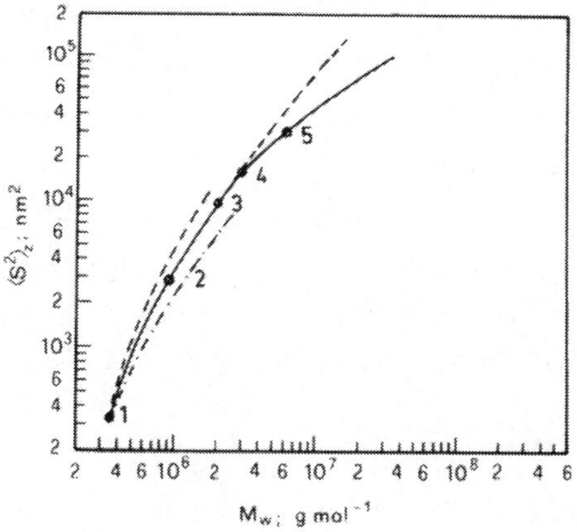

Figure 4: $\langle s^2 \rangle_z$ as a function of M_w

Curve 1 in Figure 3 represents the scattering behaviour of fibrinogen. Curve 2 exhibits the scattering behaviour of a rod-like fibrin with a length of approximately 250 nm. Typical for such rod-like structures is that a parallel asymptote is reached at high values of h^2, i.e. large angles of observation at high values of h^2, i.e. large angles of observation. Curve 3 shows the corresponding angular distribution of a fibrin with a length of approximately 400 nm. A parallel asymptote over a wide range of scattering angles is recognized. The data of $\langle s^2 \rangle$ and M_w resulting from the evaluation of the scattering curves are then plotted in Figure 4 in a double logarithmic scale. Whenever rod-like structures are formed, these evaluated points are lying on the steep initial part of Figure 4. The height of the parallel asymptote in Figure 3 denotes the molecular weight per unit length (M/L-value) and is a measure of the degree of lateral aggregation. It has its counterpart in different initial slopes in $\langle s^2 \rangle_z$ versus M_w (see the different dotted lines).

After various times of this rod-like aggregation process, the asymptote is superimposed by the scattering behaviour of the branched fibrin structures. In Figure 3, Curve 4 a decaying scattering curve is observable. At higher degrees of aggregation, the scattering behaviour is now governed by the particularities of the branched fibrin structures (Curve 5) and the evaluated points are lying on the flatter part of the $\langle s^2 \rangle_z$ versus M_w (Point 5). Thus, the kink region between the steep initial part and the flatter part at high molecular weights is a measure of that fibre length when first branching reactions occur (Point 4).

Figure 5 exhibits the influence of the rod length between branching points in a plot of $\langle s^2 \rangle_z$ as a function of the molecular weight of the fibrins. Thus, the longer the fibrils between branching points, the further the kink region is shifted towards higher molecular weights. Recently, a theory was developed by the present authors (25) which can quantitatively describe this aggregation process. The experimental data can now be fitted with the theory and this allows the description of the structure formation in terms of lateral aggregation and fibre length between branching points.

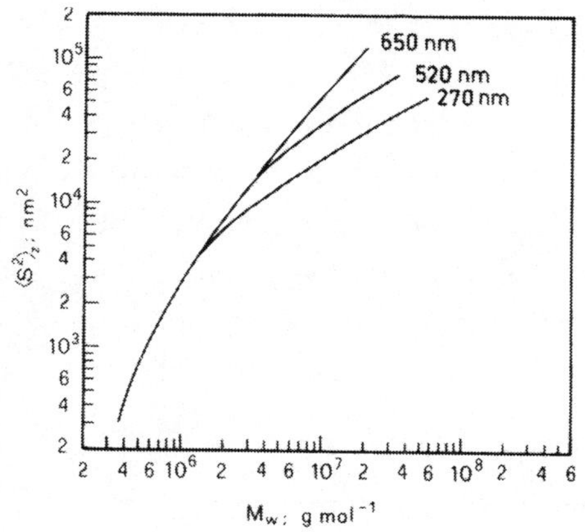

Figure 5: $\langle s^2 \rangle_z$ versus M_w. The numbers indicate the rod length between branching points

Figure 6 exhibits the experimental data utilizing human fibrinogen at pH 7.4 and 9.5; all these measurements were performed in TRIS-buffered solutions with 0.3 m KCl and 0.01 m Ca^{2+}.

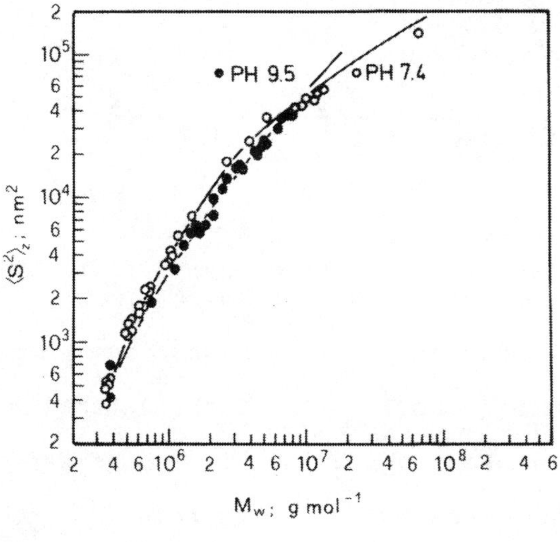

Figure 6: Aggregation experiments with human fibrinogen at pH 7.4 and pH 9.5

The fit produces a value of 525 nm for the rod length between branching points at pH 7.4. When the pH is increased to 9.5 the lateral aggregation becomes more pronounced than at pH 7.4. Both curves cross at a molecular weight of 7×10^6 g/mol. In other words, for pH 9.5 the kink region is shifted towards higher molecular weights. The rod length between branching points is now found to be larger than 700 nm. The increase in the lateral aggregation appears to be correlated to a decrease in the network density.

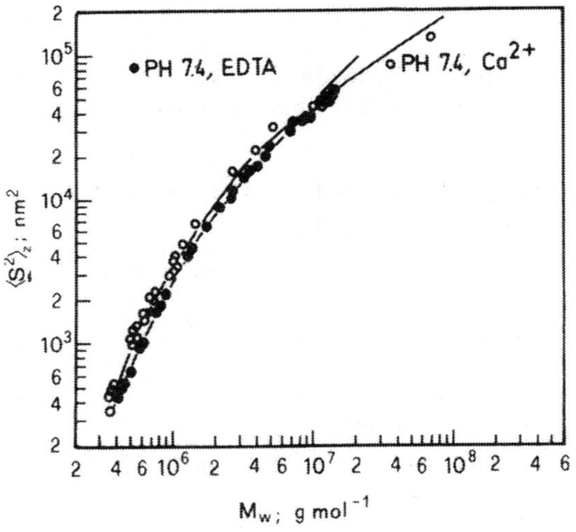

Figure 7: Aggregation experiments of human fibrinogen with Ca-ions and addition of EDTA (26)

Similar results to those at pH 9.5 are obtained when the Ca-ions are complexed by EDTA. Again, the lateral aggregation increases and branching reaction is suppressed (i.e. the fibre length was found to be larger than 700 nm). It might be suspected that the fibrinogen molecule becomes denatured at pH 9.5 so that the observed aggregation (Figure 6) represents the behaviour of a defect molecule. However, the experiments with EDTA might possibly indicate that a change has occured in the aggregation behaviour due to a change in the electrostatic fingerprint of the fibrinogen molecule.

In both cases, at high pH as well as by addition of EDTA, the number of positively charged groups is drastically reduced. These results may be interpreted such that the aggregation behaviour is strongly controlled by special thermodynamic interactions.

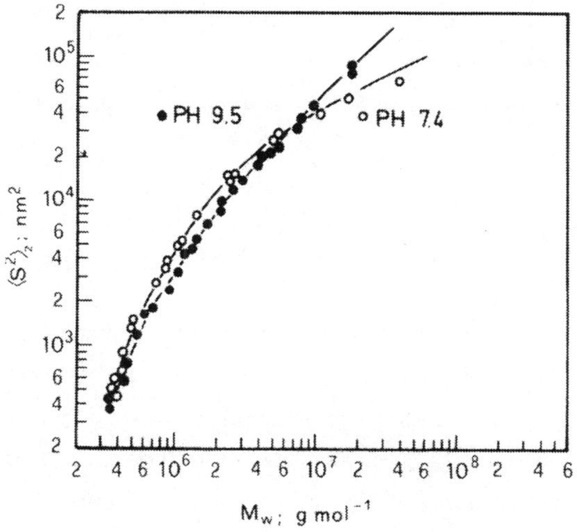

Figure 8: Aggregation experiments of bovine fibrinogen at pH 7.4 and 9.5

Figure 8 exhibits the aggregation experiments when bovine fibrinogen is used instead of human fibrinogen. Although bovine fibrinogen has a different amino acid composition(27) the aggregation behaviour is only slightly different when compared to human samples. The rod length between branching points was detected to be 500 nm at pH 7.4. When the pH is increased to the non-physiological value of 9.5, again the lateral aggregation increases and the rod length connecting branching units was 650 nm.

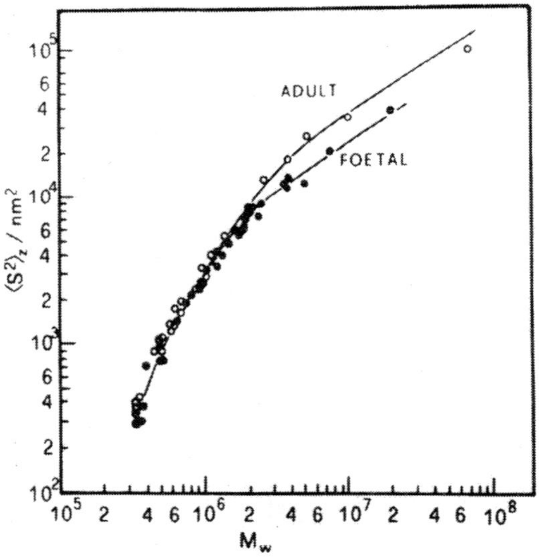

Figure 9: Aggregation experiments with foetal fibrinogen in comparison with human adult samples

Similar measurements have been performed in order to analyse the structural particularities of foetal fibrinogen (28). Morphometric analyses (29) of electron microscopic pictures revealed a higher network density in the foetal clot. The rod length between branching points was determined to be 350 nm. In addition, much thinner fibrils are found in the foetal network. The light scattering experiments during the aggregation process on soluble fibrins confirm the formation of a very dense network. The rod length connecting branching points was found to be 370 nm, compared to 525 nm for the adult samples. This is in good agreement with the EM studies. However, as can be noted in Figure 9, no differences in the lateral aggregation of the two kinds of fibrin have been observed in the present experiments. (See also Figure 6 of the preceeding contribution (8).) One reason for the differences in the final fibre thickness of the two types of fibrin may be due to the higher negative electric charge of the foetal fibrinogen. Because of electrostatic repulsion, the charge opposes the association tendency. For details see reference 28.

In conclusion, some comments pertaining to the opacity of the gels formed in the final stage is necessary. The fibrin clots were mainly translucent or transparent when the fibre length between branching points was longer than 550 nm while clots with higher branching density were generally coarse.

The authors wish to thank Deutsche Forschungsgemeinschaft for financial support.

We are indebted to Professors I.Witt, Freiburg, and R.E. Zimmermann, Münster, for supplying us with foetal and bovine fibrinogen, respectively.

References

1. Scheraga, H.A., Laskowski, M.: Advan.Protein Chem. 12, 1 (1957).
2. Doolittle, R.F.: Advan. Protein Chem. 27, 1 (1973).
3. Kay, D., Cuddigan, B.J.: Brit. J. Haematol. 13, 341 (1967).
4. Hall, C.E., Slayter, H.S.: J. Biophys. Biochem. Cytol. 5, 11 (1959).
5. Bang, N.U.: Thromb. Diath. Haemorrh. Suppl. 13, 73 (1964)
6. Köppel, G.,: Z. Zellforsch. 77, 443 (1967).
7. Müller, M., W. Burchard, W.: Biochim. Biophys. Acta 208, 537 (1978).
8. Burchard, W., Müller, M.: preceeding paper.
9. Shulman, S.: J. Amer. Chem. Soc. 75, 5846 (1953).
10. Katz, S., Gutfreund, K., Shulman, S., Ferry, J.D.: J. Amer. Chem. Soc. 74, 5706 (1952).
11. Hocking, C.S., Laskowski, M., Scheraga, N.A.: J. Amer. Chem. Soc. 74, 775 (1952).
12. Edsall, J.T.: J. Polym. Sci. 12, 253 (1954).

13. Bachmann, L., Schmitt-Fumian, W.W., Hammel, R., Lederer, K., Makromol. Chem. 176, 2603 (1975).

14. Karkow, K., Endres, G.F., Siegel, B.M., Scheraga, H.A., J. Mol. Biol. 71, 95 (1972).

15. Köppel, G.: Nature 212, 1608 (1966).

16. Marguerie, G., Stuhrmann, H.B.: J. Mol. Biol. 102, 143 (1976).

17. Serralach, E.: PhD Thesis, ETH Zürich.

18. Weisel, J.W., Phillips, G.N., Cohen, C.: Nature 289, 263 (1981).

19. Lederer, K., Hammel, R.: Makromol. Chem. 176, 2619 (1975).

20. Holtzer, A.M.: J. Polym. Sci. 17, 432 (1955).

21. Müller, M., Lasarcyk, H., Burchard, W.: Int. J. Biol. Macromol. 3, 19 (1981).

22. Neugebauer, T.: Ann. Phys. 42, 509 (1943).

23. Müller, M., Burchard, W.: Int. J. Biol. Macromol. 3, (1981) in press.

24. For more details of the technique see, e.g. R. Pecora, "Dynamic Light Scattering", Wiley, New York 1976.

25. Burchard, W., Müller, M.: Int. J. Biol. Macromol. 2, 225 (1980).

26. Bithell, T.C.: Biochem. J. 93 431 (1964).

27. Witt, I.: Biochemie der Blutgerinnung und Fibrinolyse, Weinheim 1975.

28. Müller, M., Burchard, W., Witt, I.: Thrombosis Research in press.

29, Tesch, R., Tolp, R., Witt, I.: Thrombosis Research 16, 239 (1979).

30. Witt, I., Tesch, T.: In: Fibrinogen, Fibrin und Fibrinkleber, Kl. Schimpf (ed.), Schattauer Verlag, Stuttgart-New York, 1981, pp. 31-37.

TEMPERATURE EFFECTS ON FIBRIN(OGEN)-FIBRIN SOLUTIONS
(Preliminary results)

Pierre Wiltzius and Werner Känzig
Laboratorium für Festkörperphysik, ETH
CH-8093 Zürich, Switzerland

Victor Hofmann
Medizinische Klinik, Universitätsspital
CH-8091 Zürich, Switzerland

P.W. Straub
Thrombose-Laboratorium, Medizinische Universitätsklinik,
Inselspital, CH-3010 Bern, Switzerland

In case of intravascular coagulation and often also in patients with other thrombo-embolic phenomena, the intravascular formation of thrombin results in so-called "circulating soluble fibrin", which may be detected using the alcohol gelation test or more cumbersome chromatographic methods. It has been suggested that circulating fibrin corresponds either to macromolecular complexes of fibrinogen with fibrin (1,2), due to the fibrin-solubilizing effect of fibrinogen, or to complexes of fibrinogen with proteolytic fibrin degradation products (3). The question whether these in vitro observed fibrinogen-fibrin complexes exist also in vivo is still not answered.

Light scattering methods (static and dynamic) are useful for the investigation of the fibrinogen to fibrin transition (4) and have been applied in this work to the study of the temperature dependence of complex formation.

Fibrinogen - Recent Biochemical and Medical Aspects

0.025 NIH units/ml of thrombin (Hoffmann LaRoche, Basel) were added at 20°C to 1 ml of human fibrinogen (Imco, Stockholm) with a clottability greater than 95 % and a concentration of 2 mg/ml. The activity of thrombin was quenched after a certain incubation time (varying between 10 and 45 minutes for different runs). After the check that the aggregation process was stopped at 20°C the temperature variation was initiated. The temperature dependence of the mean hydrodynamic radii (obtained from the analysis of the dynamic light scattering measurements) is plotted in the figure below.

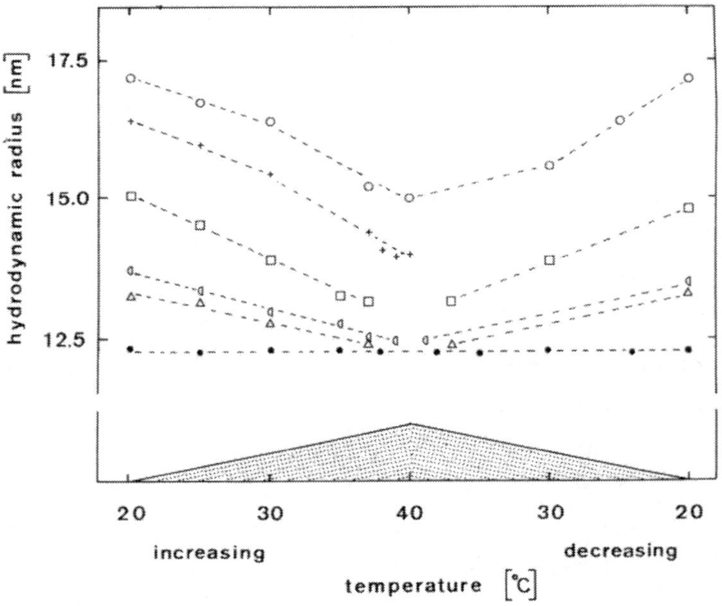

Figure: Temperature dependence of the apparent mean hydrodynamic radius of a fibrinogen solution (●) and of fibrin(ogen)-fibrin solutions with various incubation periods :
△ 10 min, ◐ 15 min, □ 27 min,
+ 37 min, ○ 45 min.

These measurements (as well as the scattered intensity measurements) suggest a reversible dissociation-association process of the oligomers formed after thrombin addition at 20°C. A solution of fibrinogen monomers shows no significant temperature dependence between 20° and 40°C.

A detailed analysis of the data shows that when less than about 20 % of fibrinogen is used up to form polymers full dissociation into monomers is reached at 37°C.

These experiments suggest that the formation of fibrin(ogen)-fibrin complexes in vitro is temperature dependent and that small concentrations of fibrin devoid of fibrinopeptides A exist in vivo at body temperature. Our conclusions are compatible with the findings of other authors (5,6).

References

1. Shainoff, J.R. & Page, I.H.: J. of Exp. Medicine 116, 687 (1962).

2. Saski, T., Page, I.H. & Shainoff, J.R.: Science 152, 1069-1071 (1966).

3. Lipinski, B., Wegrzynowicz, Z., Budzynski, A.Z., Kopec, M., Latallo, Z.S. & Kowalski, E.: Thrombos. Diath. Haemorrh. 17, 65 (1967).

4. Wiltzius, P.: Thesis, ETH Zürich, 6764 (1981)

5. Blaettler, W., Straub, P.W. & Peyer, A.: Thromb. Res. 4, 787-801 (1974).

6. Müller-Berghaus, G., Mahn, I. & Krell, W.: Thromb. Res. 14, 561-572 (1979).

Supported by the Swiss National Science Foundation 3.938-0.78

HYDRODYNAMIC CHARACTERIZATION OF PLASMIN DEGRADATION PRODUCTS FROM HUMAN FIBRINOGEN

Alfons C.M. van der Drift and Aldi Poppema
Medical Biological Laboratory, Rijswijk Z.H., The Netherlands

Frits Haverkate and Willem Nieuwenhuizen
Gaubius Institute TNO, Leiden, The Netherlands

Introduction

Recently we have shown that the size of terminal fragments formed during degradation of human fibrinogen or fibrin in vitro by the action of plasmin, depends on the presence of Ca^{2+} ions and ligands like ethylenediamine tetraacetic acid (EDTA) or ethyleneglycol bis(β-aminoethyl ether)N,N-tetra-acetic acid (EGTA) (1,2,3,4). In the presence of Ca^{2+} ions at physiological concentration, i.e. at 1-2 mM, degradation is less extensive than in the absence of Ca^{2+}. This is due to the specific binding of Ca^{2+} ions to fibrin(ogen) (5), causing a decreased susceptibility of the C-terminus of the γ-chain to plasmin degradation (2). On the other hand, in the presence of EDTA or EGTA the susceptibility of fibrin(ogen) to plasmin degradation is increased. There are strong indications that this effect should be ascribed to the observed specific binding of these ligands to fibrin(ogen) (3,4).

Two well-defined high molecular weight degradation products have been obtained from human fibrinogen after prolonged digestion by plasmin in vitro in the presence of Ca^{2+} at physiological concentrations. One of these homogeneous fragments is of the D-type and is called D-cate in order to emphasize that

FIGURE I

SCHEME OF PHYSICOCHEMICAL CHARACTERIZATION

Method	Quantity		Units
velocity sedimentation	→ sedimentation coefficient	$s^{o}_{T,s}$ *	S(vedberg)
	→ translational frictional coefficient	$f^{o}_{T,s}$ **	$erg.s.cm^{-2}$
equilibrium sedimentation			
	→ molecular weight	M	D(alton)
densitymetry	→ partial specific volume	$\bar{v}$	cm^3g^{-1}
viscometry	→ intrinsic viscosity	$[\eta]$	cm^3g^{-1}
UV- absorption spectroscopy	→ specific extinction	E_λ	cm^2g^{-1}

* $s^{o}_{T,s}$ denotes the sedimentation coefficient extrapolated to zero protein concentration in solvent s at temperature T.

** The translational diffusion coefficient can be calculated according to $D^{o}_{T,s} = \frac{kT}{f^{o}_{T,s}}$ (cm^2s^{-1}) where k is the Boltzmann constant.

it is a terminal product obtained in the presence of Ca^{2+}. This fragment contains the intact C-terminus of the fibrinogen chain. It has anticlotting properties(6). D-cate binds one Ca^{2+} ion per molecule (5) and this binding protects it from further degradation by plasmin (1,5). These properties emphasize the physiological significance of this region of the fibrinogen molecule. Besides, this part of the fibrinogen molecule is likely to be involved in the binding of EDTA(3,4).

The other fragment is of the E-type since it contains the central disulfide knot of the fibrinogen molecule. This so-called E-cate formed under the same conditions as D-cate has no anticlotting activity (6), does not bind Ca^{2+} ions(5) and, presumably, is not involved in the binding of the EDTA.

The aim of the present study was to characterize D-cate and E-cate by means of hydrodynamic methods and to investigate whether interactions with Ca^{2+} or EDTA give rise to conformational changes which can be detected from the hydrodynamic behaviour.

Materials and Methods

Pure D-cate and E-cate, free of Ca^{2+}, were obtained as described elsewhere(7). The methods used for the physico-chemical characterization and the quantities derived are summarized in fig. 1. Sendimentation experiments were carried out in a Beckman Model E analytical ultracentrifuge equiped with a temperature regulator and a UV-absorption scanning system, combined with mirror optics. Density measurements were performed with a Paar Digital Precision Density Meter (model DMA 02 D.), viscostiy measurements with an Ostwald viscometer and UV-absorption measurements with a double beam Beckman Model 25 spectrophotometer. All measurements were

TABLE 1

PHYSICOCHEMICAL CHARACTERISTICS OF THE PLASMIN DEGRADATION PRODUCTS E-CATE AND D-CATE FROM HUMAN FIBRINOGEN*

Quantity**	E-cate	D-cate
$s^o_{20,b}$	3.00 ± 0.06	4.73 ± 0.16
$f^o_{20,b} \times 10^7$ ***	0.65 ± 0.06	0.83 ± 0.07
M	41,700 ± 3,600	87,000 ± 5,700
$\bar{v}$	0.713 ± 0.008	0.722 ± 0.011
$[\eta]$	6.2 ± 0.8	6.3 ± 0.7
E_{280}	1020 ± 50	1900 ± 100

* measured in standard buffer i.e. 0.05 M Tris + 0.15 M NaCl, pH = 7.4 at 20°. Errors are standard errors of the mean.

** For units, see Fig. 1.

*** $D^o_{20,b} \times 10^7 = 6.2 \pm 0.6$ and 4.9 ± 0.4 $(cm^2\ s^{-1})$

done at 20.0°C.

Results and Discussion

The values of the various quantities determined for D-cate and E-cate have been compiled in Tabel 1. Comparison with literature data obtained with less homogeneous D- and E- fragments prepared at less defined Ca^{2+} concentrations shows minor differences in the values of the sedimentation coefficient and specific extinction(8,9). The molecular weight of D-cate is consistent with the value of 93,000 obtained from gel-electrophoresis(1,7) and falls within the range of 81,000-94,000 reported for D-fragments (8,9). The molecular weight of E-cate appears to be somewhat lower than some of the values of 45,000-62,000 reported by others for E-fragments(8,9,10). The values of the partial specific volume of D-cate and E-cate do not differ significantly from the values 0.71-0.72 reported for intact fibrinogen(11,12,13). This points to a similar relative aminoacid composition for D-cate, E-cate and fibrinogen.

From the data in Table 1, one can estimate the dimensions and the amount of hydration of a rigid, impenetrable model particle which exhibits the same hydrodynamic behaviour as the fragments. Two approaches are available each based on particular presuppositions with regard to the shape and effective hydrodynamic volume of the solvated particle(13,14,15,16). The method of Oncley uses either the sedimentation or the viscosity data(13,14). Calculations according to this method have been performed for the extreme cases: (1) the model particle is not hydrated and has miximal asymmetry and (2) the model particle is maximally hydrated and is spherically symmetric. It is generally assumed that the most probable shape and hydration of a molecule will lie between these extreme cases (13). Table 2 only shows the results for prolate ellipsoids,

TABLE 2

ESTIMATED SHAPE AND HYDRATION OF THE PLASMIN DEGRADATION PRODUCTS E-CATE AND D-CATE FROM HUMAN FIBRINOGEN IN AQUEOUS SOLUTION*

Method of calculation	Assumptions		Data used	E-cate			D-cate		
	δ	shape		δ	2a	2b	δ	2a	2b
Oncley (1941)	0	prolate ellips.	sedimenta-	0	19	2	0	24	3
	maximal	sphere	tion	1.5	7	7	1.6	9	9
	0	prolate ellips.	viscosity	0	17	2	0	22	3
	maximal	sphere		1.8	7	7	1.8	9	9
Scheraga-Mandelkern (1953)	-	-	sedimentation + viscosity	0.8	13	4	0.7	18	5

* In standard buffer, i.e. 0.05 M Tris + 0.15 M NaCl, pH = 7.4, at 20°.
δ denotes the hydration in gram H_2O/gram dry protein; 2a and 2b denote the major and minor axis of the ellipsoid respectively, expressed in nm.

but similar results were obtained for oblate ellipsoids. It is seen that the sedimentation and viscosity data yielded essentially the same limiting values. The second approach, according to Scheraga and Mandelkern, uses both the sedimentation and viscosity data (15,16). In this case it is assumed that the same hydrodynamic volume is effective in sedimentation and viscosity and that this quantity is composed of additive contributions of the particle and bound solvent. The dimensions and hydration calculated according to this approach (Table 2) appear to approximate very well the most probable values expected from the results obtained with the Oncley method. Because for D-cate and E-cate the value of the Scheraga-Mandelkern β-function is 2.19×10^6 and 2.18×10^6 respectively, an oblate ellipsoid can be ruled out in this case (15,16).

On the basis of the estimates in Table 2 we may assume that D-cate and E-cate in solution behave as rigid prolate ellipsoidal particles with a hydration of about 0.7 - 0.8 g H_2O/g protein and an axial ratio of about 3.5. Electron microscopic investigations on fibrinogen and D- and E-fragments reveal compact globular structures for the regions from which D-cate and E-cate are derived (17,18,19,20) and the dimensions estimated for anhydrous D- and E-fragments are compatible with the Scheraga-Mandelkern data in Table 2 (19).

In order to investigate whether binding of Ca^{2+} or EDTA influences the hydrodynamic behaviour, velocity sedimentation experiments have also been performed in the presence of 20 mM Ca^{2+} or 20 mM EDTA. Under these conditions the binding sites involved may be assumed to be saturated (5). The presence of these ligands did not have a significant effect as can be seen in Table 3. Furthermore it was found that the partial specific volume was not affected either by these additions. Consequently, it appears that binding of Ca^{2+} or

TABLE 3

SEDIMENTATION COEFFICIENTS* OF THE PLASMIN DEGRADATION PRODUCTS E-CATE AND D-CATE FROM HUMAN FIBRINOGEN

	E-cate	D-cate
buffer	3.00 ± 0.06	4.73 ± 0.16
buffer + 20 mM Ca^{2+}	2.95 ± 0.06	4.84 ± 0.06
buffer + 20 mM EDTA	2.85 ± 0.06	4.68 ± 0.08

* corrected to standard conditions in pure buffer, i.e. 0.05 M Tris + 0.15 M NaCl, pH = 7.4, 20° and extrapolated to zero protein concentration. For correction to standard conditions in water the values have to be multiplied by 1.055.

EDTA does not have an appreciable effect on the overall conformation of D-cate and E-cate under the conditions employed in this investigation, i.e. they behave like the rigid prolate ellipsoidal particles described above. Also for highly purified human fibrinogen binding of Ca^{2+} or EDTA was found to have no significant effect on the sedimentation rates (results not published). This is in agreement with the results of Endres who also found a neglegible effect of EDTA on the sedimentation of fibrinogen(21).

The size and shape of fibrinogen have been the objects of numerous investigations. There exists some controversy between the results obtained with different methods(11). In general hydrodynamic studies indicate an elongated rod-like structure with a length between 60-110 nm and a diameter of 4-5 nm (11,13,22), whereas other methods like low angle X-ray scattering (23), neutron small angle scattering (24) and electronmicroscopy (17,20,25,26) point at a more compact particle with a length of 40-50 nm and a diameter of 8-10 nm. Strictly speaking the data of Table 2 do not allow an unambiguous discrimination between these results as they more or less may fit either model. However, the estimates resulting from the Scheraga-Mandelkern calculation are consistent with a hydrodynamic rod-like model if we assume that the two fragments have retained the conformation they possess in the intact molecule and that the long axis of the fragments is approximately coincident with the long axis of complete fibrinogen. Each fibrinogen molecule (M=340,000) comprises two D-cate and one E-cate fragments. These fragments have a total molecular weight of about 220,000, which corresponds to about 65 percent of the total mass of fibrinogen. Under the above assumptions a maximal total length of about 49 nm would be obtained in case of a linear end-to-end arrangement of the three fragments. On basis of this reasoning a maximum length of about 76 nm would follow for a rod shaped fibrinogen molecule with an approximately uniform diameter of 4-5 nm. According to this estimation the present data seem to favour the results of hydrodynamic investigations of intact fibrinogen rather than the compact structure indicated by other methods. However, to what extent the dimensions of this hydrodynamic model correspond with the actual shape and hydration of the molecule in solution remains unclear as yet (11,16).

References

1. Haverkate, F., Timan, G.: Thromb. Res. 10, 803-812 (1977).
2. Haverkate, F., Nieuwenhuizen, W., van der Drift, A.C.M.: these proceedings.
3. Nieuwenhuizen, W., Vermond, A., Haverkate, F.: these proceedings.
4. Nieuwenhuizen, W., Vermond, A., Haverkate, F.: Biochim. Biophys.Acta: 667, 321-327 (1981).
5. Nieuwenhuizen, W., Vermond, A., Nooijen, W.J., Haverkate, F.: FEBS-Letters 98, 257-259 (1979).
6. Haverkate, F., Timan, G., Nieuwenhuizen, W.: Europ.J.Clin. Invest. 9, 253-255 (1979).
7. Van Ruijven-Vermeer, I.A.M., Nieuwenhuizen. W., Haverkate, F., Timan, G.: Hoppe-Seyler's Z. physiol.Chem. 360, 633-637 (1979).
8. Marder, V.J., Shulman, N.R.: J. Biol.Chem. 244, 2111-2119 (1969).
9. Marder, V.J., Budzynski, A.Z.: Progress in Hemostasis and Thrombosis, Vol. II, p. 141-174 (1974).
10. Slade, C.L., Pizzo, S.V., Taylor Jr., L.M., Steinman, H.M., McKee, P.A., J. Biol.Chem. 251, 1591-1596 (1976).
11. Doolittle, R.F.: Adv. in Prot.Chem. 27, 1-109 (1973).
12. Scheraga, H.A., Laskowski Jr., M.: Adv. in Prot.Chem. 12, 1-131 (1957).
13. Tanford, C.: Physical Chemistry of Macromolecules, John Wiley, New York (1967).

14. Oncley, J.L.: Ann. N.Y.Acad.Sci. 41, 121-150 (1941).

15. Scheraga, H.A., Mandelkern, L.: J.Am.Chem.Soc. 75, 179-184 (1953).

16. Yang, J.T.: Adv. in Prot.Chem. 16, 323-400 (1961).

17. Hall, C.E., Slayter, H.S.: J.Biophys.Biochem.Cytol. 5, 11-15 (1959).

18. Fowler, W.E., Erickson, H.P.: J.Mol.Biol. 134, 241-249 (1979).

19. Fowler, W.E., Fretto, L.J., Erickson, H.P., McKee, P.A.: J.Clin.Invest. 66, 50-56 (1980).

20. Weisel, J.W., Phillips Jr., G.N., Cohen, C.: Nature 289, 263-267 (1981).

21. Endres, G.E., Scheraga, H.A.: Arch.Bioch.Biophys. 144, 519-528 (1971).

22. Mueller, M., Burchard, W.: Biochim. et Biophys Acta 537, 208-225 (1978).

23. Lederer, K., Hammel, R.,: Makromol.Chemie 176, 2619-2639 (1975).

24. Marguerie, G., Stuhrmann, H.B.: J.Mol.Biol. 102, 143-156 (1976).

25. Bachman, L., Schmitt-Fumian, W.W., Hammel, R., Lederer, K.: Makromol.Chem. 176, 2603-2618 (1975).

26. Weisel, J.W., Warren, S.G., Cohen, C.: J.Mol.Biol. 126, 159-183 (1978).

THE KINETICS OF FIBRIN FORMATION

R. Eberhard Zimmermann

Physiologisches Institut der Universität Münster, Abt. Molekularbiologie der Gerinnung, Robert Koch-Str. 28, D-4400 Münster, FRG

Within the plasmatic compartment of the circulatory system, the molecular forms of fibrin (des-AA or des-AABB fibrin monomers) are solubilized through the formation of complexes with fibrinogen and/or other plasma proteins. The conversion of fibrinogen to fibrin is caused by the proteolytic removal of fibrinopeptides A and B from the N-terminal part of the fibrinogen molecule. This event induces the conformational exposition of a set of binding sites, one located in the amino-terminal disulfide knot and the other associated with the D-region of fibrinogen (1). Hydrogen bonds as well as electrostatic bonds are thought to be involved in the formation of soluble and insoluble fibrin.

The rate of blood flow has a strong influence on the structure and the characteristics of thrombus formation. The embolic material is formed in veins. Shock and acidosis potentiate the intravascular coagulation, leading to the precipitation of fibrin, whereby α -receptors of the vascular wall play a key role for fibrin elimination. Since the circulatory system is functionally interconnected with the lymphatic vessels, the exchange of fibrinogen, fibrin monomers, and split products with the extravascular space is permanent and includes possible extravascular/intravascular transport mechanisms (2). The endothelial lining of the vessel wall contributes to normal blood flow by prohibiting platelet activation through its inert surface and by the synthesis of certain prostaglandins, plasminogen activator and factor VIII. In the course of blood coagulation, thrombin (EC 3.4.21.5) generated by cleavage of prothrombin stimulates not only the conver-

Fibrinogen - Recent Biochemical and Medical Aspects

sion of fibrinogen to fibrin but also aggregates platelets. In this way, fibrinogen receptors are exposed (3, 4). The structural parts of fibrinogen which may interact with platelets in the presence of ADP, collagen, and thrombin are likely to be found in the N-terminal region.

In contrast to the physiological conditions prevailing during hemostasis, in vitro fibrin formation occurs in the absence of a number of important parameters such as blood flow, contact with the vessel wall, and participation of cellular particles. Therefore, measurements of coagulation time by the methods for prothrombin time, activated partial thromboplastin time, and thrombin time are unreal and include several standardized errors inherent in the method chosen and the measuring devices.

The classical view of the molecular events occurring during fibrin formation has been quantitated in a recent study using light scattering measuring devices (5). After the removal of fibrinopeptide A, activated fibrin monomers polymerize to form protofibrils, which in turn consist of monomers aligned end-to-end in two parallel strands (fibrin I). Fibrinopeptide B can be removed from the growing protofibrils causing an increase in the rate of lateral association which normally depends upon the length of the fibrin I polymer (fibrin II). Blood flow and the presence of plasmin may affect the amount of fibrin II formed (6). The resultant three-dimensional gel network triggers mechanical or optical measuring devices and serves in this way for the determination of the clotting or coagulation time.

In the early stages of this conversion, soluble fibrin polymers are formed containing des-AA fibrin and fibrinogen. Since the optical density of the solution is not drastically changed, this process

cannot be recorded by commonly available clotting timers. Furthermore, the sedimentation coefficient is also insensitive to the length of rod-like polymers, so that oligomers cannot be distinguished until lateral aggregation of protofibrils occurs. (Fig. 1)
Oligomers are only demonstrable by polyacrylamide gel electrophoresis without disulfide reduction. Their presence explains the concurrent viscosity changes of the fibrinogen solution during this time. As soon as the protofibrils associate laterally to form fibers, the fibers are joined together in an irregular fashion, forming gels or clots, which have the properties of solids and can easily be detected by mechan-

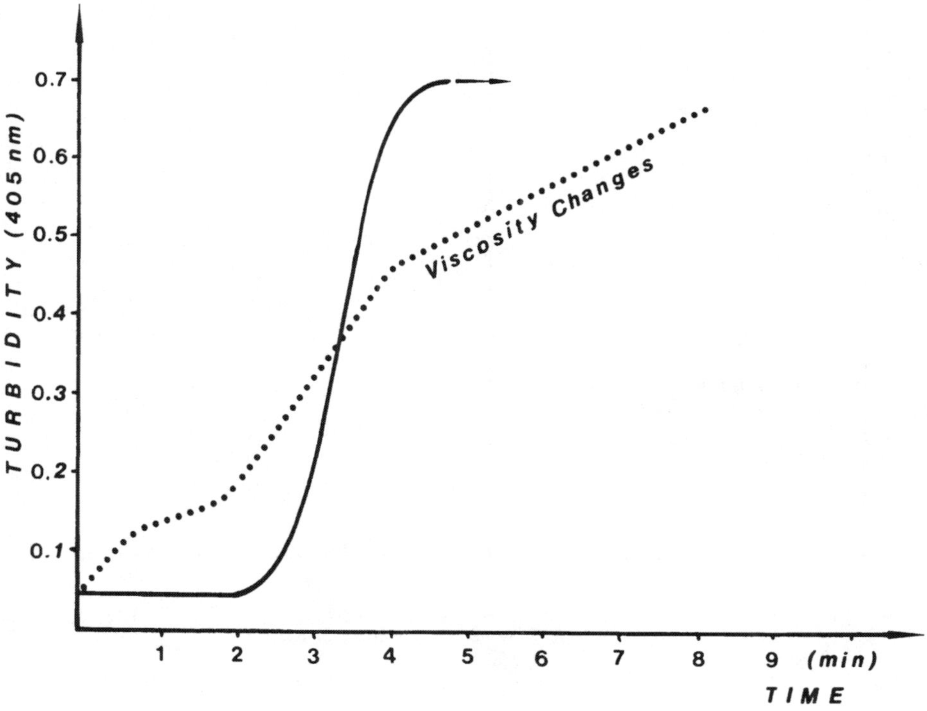

Fig. 1: Registration of turbidity and viscosity changes during fibrinogen/fibrin conversion.
(3 mg human fibrinogen/ml, 0.1 M Tris-HCl buffer, pH 7.4)

ical or optical measuring devices. At the beginning of clotting, only 3.5 to 4 % of the plasmatic fibrinogen is converted to fibrin monomers depending upon the ionic strength and the pH of the solution. At high pH (8.5 - 9) and high ionic strength (0.3-0.45) the formed clots are highly transparent and have very little side-to-side aggregation. Under these circumstances, optical measuring devices fail to indicate the gelation of the fibrinogen/fibrin solution. With the increase of the albumin/globulin concentration in the sample, the

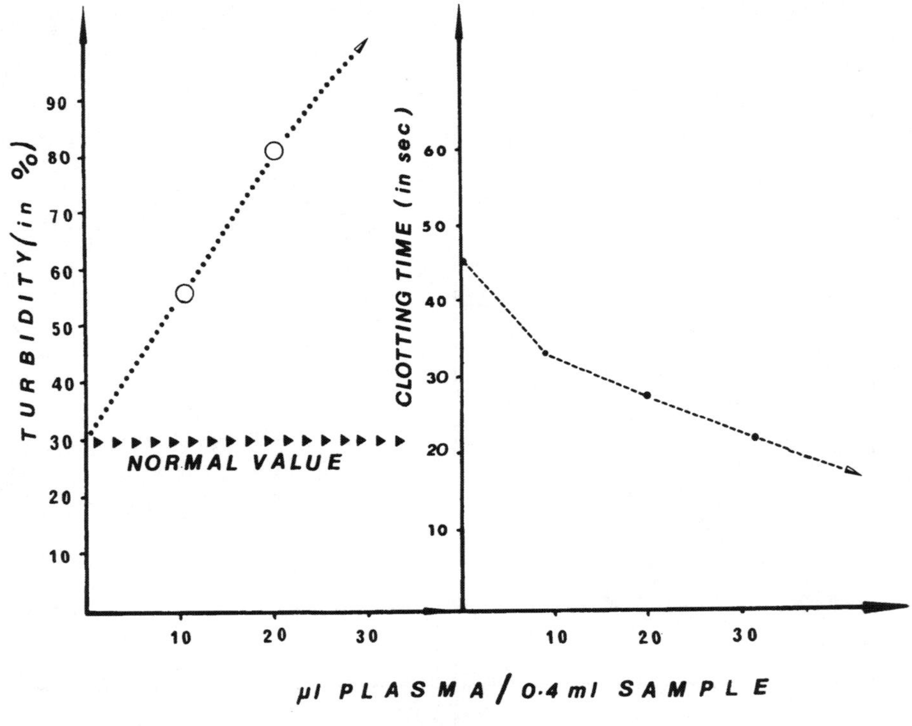

Fig. 2: Influence of plasma proteins upon the turbidity of a clotting fibrinogen solution.
(3 mg human fibrinogen/ml, 0.1 M Tris-HCl buffer, pH 7.4)

turbidity of the fibrinogen/fibrin solution drastically increases and fibrin formation can now be measured. (Fig. 2)
"Fine" clots have a low light scattering intensity, and for this reason, they easily escape routine detection. Additives like glycerin (to stabilize the thrombin solution) prevent the lateral aggregation, whereas heparin or dextrose produce coarser fibrin fibers with larger cross-sectional areas. Under standard conditions of pH and ionic strength, clotting time depends directly upon the amount of enzyme available

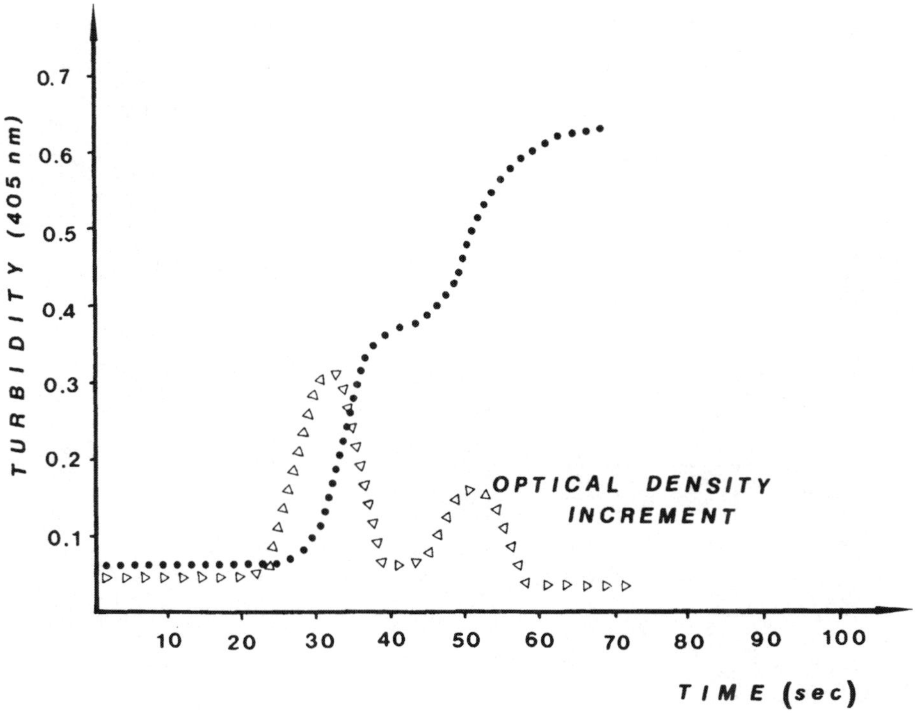

Fig. 3: Biphasic clotting mechanism of a plasma sample during PTT registration. Comparison of turbidity changes with the optical density increment.

for converting the substrate fibrinogen to fibrin. As soon as the fibrin monomer concentration exceeds 5% of the original fibrinogen concentration the lag period ends. The lateral association of protofibrils amplifies the turbidity many times so that the clotting time can easily be measured. The permanent recording of the optical changes during the process of fibrin formation demonstrates different rates for polymerization and subsequent fiber growth. (Fig. 3)

By increasing the amplitude of the optical density increment, the biphasic nature of fibrin network formation can be demonstrated. This behaviour may be attributed to a structural readjustment during gelation, indicating the differences between fiber growth and fibrin network formation. The same structural rearrangements of fibrin gels are causing continuing changes of the viscosity for some time after the optical density has reached its final level.

References

1. Blombäck, B., Hessel, B. , Hogg, D., Therkildsen: Nature (Lond.) 275, 501-505 (1978).

2. Rüegg, R., Straub, P. W.: J. Lab. Clin. Med. 95, 842-856 (1980).

3. Niewiarowski, St., Budzynski, A. Z., Norinelli, T. A., Brudzynski, T. M., Stewart, C. J.: J. Biol. Chem. 256, 917-925 (1981).

4. Marguerie, G. A., Plow, E. F.: Biochemistry 20, 1074-1080 (1981).

5. Hantgan, R. R., Hermans, J.: J. Biol. Chem. 254, 11272-11281 (1979).

6. Nelb, G. W., Kamykowski, G. W., Ferry, J. D.: J. Biol. Chem. 255, 6398-6402 (1980).

II. BIOCHEMICAL ASPECTS

STRUCTURE - FUNCTION - EVOLUTION RELATIONSHIP IN FIBRINOGEN

Agnes Henschen, Friedrich Lottspeich, Maria Kehl, Chris Southan and John Lucas

Max-Planck-Institut für Biochemie
D-8033 Martinsried bei München, Germany

Introduction

The fibrinogen molecule is composed of three pairs of non-identical peptide chains, Aα, Bβ and γ, and the structure can therefore be described as $(A\alpha, B\beta, \gamma)_2$. Thrombin cleaves two of the peptide chain pairs within the N-terminal parts of the chains (see Figures 1 and 2), whereby the fibrinopeptides A and B are released and fibrin is formed. Accordingly, the structure of fibrin is $(\alpha, \beta, \gamma)_2$. The complete amino acid sequences of all three human peptide chains, containing 610, 461 and 411 residues respectively, have recently been elucidated (1, 2, 3, 4). The connections formed by the 29 sulfur bridges have been established (1, 5). Partial sequences, mainly very short sections, from many animal fibrinogens have been published. Thus, fibrinopeptide sequences from over 40 mammals are known (6). The only complete animal peptide chain sequence is that of the bovine Bβ-chain (7).

The amino acid sequence information can be used to study the relationships between structure, function and evolution. Thus, the likelihood that all three peptide chains in fibrinogen have a common evolutionary ancestor may be tested. Furthermore, the evolutionary stability of different regions within the fibrinogen molecule may be evaluated as an indicator of their importance for the function. Finally, the possible relatedness

Fibrinogen - Recent Biochemical and Medical Aspects

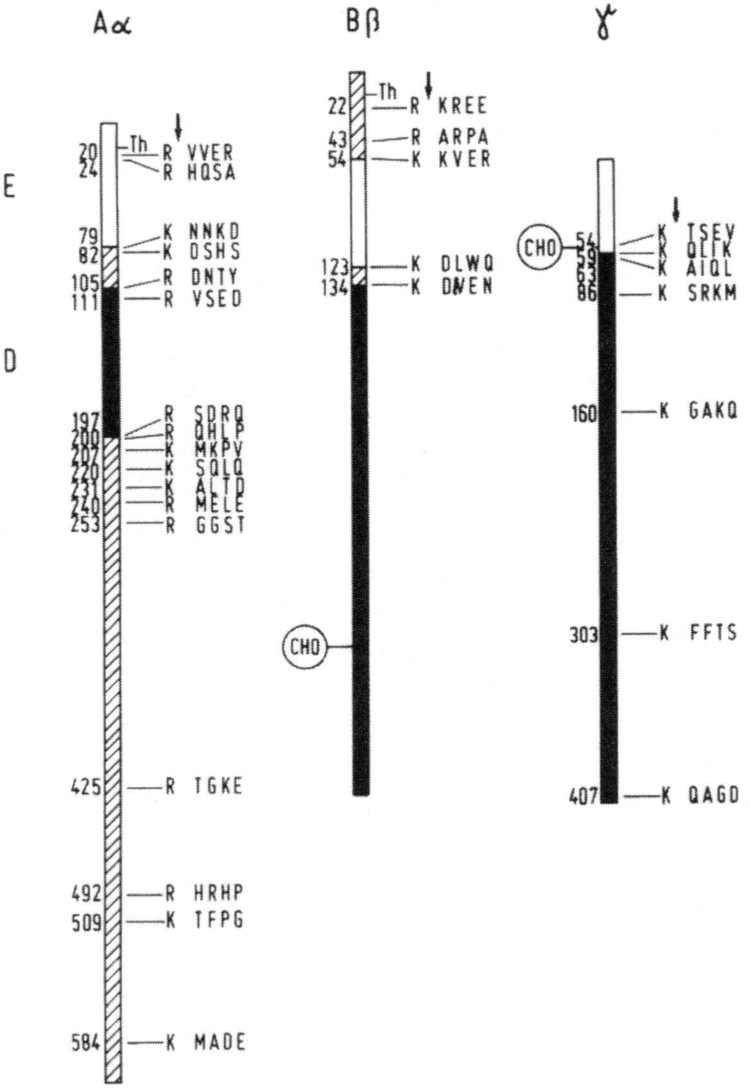

Figure 1. Plasmin cleavage sites in the three chains of human fibrinogen. The chains have been aligned according to homology, the N-termini being at the top. The arrows show the cleavages in the indicated sequences, the numbers give the positions of the new N-terminal residues. Plasmic fragment E is marked white and fragment D black. Th denotes a thrombin cleavage site and CHO a carbohydrate side chain.

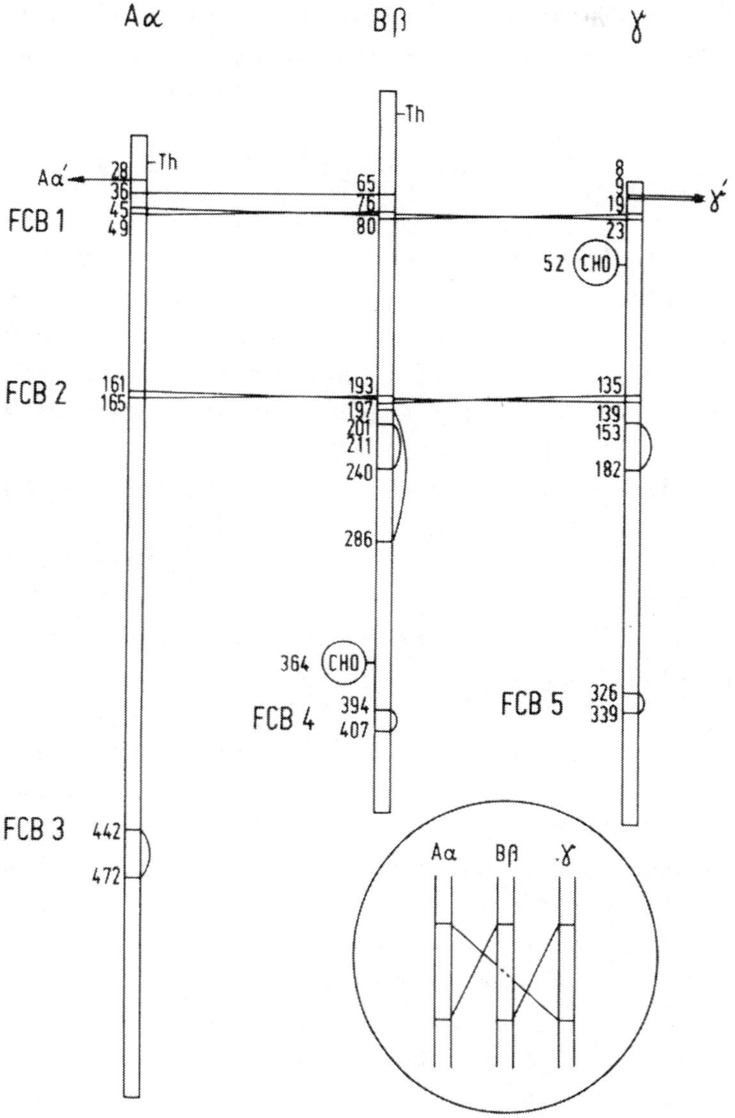

Figure 2. Disulfide bridges in human fibrinogen. The chains have been aligned according to homology, the N-termini being at the top. The numbers give the positions of the half-cystine residues. Aα' and γ' indicate the corresponding chain in the other half of the molecule. FCB1-FCB5 denote the disulfide knots obtained on cyanogen bromide cleavage, Th a thrombin cleavage site, and CHO a carbohydrate side chain. Within the circle the triple-connection is shown in an enlarged form.

of fibrinogen to other proteins may be examined. Several searches for these kinds of relationships are presented below.

Comparisons between the Three Human Peptide Chains

One may speculate on the evolutionary origin of the three different peptide chains in fibrinogen. There are, in fact, many good reasons to believe that there once existed a primitive fibrinogen with only one type of peptide chain and that the present three chains all are descendants of this chain. Thus, the highly specific proteolytic enzyme thrombin cleaves very selectively two pairs of chains, Aα and Bβ (Figures land 2). Furthermore, the somewhat less specific enzyme plasmin shows a clear tendancy to cleave all three chains on analogous levels, producing mainly similar size fragments, as demonstrated in Figure 1, where the chains have been aligned according to homology (see below).

The most obvious similarity between all three chains lies, however, in the distribution of the half-cystine residues, the arrangement of the disulfide connections between them and the amino acid sequences around them. Figure 2 shows that each chain contains one N-terminal, one intermediate and one C-terminal cluster of half-cystine residues. The two first mentioned clusters have been used to align the chains, as most of the half-cystines in each chain can be brought into perfect register with those of the other chains. The high degree of symmetry in the connections between the half-cystine residues is striking, especially in the N-terminal and intermediate triple-connections and in the intermediate and C-terminal disulfide loops. The amino acid sequences within the triple-connections and the C-terminal disulfide loops are given in Figure 3 and show very clear similarities.

Aα	45	C P S G C	161	C R G S C
Bβ	76	C P T G C	193	C R T P C
γ	19	C P T T C	135	C Q E P C

Aα	442	C S K T V T K T V I G P D G H K E V T K E V V T S E D G S D C
Bβ	394	C S K E D G G G W W Y N R C
γ	326	C A E Q D G S G W W M N K C

Figure 3. Amino acid sequences within the triple-disulfide-connections and the C-terminal disulfide loops of human fibrinogen. The numbers indicate the positions of the most N-terminal residues.

When the complete sequences of all three chains are aligned, a remarkable homology between the Bβ- and γ-chain can be observed, i.e. a large number of positions in these two chains have the identical amino acid residue (8). However, the homology is only present when the chains are shifted with respect to each other as indicated in Figures 1 and 2, i.e. when the half-cystine residues are in register. The Bβ- and γ-chain are sufficiently similar to make the construction of a Bβ-γ-"average"-chain, as shown in Figure 4, meaningful. In order to obtain the best fit, some deletions and insertions have to be allowed for. Identical residues in corresponding positions are found in 33% of the positions, when the long N-terminal extension of the Bβ-chain and the short C-terminal extension of the γ-chain are disregarded. However, the identical position are unevenly distributed, the C-terminal two thirds of the chains being much more similar. A detailed comparison between the chains can be obtained by calculating the number of identical residues in a 20-residue segment, which is shifted along the chain sequences. The lower heavy line in Figure 5 shows the result. High degrees of identity are found around half-cystine residues. However, for the presence of most of the highly identical regions no explanation is yet available.

```
  1  - - - - - - - - - - - - - - - - - - - - - - - - - - - - - -
 31  - - - - - - - - - - - - - - - - - - - - - - - - - - - + + +
 61  + + + + C + + + D + + + G + + C P T + C + + + + + L + + + +
 91  + + + + + + + + + L + + + + + + V + + + + - S + + + + + + +
121  + + + L + + + + + + + + K + N + + + + + + + + S + + + + + +
151  + + + + E + + + + + + + + + + R + L + + I + + + + + + K I +
181  + L + + + V + + + + + + C + + P C + + + + + I + + + + G K +
211  C + + I + + K G + + + S + + Y + I + P + + + + + + + + V Y C
241  + + + + + + + G W T V + Q + R + D G S V D F + + + W + + Y K
271  + G F G + + + + + - - G + + - - - - - - E + W L G N + K I +
301  + + - T - + + + P + + L + + E + E D W + G + + + + A + Y + +
331  F + V + + E A + K Y + + + + + + + - + G + A G + A + + + + +
361  + + + + + + + + + + T + H N G M + F S T + D + D N D - - - +
391  + + - + + - C + + + D G + G W W + N + C H A + + + N G + Y Y
421  + G G + Y + + + + + + + G + D + G + + W + + W K + + W Y S M
451  + K + + M K I + P F + + + + - - - - - - - - - - - - - - - -
481  - -
```

Figure 4. The Bβ-γ-"average" sequence in human fibrinogen. - Denotes insertion/deletion, + nonidentical residue.

The alignment of Bβ- and γ-chain might offer an explanation for the absence of a fibrinopeptide C, the γ-chain being "too short" at the N-terminus. In a similar way, the Bβ-chain may be regarded as being "too short" at the C-terminus to include a cross-linking site corresponding to the one in the γ-chain.

When the Aα-chain is aligned with either Bβ- or γ-chain only a low degree of homology can be detected. In fact, the homology is often too weak to allow a unique correct alignment. Only 12% of the positions show identity, even when regions with differences in chain length are disregarded. The Aα-chain shows better agreement with the other chains in the N-terminal half

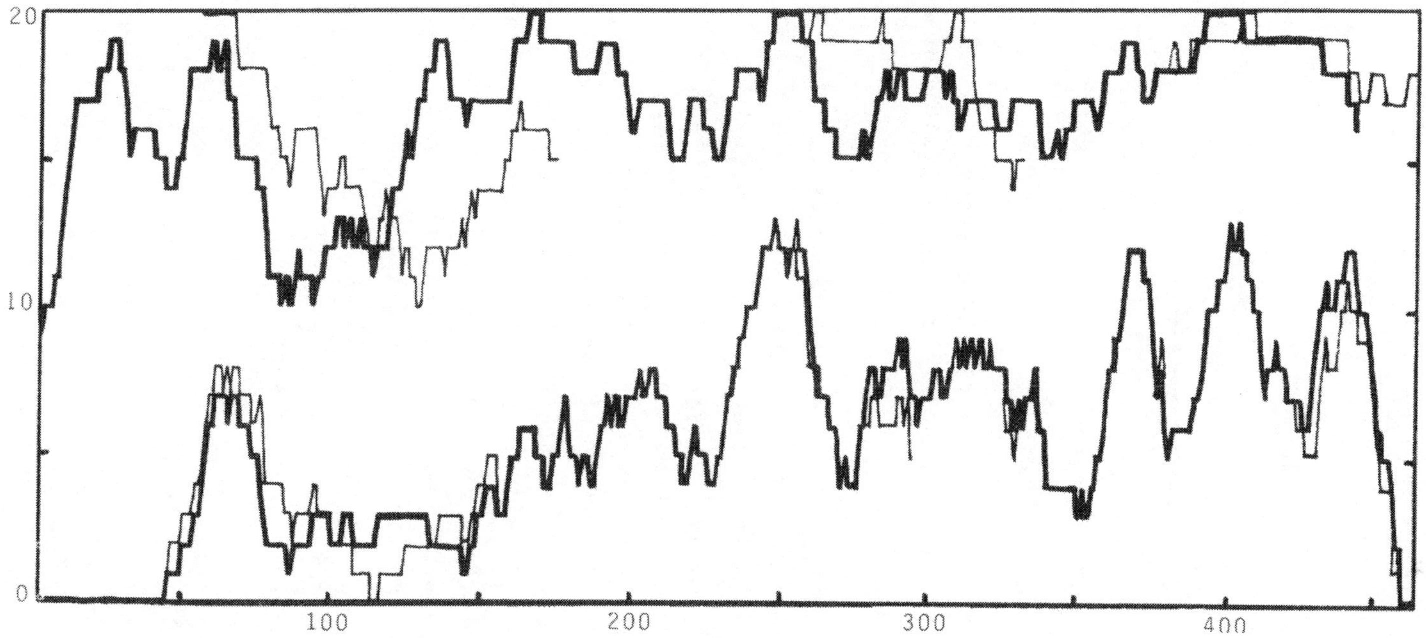

Figure 5. Number of identical residues in 20-residue segments of aligned human and bovine Bβ- and γ-chains. The lower heavy line is human Bβ- against γ-chain, the lower light line bovine Bβ- against γ-chain, the upper heavy line human against bovine Bβ-chain, and the upper light line human against bovine γ-chain.

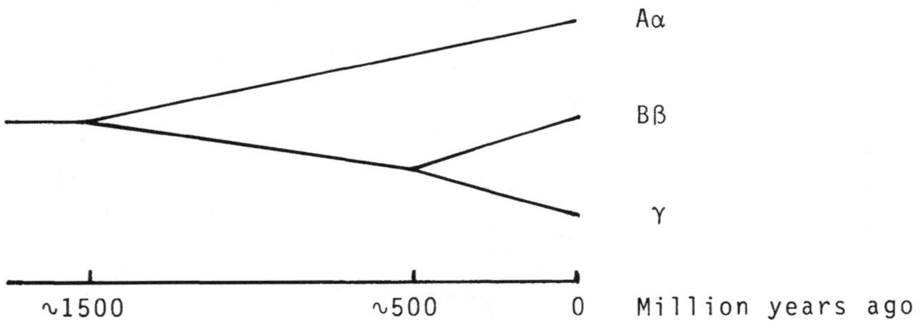

Figure 6. Evolution of the three fibrinogen chains.

of the sequence. Here, at least, the degree of homology is higher than the one expected from a comparison between randomised sequences.

The general interpretation of the findings is that all three chains have shared an evolutionary ancestor, but the γ- and Bβ- chains have a more recent common ancestor than all three chains together (8). This hypothesis is illustrated by Figure 6. The γ- and Bβ-chain must have diverged during evolution before the vertebrates appeared, as even primitive vertebrates seem to have three different peptide chains. This would place the time for the divergence to at least 500 million years ago. As the Aα-chain is much less similar to the other chains, the ancestor of the Aα-chain might have separated from the common Bβ-chain ancestor 1 500 million years ago, which then would be at about the same time as thrombin and trypsin separated from each other (6). This would mean that the highly specific enzyme thrombin and its substrate, the primitive Aα-chain, could have evolved together in a process of continuous mutual adaptation.

The original fibrinogen might have been a hexamer of identical ancestral Aα-chains. As a result of two gene duplications and many mutations four of the chains would have been replaced, γ 148 I T G K D C Q D I A N K G A K Q S G L Y

γ 392 L T I G E G Q Q H H L G G A K Q A G D V

Figure 7. Possible internal sequence repeat in the human γ-chain (9). The crosslinking sites at the C-terminus are marked by dots. The numbers indicate the positions of the most N-terminal residues.

first by the common Bβ-γ-chain ancestor and then by two Bβ- and two γ-chains.

Internal Repeats in the Human Peptide Chains

Comparing different sections within each of the human peptide chains a number of internal repeats may be observed. These sequence repeats can be interpreted as being the still noticable remnants of gene duplications of the type which leads to chain elongation. In all three chains the characteristic amino acid sequence Cys-X-X-X-Cys is found twice, once in the N-terminal part and once in the middle part of the chain, as shown in the upper half of Figure 3. As the degree of similarity is higher within each group than within each chain it seems likely that the gene duplication leading to chain elongation occured before the one leading to three different chains.

Additional internal sequence repeats are found in the γ- and in the Aα-chain. In the γ-chain a sequence, similar to the one at the C-terminus and containing the crosslinking residues glutamine and lysine, is also present in the middle of the chain (9), as indicated in Figure 7. The sequence from the middle part contains the residues glutamine and lysine in the corresponding positions. A very obvious repetitive sequence, shown in Figure 8, occupies the middle part of the Aα-chain (2).

Keratin	81	G G S I G Y G Q V G S S G
Aα	253	G G S T S Y G - T G S E T
	265	E S P R N P S S A G S W N
	278	S G G S G P G G T G N R N
	291	P G S S G T G G T A T W K
	304	P G S S G P G S T G S W N
	317	S G S S G T G S T G N Q N
	330	P G S P R P G S T G T W N
	343	P G S S E R G S A G H W T
	356	S E S S V S G S T G Q W H
Aα "average"		(P)G S S(G P)G S T G + W N

Figure 8. Internal sequence repeats in the human Aα-chain and homology with a keratin sequence (10). Only residues identical to the keratin sequence are underlined. The numbers indicate the positions of the most N-terminal residues.

Aα	266	S P R N P S S A G
Aα	397	V S G N V S P G T
		+ (over N)
Bβ	361	M G E N R T M T I
		+ (over N)
γ	49	Q V E N K T S E V

Figure 9. Sequence coding for N-glycosidic linkage in human fibrinogen. The + over N marks a carbohydrate side chain. The numbers indicate the positions of the most N-terminal residues.

A 13-residue segment is repeated nine times and over 60% of the positions show identity with an "average" sequence.

Carbohydrate Attachment Sites in the Human Peptide Chains

The characteristic sequence, Asn-X-Ser/Thr, coding for N-glycosidic linkage, occurs four times in the human fibrinogen structure (Figure 9). However, only two of these sequences do indeed carry carbohydrate side chains, i.e. those in the Bβ- and γ-chain (11,12). In the Aα-chain sequences the neighbouring proline residues might interfere with the carbohydrate attachment.

Comparisons between Peptide Chains from Different Species

Information about amino acid sequences in fibrinogens from other species than man has been very limited, with the exception of those in fibrinopeptides (6). The extreme variability in the fibrinopeptide sequences is illustrated by Figure 10, the fibrinopeptides being to the left of the arrows indicating thrombin cleavage. In almost every position many variants seem to be compatible with a functioning fibrinogen molecule. In contrast, only limited variability is found on the fibrin side (Figure 10, to the right of the arrows) of the thrombin cleavage sites in the Aα- and Bβ-chain and even lower variability in the N-terminus of the γ-chain (Figure 10, bottom), the low variability obviously pointing to high functional importance. It will in the future be of great interest to compare the "allowed" variability in the mammalian fibrinogen sequences with the "not allowed" variability in the genetically abnormal human fibrinogen sequences (cf. other communication in this volume).

Recently, a considerable part of the bovine fibrinogen sequence has become available. Thus, 85% of the γ-chain and, partly deduced from CDNA analysis, the total Bβ-chain sequences are now known (7,13,14,15). It is therefore now possible to make a de-

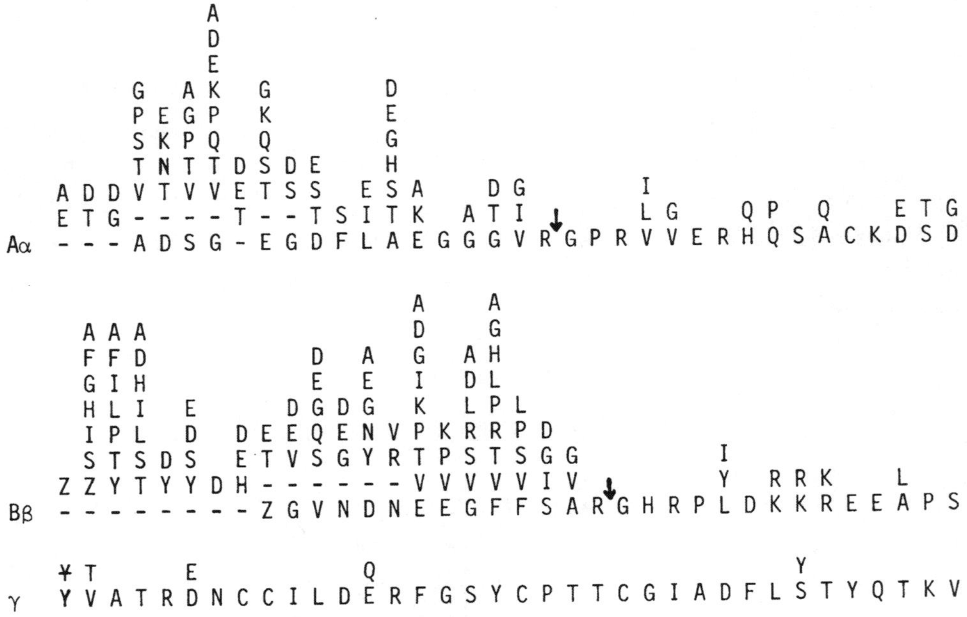

Figure 10. N-Terminal amino acid sequence variations in mammalian fibrinogens (cf. 6). The bottom line for each chain represents the human sequence, the arrows denote the thrombin cleavage sites, - denote deletion (i.e. shorter sequence), ¥ denotes a blocked Tyr. Fibrinopeptide positions represent about 40 species, fibrin positions only 5-9 species.

tailed comparison between these two chains and the corresponding human chains. The result is shown in Figure 5. It is striking that a very similar pattern is obtained when bovine Bβ- and γ-chain are aligned (Figure 5, lower light and heavy lines). This may be taken to indicate that the patterns illustrate significant properties of the sequences, even though the meaning of the patterns still is unkown.

Aligning Bβ- or γ-chain from the two species results in the two additional patterns depicted in Figure 5 (upper heavy and

```
κ   11  C E K D E R F F D D K I A K Y I P I Q Y V L S R Y
γ    9  C I L D E R F G S Y C P T T C - G I A D F L S T Y

κ   34  R Y P S Y G L N Y Y Q Q K P V A L
γ  108  R Y L Q E I Y N S N N Q K I V N L

κ  136  I E A V E S T V A T L E A S P E V I E S P P E I N T V Q
γ  121  I V N L K E K V A Q L E A - Q C Q - - - E P G Q D T V Q

κ   86  K S C Q A Q P T T M A R H P H P H L S F M A I P P K K N Q
γ  151  K D C Q D I A N K G A K Q S - - G L Y F I - K P L K A N Q

κ  123  N T I A S G E P T - - - S - T P T T E
γ  207  N W I Q Y K E G F G H L S P T G T T E
```

Figure 11. Homology between sequences in bovine κ-casein and human fibrinogen γ-chain (16). - denotes deletion. The numbers indicate the positions of the most N-terminal residues.

light lines). It may be noticed that most maxima and minima of these curves coincide with those of the Bβ-γ-alignment curves (15). This would mean that those sections of the chains which were important for the function when Bβ- and γ-chain diverged during evolution still are important and therefore conserved. Occasionally, a section within one chain has stayed more constant than the corresponding section in the other chain, which then would indicate a specialised function of the conserved structure.

Comparison between Fibrinogen and Some Other Proteins

The sequences in fibrinogen have been compared with those in other proteins to find signs of evolutionary or function-de-

```
                                    x   y   z    -y  -x    -z

Calm.           + E X + + X X + + X D + O + O G - + I O + + E X + + X X + + X
        ↓
γ 301 D K F F T S H N G M Q F S T W D N D N D K E E G N C A E Q D G S G W W M
      +
Bβ364 N R T M T I H N G M F F S T Y D R D N D G - W L T S - D P R K Q C S K E
```

Figure 12. Homology between calcium-binding sequence in calmodulin (17) and suggested calcium-binding sequences in human γ- and Bβ-chain. Calm. denotes calmodulin, the arrow the calcium-protected plasmin cleavage site, x — -z denote calcium ligands, X hydrophobic residues, O oxygen-containing ligand residues, - deletion, + arbitrary residue, + over N carbohydrate side chain. Underlined are residues in agreement with calmodulin "test sequence" (17). The numbers indicate the positions of the most N-terminal residues.

pendent relationships. Only in few cases have sequence similarities been found. The possible structural relatedness of κ-casein and the γ-chain is illustrated by Figure 11. A total of 80% of the κ-casein sequence has counterparts in the γ-chain sequence, and more than 35% of the corresponding positions are occupied by identical residues (16). The repetitive Aα-chain sequence, shown in Figure 8, is to 33% homologous with a keratin sequence (10). However, the significance of these puzzling observations is not yet clear.

Finally, the structural similarity between calcium-binding sequences, e.g. in calmodulin (17), and sequences in the C-terminal parts of the γ-chain and Bβ-chain is demonstrated in Figure 12. Possible ligands for calcium-binding seem to be present in both fibrinogen chains. It has been known for some time that fibrinogen contains calcium binding sites (18), but their exact positions were so far unknown. An important fact is that the γ-chain Lys-Phe bond shown is protected against plasmin cleavage by calcium ions. The corresponding Bβ-chain bond might independently be protected by the carbohydrate side chain attached to the neighbour residue.

It may be concluded that the fibrinogen sequence, apart from its interest per se, also contains a wealth of information, partly puzzling, partly clear.

References

1. Henschen, A.: Hämostaseologie 1, 30-40 (1981).
2. Henschen, A., Lottspeich, F., Hessel, B.: Hoppe-Seyler's Z. Physiol. Chem. 360, 1951-1956 (1979).
3. Henschen, A., Lottspeich, F.: Hoppe-Seyler's Z. Physiol. Chem. 358, 1643-1646 (1977).
4. Lottspeich, F., Henschen, A.: Hoppe-Seyler's Z.Physiol. Chem. 358, 935-938 (1977).
5. Henschen, A.: Hoppe-Seyler's Z. Physiol. Chem. 359, 1757-1770 (1978).
6. Dayhoff, M. O.: Atlas of protein sequence and structure. National Biomedical Research Foundation, Washington (1972).
7. Chung, D., Rixon, M. W., MacGillivray, R. T. A., Davie, E. W.: Proc. Natl. Acad. Sci. U.S.A. In press (1981).
8. Henschen, A., Lottspeich. F.: Thromb. Res. 11, 869-880 (1977).
9. Lottspeich, F., Henschen, A.: Hoppe-Seyler's Z. Physiol. Chem. 358, 703-707 (1977).
10. Jollès, P., Jollès, J., Henschen, A.: J. Mol. Evol. 17, 188-189 (1981).
11. Töpfer-Petersen, E., Lottspeich, F., Henschen, A.: Hoppe-Seyler's Z. Physiol. Chem. 357, 1509-1513 (1976).
12. Blombäck, B., Gröndahl, N. J., Hessel, B., Iwanaga, S., Wallén, P.: J. Biol. Chem. 248, 5806-5820 (1973).
13. Töpfer-Petersen, E., Henschen, A.: Throm. Res. 11,881-892 (1977).
14. Timpl, R., Fietzek, P. P., van Delden, V., Landrath, G.: Hoppe-Seyler's Z. Physiol. Chem. 359, 1553-1560 (1978).
15. Henschen, A., Lottspeich, F., Töpfer-Petersen, E., Kehl, M., Timpl, R.: In: Protides of biological fluids (Peeters, H., ed.) 28, 47-50 (1980).

16. Jollès, P., Loucheux-Lefebvre, M. H., Henschen, A.: J. Mol. Evol. 11, 271-277 (1978).

17. Tufty, R. M., Kretsinger, R. H.: Science, 187, 167-169 (1975).

18. Haverkate, F., Timan, G.: Throm. Res. 10, 803-812 (1977).

ARTIFACTUAL HETEROGENEITY OF HUMAN FIBRINOGEN POLYPEPTIDE CHAINS CAUSED BY CARBOXYMETHYLATION

Cemal Kuyas, André Haeberli, and P. Werner Straub
Laboratory for Thrombosis Research, Department of Medicine, Inselspital, CH-3010 Bern, Switzerland.

Summary

The heterogeneity of the polypeptide chains of human fibrinogen obtained after full reduction and alkylation was compared to the one of asialofibrinogen by isoelectric focusing.
A surprisingly great number of the Bβ- and γ-chain variants was observed, and even more striking no decrease of the number of Bβ- and γ-chain variants was detectable in asialofibrinogen, although the two chain types contain heterogeneous subfractions with respect to sialic acid each.
However, when the alkylation step was ommitted the isoelectric focusing pattern revealed 1) a decreased number of Bβ- and γ-bands, namely three main bands for each, and 2) a reduction of the number to two main distinct bands for each type in asialofibrinogen.
It is concluded that the alkylation of reduced fibrinogen gives rise to artifactual charge heterogeneities and therefore hides the decrease of the polypeptide chain heterogeneity of human asialofibrinogen as compared to native fibrinogen.

Fibrinogen - Recent Biochemical and Medical Aspects

Introduction

Human fibrinogen consists of three pairs of polypeptide chains, namely Aα, Bβ, and γ. It is a glycoprotein with a sialic acid content of 6 residues per mole fibrinogen (1, 2). The sugar moieties are located in the Bβ- and γ-chains, whereas the Aα-chains do not contain any carbohydrate (3). A heterogeneity associated with different sialic acid contents of the variants was reported by Gati and Straub (4), who were able to separate both the Bβ- and the γ-chains of SCm-fibrinogen into two main populations, one containing close to two, the other close to one mole of sialic acid per mole of chain.
During the investigation of the contribution of these differences in sialic acid content to the heterogeneity of the Bβ- and γ-polypeptide chains, we became aware that the alkylation of fully reduced fibrinogen was disturbing the analysis of the charge heterogeneities of the polypeptide chains studied by isoelectric focusing. This report presents results on artifactual heterogeneities produced by carboxymethylation.

Material and Methods

Material

Human fibrinogen was obtained from AB Kabi, Stockholm (Grade L) or was isolated by fractionated ethanol precipitation (5). The clottability was 95%. All reagents used were of analytical grade, and all urea solutions were freshly deionized with Amberlite MB-3 (Serva, Heidelberg FRG) prior to use. The urea solutions were also checked for the presence of cyanate ions (6). Neuraminidase from clostridium perfringens was obtained from Worthington, Freehold, N.J., USA, neuraminidase from vibrio cholerae from Behring-Werke, Marbürg FRG.

Preparation of asialofibrinogen: 100 mg human fibrinogen, dissolved in 9.5 ml 0.3 M NaCl adjusted to pH 6.5 was incubated with 0.5 U neuraminidase from vibrio cholerae, or clostridium perfringens for 24 h at 37°C. When neuraminidase from vibrio cholerae was used, 5 mM $CaCl_2$ was added. The incubation mixture was dialized overnight at 4°C against 0.15 M Tris-HCl buffer, pH 8.5, containing 8 M urea. The clottability of the asialofibrinogen was 90% or greater. Polyacrylamide gel electrophoresis of asialofibrinogen in presence of sodium dodecyl sulfate confirmed the integrity of the polypeptide chains.

Reduced and alkylated fibrinogen: Reduction and alkylation was carried out according to Murano et al. (7). Human fibrinogen or asialofibrinogen (100 mg) was dissolved in 10 ml 0.15 M Tris-HCl buffer, pH 8.5, 8 M urea, and was reduced with 90 mg dithiothreitol (30 x molar excess) (Calbiochem) for 15 min. at room temperature. Alkylation was carried out immediately thereafter with 300 mg iodoacetic acid (90 x molar excess) for 20 min. in a light-protected vessel. Excess reactants were removed by gel filtration on a Sephadex G-25 column (2.5 x 30 cm) equilibrated with 0.15 M Tris-HCl buffer, pH 8.5, 8 M urea. The fractions containing fibrinogen were pooled and dialyzed against the elution buffer in the cold prior to isoelectric focusing.

Reduced fibrinogen: Fibrinogen and its asialoderivative were reduced with dithiothreitol (2 mg per mg protein, 60 x molar excess) for 20 min. at 37°C prior to isoelectric focusing.

Isoelectric focusing: Isoelectric focusing was carried out in highly crosslinked gels. DATD (N,N'-diallyltartardiamide) was used as crosslinking agent (8). The final gel concentration of the gels was T = 7%, C = 15%. The isoelectric focusing gels contained 7 M urea, 1% Triton X-100. The pH gradient of 4 - 9 was obtained by mixing equal amounts of Pharmalyte pH 4 - 6.5 and pH 6.5 - 9 (Pharmacia, Uppsala, Sweden). The anolyte was 10 mM glutamic acid and the catholyte 10 mM ethanolamine. The focusing runs were carried out in gel rods (0.5 x 8 cm) with 30 V/gel at 8°C, for 12 hours. The Pharmalyte concentration was 2.7%. The gels were chemically polymerized with K-peroxosulfate. Before sample application, the gels were prerun until a constant voltage was obtained. The gels were stained with 0.02% Coomassie Brillant Blue solution.

Chemical determinations: Sialic acid was determined by the thiobarbituric acid assay according to Aminoff (9) after release of sialic acid with 0.1 M H_2SO_4 at 80°C for 60 min. As standard N-acetylneuraminic acid with 98% purity (Sigma) was used. Protein concentrations were determined with a modified Lowry-Folin assay (10) using human fibrinogen as standard.

Results and Discussion

The polypeptide chains of SCm-fibrinogen were focused in a pH-gradient 4 - 9 in presence of 7 M urea. SCm-fibrinogen (Fig. 1, gel No. 3) showed five main γ- and five main Bβ-bands, with apparent isoelectric points of 4.8 - 5.1 for the SCm-γ-chains and 5.2 - 5.4 for the SCm-Bβ-chains. The SCm-Aα-chains were not well separated with apparent isoelectric points of 5.7 - 6.2. These are all apparent isoelectric points because urea shifts the dissociation constants of acids and bases leading to higher pK values compared to the values in absence of urea (11). The isoelectric points of the carboxymethylated asialo-γ- and asialo-Bβ-chains were shifted to a more alkaline pH. At this point we made two rather unexpected observations. First, the number of Bβ- and γ-chain variants of carboxymethylated normal human fibrinogen was rather large, and second, no de-

crease in band number was observed when the carboxymethylated Bβ- and γ-chains of asialofibrinogen were compared to the ones of normal fibrinogen (Fig. 1, gel No 1, 2). This was somewhat surprising since we expected that the almost complete removal of sialic acid (more than 90%) from the Bβ- and γ-polypeptide chains would lead to more homogeneity for this very reason that both, Bβ- and γ-chain populations are heterogeneous in respect to sialic acid as shown by Gati and Straub (4). We therefore had to assume that side reactions could possibly occur during the reduction and/or the alkylation step. Alkylation of methionine, histidine and ε-amino groups of lysine as well as the alkylation of half-cystine has been described (12, 13). These reactions could well produce artifactual charge heterogeneities. Moreover, the alkylation of all half-cystines introduces a considerable number of carboxyl groups, namely 10 into the γ-chain and 11 into the Bβ-chain (14-16). This might well veil the reduction of charge heterogeneity after the removal of one or two sialic acid residues.
Indeed, when reduced but not alkylated fibrinogen was focused in the same pH-gradient, only three main Bβ-bands and three main γ-bands with more alkaline isoelectric points compared to the reduced and alkylated γ- and Bβ-chains of fibrinogen were detected. The γ-chains were focusing at an apparent pH 6.0 - 6.3 and the Bβ-chains at pH 7 - 7.5. In reduced asialofibrinogen only two distinct Bβ- and γ-variants could be observed in isoelectric focusing gels (Fig. 1, gel No 2). The removal of sialic acid was again accompanied by a shift of the apparent isoelectric points for the asialo-γ-chains to 6.3 - 6.5 and for the asialo-Bβ-chains to 7.5 - 7.7. This decrease in the number of Bβ- and γ-variants between reduced and alkylated versus merely reduced fibrinogen strengthened our assumption that undesired side reactions occured. Using iodoacetamide, an uncharged alkylating reagent, the same variant pattern as with iodoacetic acid was obtained, thus excluding that the artifac-

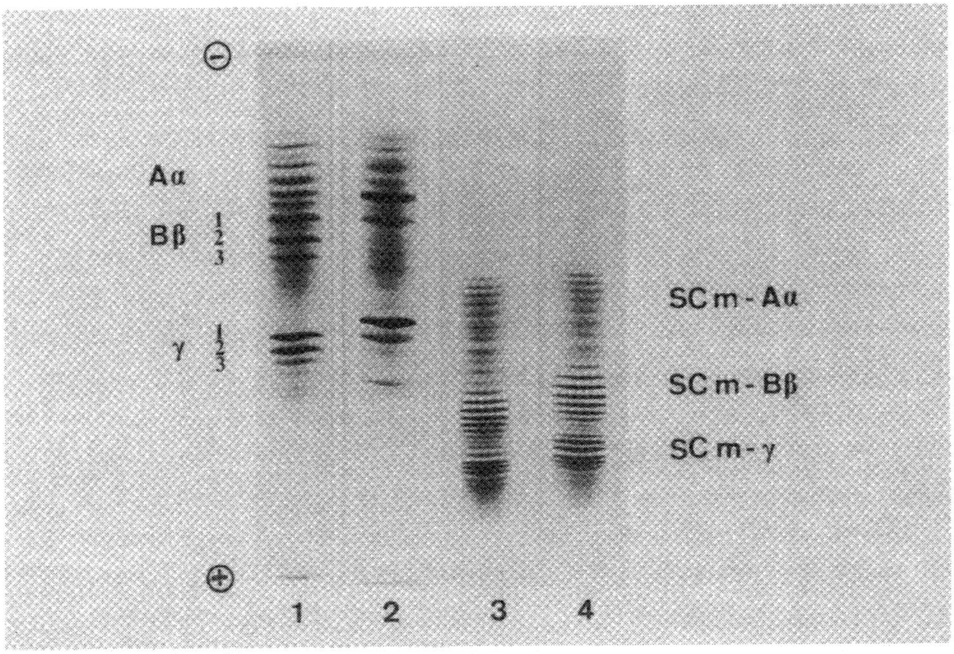

Fig. 1 Isoelectric focusing on polyacrylamide gels of reduced but not alkylated fibrinogen (gel No 1), reduced asialofibrinogen (gel No 2), SCm-fibrinogen (gel No 3), and SCm-asialofibrinogen (gel No 4) in presence of 7 M urea and 1% Triton X-100. pH-gradient 4 - 9, the cathode is indicated. The bands of reduced fibrinogen were designated with arabic numbers according to IUPAC - IUB Commission of Biochemical Nomenclature (21).

tual heterogeneity was not only the effect of the carboxyl group of iodoacetic acid.

Several authors have reported heterogeneities of S-sulfo- and SCm-fibrinogen polypeptide chains upon isoelectric focusing in polyacrylamide gels (17 - 19). Four to five γ- and Bβ-bands and several Aα-bands were described. Weinstein et al. (20) reported that these variants were of the same molecular size as analyzed by two dimensional electrophoresis. Our data on SCm-fibrinogen are in good agreement with these findings. But only

the fact that in reduced fibrinogen fewer bands are observed when compared with SCm-fibrinogen points to artificially produced heterogeneities. Furthermore the effect of the removal sialic acid was only visible when the alkylation step was omitted. These two observation lead to the conclusion that alkylation with iodoacetic acid may cause artifactual heterogeneities. Since the omission of the alkylation step may facilitate the reoxidation of the sulfhydryl groups to cystines during the isoelectric focusing run, reduced fibrinogen samples were run in the presence of excess amounts of dithiothreitol in the gel and in its complete absence. No differences were observed indicating that reoxidation was not disturbing our analysis. In addition, reduced fibrinogen was analyzed by two-dimensional electrophoresis, with polyacrylamide gel electrophoresis in presence of sodium dodecyl sulfate beeing in the second dimension. If reoxidation would have occured abnormaly high molecular weight compounds should have been demonstrable. All experiments were negative in this respect.
Furthermore, carbamylation a well known side reaction occuring in the presence of cyanate ions, could not be the reason for the great number of bands, since there were no cyanate ions measurably present in the urea solutions used.
In conclusion it is suggested that studies for charge heterogeneities of fibrinogen polypeptide chains should be carried out on reduced material, the alkylation step beeing omitted.

Acknowledgements

This work was supported by the Swiss National Research Foundation, Grant No 3.9.3.7-0.78

Literature

1. Martinez, J., Palascak, J.E., and Peters, C.: J. Lab. Clin. Med. 89, 367-377 (1977).

2. Gralnick, H.R., Givelber, H., and Abrams, E.: N. Engl. J. Med. 299, 221-226 (1978).

3. Gaffney, P. J.: Biochim. Biophys. Acta 263, 453-458 (1972).

4. Gati, W.P., and Straub, P.W.: J. Biol. Chem. 253, 1315-1321 (1978).

5. Blombäck, M.: Ark. Kemi 12, 387-396 (1958).

6. Werner, E.A.: J. Chem. Soc. 123, 2577 (1923).

7. Murano, G., Wiman, B., Blombäck, M., and Blombäck, B.: FEBS Lett. 14, 37-41 (1971).

8. Baumann, G., and Chrambach, A.: Anal. Biochem. 70, 32-38 (1976).

9. Aminoff, D.: Biochem. J. 81, 384-392 (1961).

10. Hess, H.H., Lees, M.B., and Derr, J.E.: Anal. Biochem. 85, 295-300 (1978).

11. Gelsema, W.J., DeLigny, C.L., and Van Der Veen, N.G.: J. Chromatogr. 171, 171-181 (1979).

12. Gurd, F.R.N. in Methods in Enzymology (Hirs, C.H.W., and Timasheff, S.N., eds), Vol XXV, p. 424-438, Academic Press (1972).

13. Plow, E.F.: Eur. J. Biochem. 80, 55-64 (1977).

14. Henschen, A., Lottspeich, F., Töpfer-Petersen, E., and Warbinek, R.: Thrombos. Haemostas. 41, 662-669 (1979).

15. Henschen, A., and Lottspeich, F.: Hoppe Seyler's Z. Physiol. Chem. 358, 1643-1646 (1977).

16. Watt, K.W.K., Tagaki, T., and Doolittle, R.F.: Biochemistry 18, 68-76 (1979).

17. Gaffney, P.J.: Nature 230, 54-56 (1971).

18. Galanakis, D.K., Mosesson, M.W., and Stathakis, N.E.: J. Lab. Clin. Med. 92, 376-386 (1978).

19. Stathakis, N.E., Mosesson, M.W., Galanakis, D.K., and Ménaché, D.: Thrombos. Res. 13, 467-475 (1978).

20. Weinstein, M.J., and Deykin, D.: Thrombos.Res. 13, 361-377 (1978).

21. IUPAC - IUB Commission on Biochemical Nomenclature: J. Biol. Chem. 246, 6127-6128 (1971).

INFLUENCE OF SIALIC ACID ON THE ADULT AND FETAL FIBRINOGEN-FIBRIN TRANSITION

Irene Witt, Hans H. Seydewitz, Volker Dahlacker

Biochemisches Labor der Universitäts-Kinderklinik Freiburg i.Br.
D-7800 Freiburg i.Br., Germany

Introduction

The human newborn synthesizes a molecular variant of fibrinogen, called fetal fibrinogen (1). This fetal fibrinogen is presumed to cause some of the special features of neonatal coagulation.

The highly purified fetal protein is characterized by several functional and biochemical differences as compared to the adult one. The most important of these features are:

a twofold higher content of organically bound phosphorus (2,3),
a higher affinity to thrombin at pH 7.0 (3,4),
a remarkably pronlonged clotting time in the alkaline pH region (5),
a higher negative charge (concluded from the behaviour in DEAE-Cellulose-chromatography) (5),
a delayed fibrin monomer aggregation (4,6) and
a different fibrin network as revealed by morphometric analyses of electron microscopic pictures (7).

The fibre segments in the fetal fibrin are nearly half as thick and long as those of the adult and the network is more compact. There is strong evidence that the differences between adult and fetal fibrinogen are caused by a different primary structure. From fingerprint analyses of peptides we assume that the special amino acid sequence of the fetal protein is located in the Aα-chain, but the special sequence has not yet been found.

Fibrinogen - Recent Biochemical and Medical Aspects

The possibility should be taken into consideration that constituents of the fibrinogen molecule which are incorporated after the translational step, namely the phosporus and the carbohydrate residues, provide the molecular basis of the fetal features. The question as to the importance of phosphorus has already been investigated several years ago. Fetal fibrinogen contains twice as much organically bound phosphorus as the adult protein. When the phosphorus is split off by phosphatase no alterations of the special fetal features can be observed (3). This fact rules out that the higher phosphorus content of the fetal protein can be responsible for its characteristic properties.

It remains the question as to the importance of the carbohydrate residues for the funçtional properties of fibrinogen and especially for the fetal variant. The investigations of Töpfer-Petersen (8) have shown that there are two carbohydrate side chains bound in each fibrinogen molecule, one at the β-chain and one at the γ-chain. Sialic acid is always on the exposed end of the carbohydrate chain. When the sialic acid is split off by neuraminidase the transition of fibrinogen to fibrin induced by thrombin is remarkably enhanced (9,10).

Material and Methods

Materials

Neuraminidase of vibrio comma (cholerae) and bovine thrombin were purchased from Behringwerke AG, Marburg.

Purification of fibrinogen

Human umbilical cord blood was collected in siliconized 20 ml- centrifuge tubes containing 5 ml of ACD-stabilisator and 1% w/v- ε -amino caproic acid. The plasma was obtained within 24 hrs after collection by centrifugation at 2,500 x g for 20 min and stored at 4°C. Combined plasma samples were centrifuged at 10,000 x g for 15 min immediately before preparation of the fibrinogen. The preparation of adult and fetal fibrinogen was carried out by the method of Osbahr et al. (11). The purity of the fibrinogen was controlled by measuring the percentage of clottable protein by thrombin and by polyacrylamide-gel-electrophoresis (PAGE) of samples reduced with dithio-erythritol. The purity was 96 - 98 % in the clotting test and in PAGE less than 3 % of contaminating protein was seen.

Determination of N-acetyl neuraminic acid

The determination of the N-acetyl neuraminic acid content was done after 1 hr incubation at 80°C with 0.2 N H_2SO_4 by the method of Aminoff (12).

Desialylation of fibrinogen

Fibrinogen solutions (8 mg/ml) in 0.3 M KCl were incubated at 37° C with neuraminidase (0.02 units per mg protein) for 150 min.

Clotting time

The thrombin clotting time was determined by using fibrinogen solutions of 1.5 mg/ml in 0.3 M KCl. The pH was adjusted to 9.0 or 7.0 by adding 0.02 M NaOH or 0.02 M HCl, respectively.
The reaction was carried out at 37° C. To 0.1 ml fibrinogen solution 0.1 ml thrombin solution (120 N.I.H. units/100 ml) was added and the mixture manually stirred with a platinum loop until a fibrin clot appeared. Each measurement was performed in triplicate.

Results and Discussion

In Fig. 1 the rate of sialic acid release under the above described conditions is shown. About 80 % of the sialic acid is split off during 150 min of incubation with neuraminidase. The splitting rate may be somewhat faster for the fetal than for the adult fibrinogen.

During the release of the N-acetyl neuraminic acid the clotting time is shortened in the order of 40%. This effect is more pronounced in fetal fibrinogen as compared to adult fibrinogen (Fig.2).

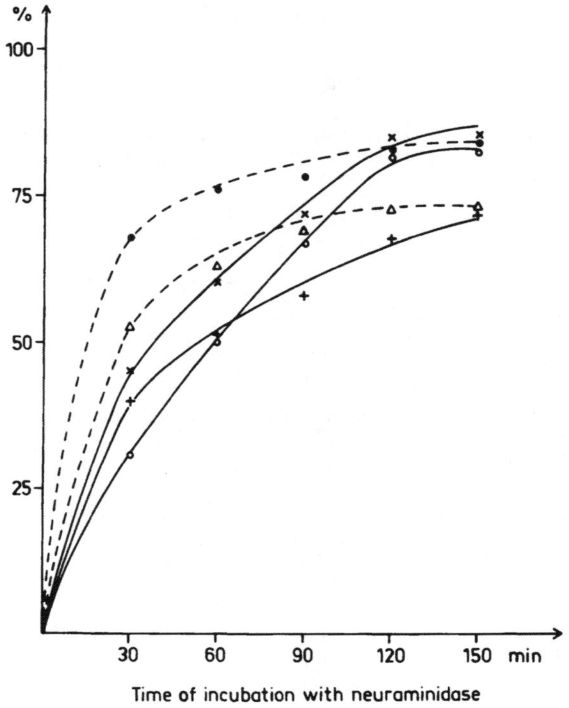

Fig.1
Relative amount of N-acetyl neuraminic acid released from adult (——) and fetal (---) fibrinogen by neuraminidase at 37°C and pH 7.0

Recently some light scattering investigations on our preparations of adult and fetal fibrinogen and asialo-fibrinogen were carried out by Müller and Burchard at the Makromolekulares Institut der Universität Freiburg (for details of the method see (13) and this issue).
In the early stages of the thrombin-induced fibrinogen-fibrin transition one can extract graphs like those shown in fig. 3. The initial part represents the rod-like fibrin growth and the flat part at high molecular weights is significant for branched structures.

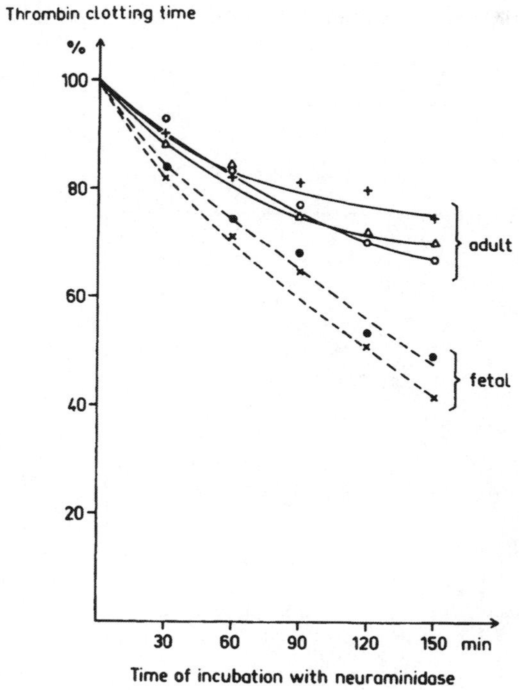

Fig.2
Relative thrombin clotting time of adult and fetal fibrinogen after incubation with neuraminidase at 37°C and pH 7.0

From a quantitative analysis of these curves it is possible to compute the mean fibre length of the fibrin segments between two branching points.

These measurements confirm nicely our former electron microscopic investigations (7). By both methods the fibrin network is found to be considerably denser when formed from fetal than from adult fibrinogen (Tab. 1).

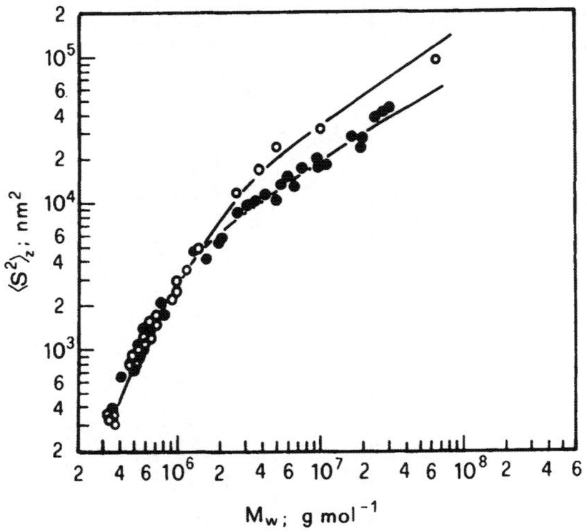

Fig.3
Dependence of the mean square radius of gyration $\langle S^2 \rangle_z$ on the molecular weight M_w for adult -0-0- and adult asialo -●-●- soluble fibrin

Tab. 1

Length parameters of fibrin as determined by light scattering and electron microscopy.

	Light scattering Mean length of fibrin fibre segments	Electron microscopy Mean distance between branching points in the fibrin network
Fetal Fibrin	370 nm	350 nm
Adult Fibrin	525 nm	520 nm
Fetal Asialofibrin	265 nm	-
Adult Asialofibrin	265 nm	-

Asialo-fibrinogen forms fibrin with a higher branching density than intact fibrinogen. This effect is more drastic in the case of adult fibrinogen. The consequence is that fetal and adult fibrinogen in the asialo-form cannot be distinguished by their fibrin fibre lenghts. We do not yet have electron microscopic pictures of asialo-fibrin.

Fetal fibrinogen has a remarkably longer thrombin clotting time at pH 9.0 compared to that of pH 7.0. Adult fibrinogen shows only minor differences in clotting time at these two pH values. When the clotting time of asialo-fibrinogen is measured at pH 7.0 and 9.0 the characteristic prolongation in clotting time at pH 9.0 of fetal fibrinogen cannot be found (Tab. 2).

Tab. 2

Thrombin clotting time of adult and fetal fibrinogen and asialo fibrinogen at different pH values

	pH 7.0	pH 9.0	Ratio pH 9.0/pH 7.0
Fetal Fibrinogen	40 sec	145 sec	3.6
Adult Fibrinogen	30 sec	55 sec	1.8
Fetal Asialofibrinogen	22 sec	27 sec	1.2
Adult Asialofibrinogen	20 sec	30 sec	1.5

These two observations indicate that some differences between adult and fetal fibrinogen are not longer apparent áfter the release of sialic acid. One might conclude that the difference between the two proteins is caused by a different sialic acid content. But careful measurements of the sialic acid content of adult and fetal fibrinogen have shown that there are no differences in the total sialic acid content (14). Adult fibrinogen contains 20.8 ± 3.3 nanomoles/mg protein and fetal fibrinogen 22.2 ± 5.7 nanomoles/mg, that is about 7 moles sialic acid per mole fibrinogen whereas in a fully sialylated fibrinogen 8 moles should be expected.

In recent measurements on isolated ß- and γ-chains of adult and fetal fibrinogen we have found that there is also no difference in the distribution of sialic acid on these two chains. In the adult as well as in the fetal fibrinogen the ratio of γ-chain to ß-chain sialic acid is 1.10.

The fact that the total content of sialic acid in adult and fetal fibrinogen is the same but the release of sialic acid brings about a nearly complete loss of the fetal-adult differences in clotting time and network density is difficult to explain in terms of a sialic acid effect alone, at least if one retains the assumption that the carbohydrate attachment points and the molecular arrangement of the carbohydrate side chains are the same. It is more likely that there is a difference in the primary structure and that the influence of this special structure on the fibrinogen-fibrin transition heavily depends on the presence of N-acetyl neuraminic acid.
The question as to the physiological importance of the carbohydrate chains of the fibrinogen molecule also has not yet been answered. During the fibrinogen-fibrin transition there is no liberation of sialic acid. It is possible that the sialic acid protects the molecule from the attack of proteases. It can be presumed that during bacterial infections sialic acid could be split off by neuraminidase thus enhancing the fibrin formation by thrombin. This may be of clinical importance for example in disseminated intravascular coagulation.

Summary:

1. The N-acetyl neuraminic acid on the fibrinogen molecule influences the speed of its transition to fibrin by thrombin and the structural network of the resulting fibrin. The physiological importance of sialic acid might be thought of as "protection" against the attack of proteases, which under pathological conditions, for example bacterial infections, could be lost.

2. The characteristic differences between adult and fetal fibrinogen are significantly reduced after sialic acid release, which, however, does not imply that the differences depend on a difference in the total sialic acid content.

References

1. Künzer, W.: Klin.Wschr. 39, 536-537 (1961).

2. Witt, I., Müller, H.: Biochim. Biophys. Acta 221, 402-404 (1967).

3. Witt, I., Hasler, K.: Biochim. Biophys. Acta 271, 357-362 (1972).

4. Galanakis, D.K., Mosesson, M.W.: Blood 48, 109-118 (1976).

5. Witt, I., Müller, H., Künzer, W.: Thromb. Diathes. Haemorrh. 12, 101-109 (1969).

6. Guillin, M.C., Ménaché, D.: Thromb. Res. 3, 117-135 (1973).

7. Tesch, R., Trolp, R., Witt, I.: Thromb. Res. 16, 239-243 (1979).

8. Töpfer-Petersen, E. in: Fibrinogen, Fibrin und Fibrinkleber, K. Schimpf (Ed.), Schattauer Verlag Stuttgart-New York 1980, pp. 43-47.

9. Chandrasekhar, N., Warren, L., Osbahr, A.J., Laki, K.: Biochim. Biophys. Acta 63, 337-339 (1962).

10. Martinez, J., Palascak, J., Peters, C.: J. Lab. Clin. Med. 89, 367-377 (1977).

11. Osbahr, A.J., Gladner, J.A., Laki, K.: Biochim. Biophys. Acta 86, 535-542 (1964).

12. Aminoff, D.: Biochem. J. 81, 384-392 (1961).

13. Müller, M., Burchard, W., Witt, I.: Thromb. Res., in press.

14. Witt, I., Tesch, R. in: Fibrinogen, Fibrin und Fibrinkleber, K. Schimpf (Ed.), Schattauer Verlag Stuttgart-New York 1980, pp. 31-37.

FIBRIN FORMATION IN THE PRESENCE OF TRYPSIN

Birgit Hessel, Birger Blombäck, Sonja Söderman

Department of Blood Coagulation Research, Karolinska Institute, S-104 01 Stockholm, Sweden

In previous studies we have found that the thrombin mediated fibrinogen-fibrin transition occurs as a two-stage reaction, in which fibrinopeptides are released and two sets of polymerization domains (A/a and B/b) are activated. As a result fibrin I and fibrin II are formed (1). First fibrinopeptide A (FPA) is released resulting in activating of the A-binding domain in the NH_2-terminal part of fibrinogen (N-DSK). The active A-domain acquires affinity for a COOH-terminal domain termed a, and interaction between these sites (A and a) of neighboring molecules leads to formation of a linear polymer, fibrin I. The second stage involves the removal of fibrinopeptide B (FPB), and this leads to the activation of the B site, which also may be located in the N-DSK region (1, 2). The active B-site can interact with a complementary site, b, located on fibrin I polymers. Thus a copolymer, fibrin II polymer, is formed which has a different structure from fibrin I polymer (2). Fibrin I and fibrin II polymers cross-link [a)] into an infinite net-work structure, whereby fibrin I and fibrin II gels are formed (3).

It is well known that early plasmin attack of fibrinogen removes polypeptides from the carboxyterminal end of the Aα chain and from the aminoterminal end of the Bβ chain (4, 5, 6). In

[a)] Foot note: This process must be distinguished from FXIII induced cross-linking.

Fibrinogen - Recent Biochemical and Medical Aspects

the latter chain, the bond Bβ 42-43 is first cleaved. Nossel et al. suggested that the <u>in vivo</u> release of Bβ 1-42 by plasmin may serve as a possible regulatory mechanism in the fibrin II formation (7). Nossel suggested that after fibrin I formation, thrombin and plasmin compete for the NH_2-terminal end of the Bβ chain, and if thrombin cleaves the Bβ chain before plasmin the products are FPB and fibrin II. If plasmin cleavage precedes that of thrombin the products are Bβ 1-42 and the high molecular weight fragment X which can not form fibrin II (8).

We have studied the effect of trypsin on the fibrin formation <u>in vitro</u>. Fibrinogen, 0.45, 1.67 and 3.22 mg/ml respectively in Tris-NaCl buffer, pH 7.2, was treated with trypsin (1:1000 - 1:4000) at 37°C, for 1 to 10 minutes. After incubation soy-bean trypsin inhibitor (SBTI) was added. Controls were 0-time sample, in which the inhibitor was added before trypsin, and samples in which the inhibitor and trypsin were replaced by sodium chloride. Clotting in the trypsin treated samples was obtained by

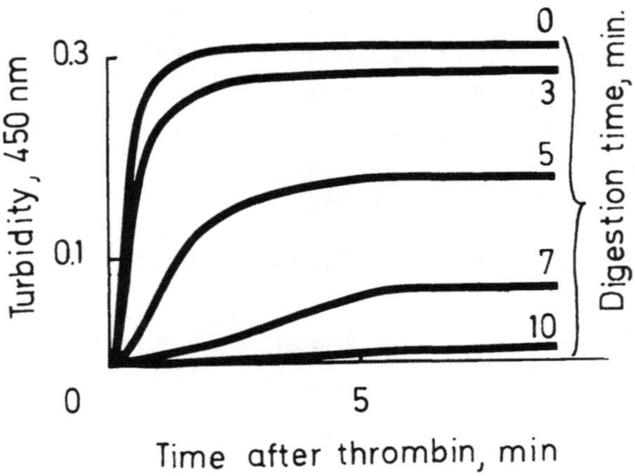

Fig. 1

Fibrinogen, 0.45 mg/ml, in tris-NaCl-buffer, pH 7.2 was treated with trypsin (Enzyme to substrate 1:2000, w/w), at 37°, for different times, whereafter SBTI was added. Thrombin was added, 2 NIH units/ml in final concentration. Incubation at 37° for 10 min. Turbidity was registrated at 450 nm in an Acta III Spectrophotometer, Beckman.

adding thrombin (1-2 NIH units/ml, 37°C for 10-15 min). The loss of clottability was determined as the decrease in turbidity at 450 nm and by analyzing the protein content in the supernatant after removal of clots. We found that during the incubation with trypsin, clottability was gradually lost and completely after 10 min. (Fig. 1).

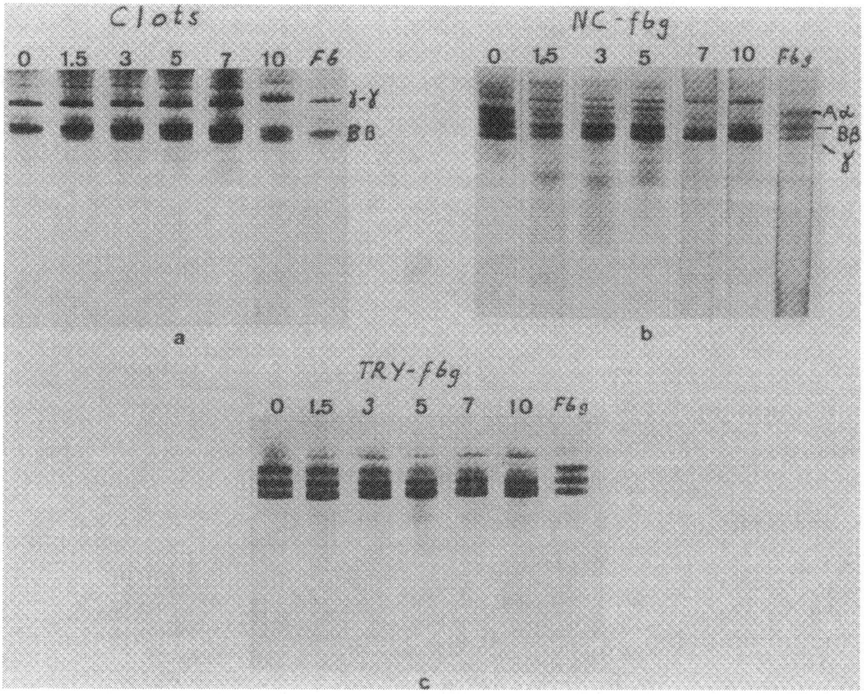

Fig. 2

SDS-gel electrophoresis, 7% gels, on samples reduced with dithiotreitol. Fibrinogen, 1.67 mg/ml, was treated with trypsin 0, 1.5, 3, 5, 7, 10 minutes. Thrombin was added, and the clots (a) removed and washed. The non-clottable fibrinogen (NC fbg), (b) in the supernatant was precipitated with ethanol. In c) the fibrinogen was treated only with trypsin (TRY fbg). All samples were dissolved in urea. Fb=cross-linked fibrin; Fbg=fibrinogen.

The clots formed were removed and washed with water. Non-clottable fibrinogen (NC fbg) in the supernatant was precipitated by adding ethanol to a concentration of 50%. NC fbg as well as the clots were dissolved in urea and subjected to SDS-gel electrophoresis on 7% gels before and after reduction. In addition, samples treated with trypsin (and SBTI) but not with thrombin (TRY-fbg), were lyophilized, dissolved in urea and electrophoresed as above. In all three cases, no appreciable degradation, compared with control samples, occured during the digestion as judged by electrophoresis of the unreduced samples.

With regard to the reduced clots (Fig. 2a) cross linking has occured in the control and digested samples. Regarding the Bβ chain it is difficult to say if it remains intact during the digestion, as a new band, increasing during the digestion appears close to the original Bβ band. Taken together with results from NH_2-terminal analysis (see below), this new band may indicate that a cleavage has occured at the NH_2-terminal part of the Bβ chain resulting in an about 4000 dalton shorter Bβ chain.

Regarding the reduced material of non-clottable fibrinogen samples (Fig. 2b), the Aα chain decreased rapidly. The Bβ chain also decreased but more slowly. The γ chain appeard to be intact. However, the γ band became less sharp. Since no low molecular weight degradation products appeared there is a possibility that the Aα and Bβ chains have been degraded to products with about the same molecular weights as the γ chain and/or to low molecular weight fragments, which penetrate the gel. A weak band corresponding to γ-γ dimer is also seen, indicating that crosslinking occurs only to a small degree. About the same profile is seen also in the samples treated only with trypsin (TRY-fbg) (Fig. 2c).

The supernatants after precipitation of NC fbg were used in radioimmunoassay (RIA) for FPA, FPB and Bβ 15-42. RIA was also used on the total clot supernatant for estimation of the COOH--terminal fragment of Aα chain, termed Hi2-DSK. This peptide constitutes the residues Aα 241-476 (9, 10). During the conditions we used cleavage of FPA and FPB occured to some degree during the tryptic digestion, but complete release was not obtained until thrombin was added (Fig. 3). FPB seems to increase during the whole digestion time, while FPA reached a plateau at about 25% of release. This is in agreement with a study reported by Bilezikian et al. (11). They found that FPB is cleaved more rapidly than FPA when they digested fibrinogen with trypsin. The most probably explanation to the low FPA-release is, that a fragment, longer than FPA (e.g. Aα1-19) has been cleaved, and that this fragment has lower immunoreactivity. Another, but rather unlikely explanation could be product inhibition. Both Bβ 15-42 and Hi2-DSK were also released by trypsin.

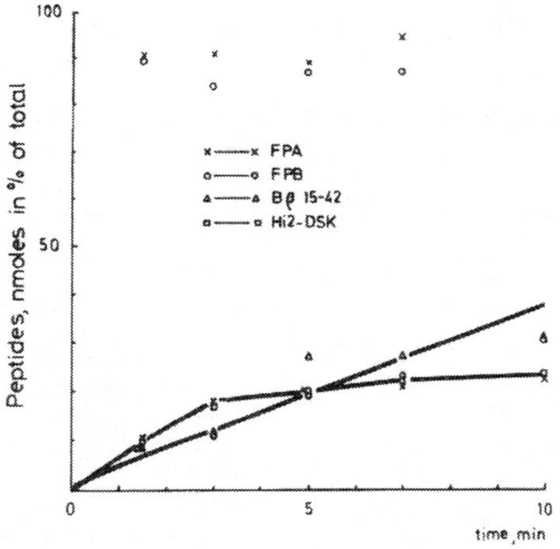

Fig. 3

Radioimmunoassays of FPA, FPB, Bβ 15-42 and Hi2-DSK after treatment of fibrinogen with trypsin. Two values are given for FPA and FPB, the higher values are obtained when thrombin has been added after incubation with trypsin.

Furthermore, we have also determined NH_2-terminal amino acids both in the clots and in NC-fbg. In the control clot we found as expected only tyrosin and glycine (ratio 1:2). During the time of digestion with trypsin alanine and valine also appeared. With regard to NC-fbg, the same NH_2-terminals were found, and in addition glutamic acid. Interestingly, glutamic acid appears at the same time and to corresponding amount as Hi2-DSK.

The above results suggest that tryptic cleavages occur in the Aα and Bβ chains of fibrinogen. To get more information regarding the bonds cleaved, N-DSK was isolated from a non clottable fibrinogen obtained after 10 minutes digestion with trypsin as well as from a control clot. New NH_2-terminals compared to the control were alanine and valine, indicating cleavage at one or more of the following bonds: Aα 19 Arg-20 Val; Bβ 42 Arg-43 Ala; 47 Lys-48 Ala; 54 Lys-55 Val; 58 Lys-59 Ala (Fig. 4). These bonds seems not to be important for the fibrin polymerization, as clots are formed even when these bonds are hydrolyzed. However, when glutamic acid appears in the non clottable fibrinogen and Hi2-DSK in the supernatant the polymerization is decreased. From the structure of Aα chain (10) and specificity of trypsin there seems to be only seven Arg/Lys-Glu bonds in the whole Aα chain which can be cleaved, and five out of these are found within Hi2-DSK (Fig. 5). Thus, one structure important for polymerization and/or clotting seems to be in the COOH--terminal part of the Aα chain.

Some other studies we have performed suggest that other factors may also participate in the <u>in vivo</u> regulation of fibrin formation. Fibrin formation in samples of whole blood taken before and after venous stasis (10 min. with a blood pressure cuff, at 100 mm Hg) were studied. The blood samples, collected in glass--tubes, were allowed to clot for various times. Appearance of activators of plasminogen in the blood was assessed by performing the euglobulin lysis test on citrated plasma samples taken

Aα 1 FPA 17 20
Ala——Arg-Glu——Arg-Val—— - - -

Bβ 1 FPB 15 43 48 55 59
0 ——Arg-Glu——Arg-Ala——Lys-Ala——Lys-Val——Lys-Ala—— - -

γ 1 63
Tyr- ——————————————————————————————Lys-Ala—— -

Fig. 4

Bonds in the NH_2-terminal part of fibrinogen (N-DSK) which may be hydrolyzed by trypsin.

171 172 407 408 421 422 427 428 457 458
Aα--- Arg-Glu——Arg-Glu——Lys-Glu——Lys-Glu——Lys-Glu——

461 462 539 540
——Lys-Glu——Lys-Glu - - -

Fig. 5

Arg-Lys/Glu bonds, in the Aα chain of fibrinogen which may be hydrolyzed during digestion with trypsin.

before and after stasis. Some changes were noted in most of the individuals who showed a positiv euglobulin lysis test. Radioimmunoassay of clot supernatants revealed small amounts of Bβ 15-42 fragments, but there was no obvious change in chain structure of this fibrin as judged by SDS-gel electrophoresis before and after reduction. Prothrombin consumption was greatly delayed after stasis and thrombin formation was also delayed. Furthermore there is a delay in fibrin formation as well as in release of FPA and FPB but the amount of fibrin eventually formed in static blood is only slightly less than in blood before stasis. It therefore seems as if regulation of fibrin formation may not only depend on plasmin, but one or several other mechanisms may also be important in vivo.

References

1. Blombäck, B., Hessel, B., Hogg, D.H., Therkildsen, L.: Nature 275, 501-505 (1978).

2. Blombäck, B., Hessel, B., Okada, M., Egberg, N.: N.Y.Acad. Sci. In print (1981).

3. Blombäck, B., Okada, M.: Submitted to Thromb. Res. (1981).

4. Budzynski, A.Z., Marder, V.J., Shainoff, J.R.: J. Biol. Chem. 249, 2294-2302 (1974).

5. Hessel, B.: Thromb. Res. 7, 75-87 (1975).

6. Takagi, T., Doolittle, R.F.: Biochemistry, 14, 940-946 (1975).

7. Nossel, H.L., Wasser, J., Kaplan, K.L., La Gamma, K.S., Yudelman, I., Canfield, R.E.: J. Clin. Invest. 64, 1371-1378 (1979).

8. Nossel, H.L.: Nature 291, 165-167 (1981).

9. Gårdlund, B., Hessel, B., Marguerie, G., Murano, G., Blombäck, B.: Eur. J. Biochem. 77, 595-610 (1977).

10. Henschen, A., Lottspeich, F.: Haematologica 65, 535-541 (1980).

11. Bilezikian, S.B., Nossel, H.L., Butler, V.P., Canfield, R.E.: J. Clin. Invest. 56, 438-445 (1975).

CHARACTERISATION OF RAT FIBRINOGEN

Maria Kehl and Agnes Henschen
Max-Planck-Institut für Biochemie,
D-8033 Martinsried bei München, German Federal Republic

Introduction

Rat fibrinogen has been widely used as a animal fibrinogen model. Thus, the synthesis and glycosylation (1), the degradation (2), the half-life (3), the effect on inflammation (4) and several other biological aspects of fibrinogen have been studied in the rat. However, the rat fibrinogen preparations described so far have differed considerably from each other and from human fibrinogen (5,6,7). Some preparations have shown only two bands in gel electrophoresis (5,6) under conditions where human preparations have shown three. Furthermore, in contrast to all species studied, rat γ-chain seemed to have a blocked N-terminus (5). Because of the controversial properties of and the wide-spread interest in rat fibrinogen it seemed worthwhile to start a structure analysis of this protein. The present report deals with the N-terminal amino acid sequences of all three rat fibrinogen chains.

Experimental and Results

N-Terminal sequence analysis of fibrinogen. The total rat fibrinogen (Koch-Light, Colnbrook, England) was subjected to a short sequenator analysis to obtain information about number and N-terminal reactivity of peptide chains. In the first cycle only alanine was present, in the second aspartic acid and threonine,

Fibrinogen - Recent Biochemical and Medical Aspects

in the third only threonine. The amino acids are those expected for the N-termini of fibrinopeptides A and B (8). No N-terminus which could have been derived from γ-chain was found.

<u>Peptide chain separation</u>. Rat fibrinogen (Koch-Light) was sulfitolysed (9) and the peptide chain mixture fractionated by CM-cellulose chromatography in a buffer containing 8 M urea, using a tris-acetate pH-gradient (10). The chromatogram is shown in Figure 1.

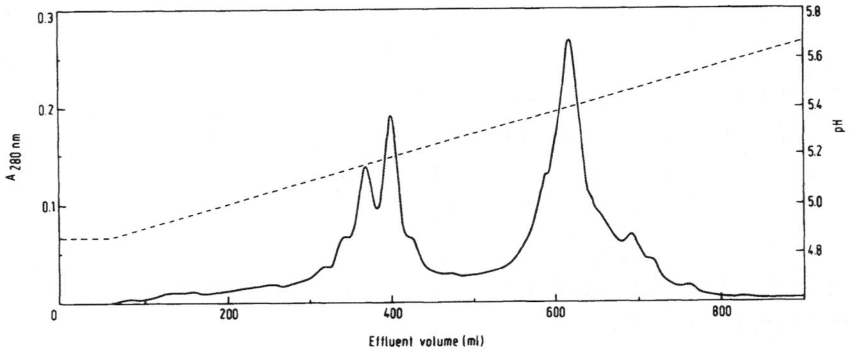

Figure 1. CM-Cellulose chromatography of S-sulfo-rat fibrinogen; the dotted line indicates the pH-gradient.

The S-sulfo-rat fibrinogen and its chromatographic fractions were analysed by SDS gel electrophoresis. The fibrinogen itself contained three bands, corresponding to γ-, Bβ-and Aα-chain. However, the Aα-chain band migrated much closer to the Bβ-chain band than in human fibrinogen. The first double-peak from the CM-cellulose column, i.e. between 300 and 450 ml effluent, contained pure γ-chain band. The second peak, i.e. between 550 and 750 ml effluent, contained in the first part mainly Bβ-chain, in the last part mainly Aα-chain band; however, a high yield of pure components was not obtained here.

N-Terminal sequence analysis of γ-chain. When the isolated γ-chain was subjected to sequenator analysis and a sensitive identification system was used a small amount of reactiv N-terminus was detected. The sequence of the first 22 amino acid residues could be determined. As demonstrated in Table 1, the sequence is highly homologous to the corresponding human sequence (10).In fact, the only difference in this part of the sequence is the valine-threonine exchange in position 2.

The nature of the group, by which the N-terminus of the γ-chain is almost completely blocked, is not yet known. An attempt to deblock the γ-chain using pyroglutamyl aminopeptidase was unsuccessful.

Table 1. N-Terminal sequence of fibrinogen γ-chain.

Rat	Y T A T R D N C C I L D E R F G S Y C P T T
Man	Y V A T R D N C C I L D E R F G S Y C P T T

Fibrinopeptide isolation. As it was difficult to prepare pure Aα- and Bβ-chain for N-terminal sequence analysis and as the fibrinopeptide sequences were only partly known the fibrinopeptides were isolated. For this purpose, rat fibrinogen (Sigma,St. Louis,U.S.A.) in 0.15 M ammonium acetate pH 8.5 was digested with 1 NIH units/ml of bovine thrombin (Behringwerke,Marburg/Lahn) for 4 h. The fibrin was removed and the supernatant liquid fractionated by high performance liquid chromatography on a RP-18 column (Hibar 250-4, LiChrosorb; Merck, Darmstadt, Germany) using a straight concentration gradient between 50 ml of 0.02 M ammonium acetate pH 6.0 with ortho-phosphoric acid / acetonitrile, 95:5 (v/v) and 50 ml of 0.05 M ammonium acetate pH 6.0 with ortho-phosphoric acid / acetonitrile, 70:30 (v/v) at a flow rate of 1.5 ml/min. The chromatogram is shown in Figure 2.

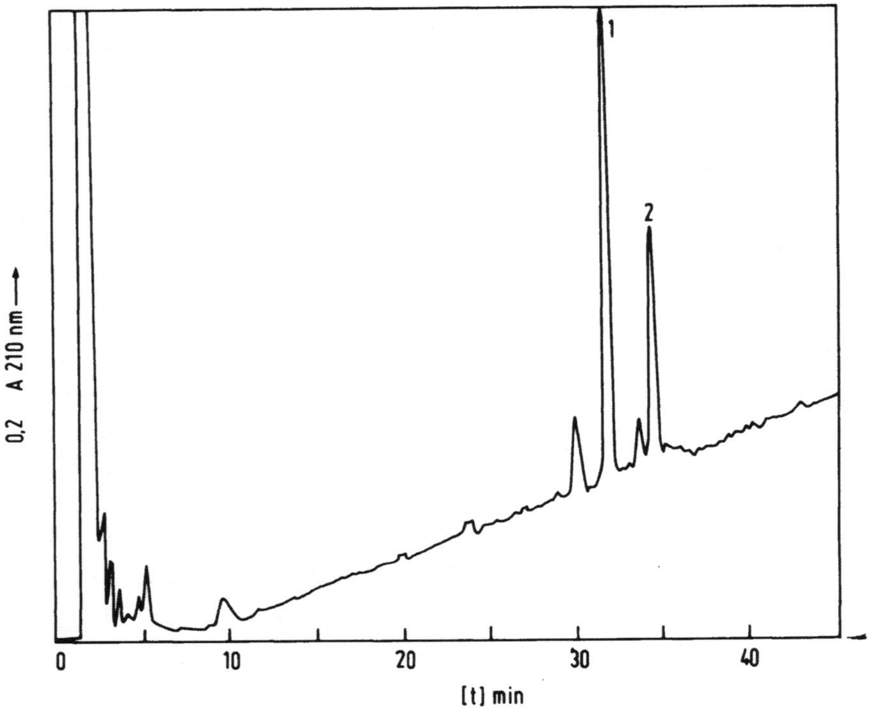

Figure 2. High performance liquid chromatography of rat fibrinopeptides A (peak 1) and B (peak 2).

The two main peak were analysed for amino acid sequence and composition. They were identified as the fibrinopeptides A and B by comparison with the partial sequences already published (8). The complete sequences of both peptides were directly determined by means of the sequenator and are presented in Table 2 together with the human fibrinopetide sequences (11). The sequences of the fibrinopeptides from the two species differ considerably from each other. However, in both species the A-peptide eluted before the B-peptide, as the A-peptides are more acidic. The minor peaks in the rat fibrinopeptide chromatography were not analysed in detail, but seemed to contain degradation products.

Table 2. Fibrinopeptide sequences.

Rat A	A D T G T T S E F I E A G G D I R
Man A	A D S G E G D F L A E G G G V R
Rat B	A T T D S D K V D L S I A R
Man B	Z G V N D N E E G F F S A R

Evolutionary Aspects

The N-terminal amino acid sequences of all three chains in rat fibrinogen are unusual. The very high mutation rate of the fibrinopeptides is well documented (12), but in their C-terminal parts these peptides show relatively little variation, the C-terminal heptapeptide sequence in almost all A-peptides being A E G G G V R . In contrast, the corresponding rat sequence is E A G G D I R. The C-terminus of the rat B-peptide differs in a similar way. The question arises if the $A\alpha$ - and $B\beta$ -chain sequences in rat are as unusual on the C-terminal side of the thrombin cleavage site as on the N-terminal side.

The N-terminus of the rat γ-chain is unusual because it is the only one out of 10 known γ-chain N-termini which is blocked (13). Apart from this, the γ-chain N-termini show very little variation; the threonine in position 2 is present not only in rat, but also in dog (14) and baboon (13) γ-chain.

This investigation was supported by a grant-in-aid from the Deutsche Forschungsgemeinschaft, which is gratefully acknowledged.

References

1. Nickerson, J.M., Fuller, G.M.: Proc. Natl.Acad.Sci.USA. 78, 303-307 (1981).
2. van Ruijven-Vermeer, I.A.M., Nieuwenhuizen, W., Haverkate, F., Timan, T.: Hoppe-Seyler's Z. Physiol. Chem. 360, 633-637 (1979).
3. Nieuwenhuizen, W., Emeis, J.J., Vermond, A., Kurver, P., van der Heide, D.: Biochem. Biophys. Res. Comm. 97, 49-55 (1980).
4. Ruhenstroth-Bauer, G., Scherer, R., Abd-El-Fattah, M., Simon, J.: This volume.
5. Bouma, III, H., Fuller, G.M.: J. Biol. Chem. 250, 4678-4683 (1975).
6. Stemberger, A., Jilek, F.: Thromb. Res. 9, 657-660 (1976).
7. van Ruijven-Vermeer, I.A.M., Nieuwenhuizen, W.: Biochem. J. 169, 653-658 (1978).
8. Blombäck, B., Blombäck, M., Gröndahl, N.J.: Acta Chem. Scand. 19, 1789-1791 (1965).
9. Henschen, A.: Arkiv Kemi 22, 1-28 (1963).
10. Henschen, A., Edman, P.: Biochim. Biophys. Acta 263, 351-367 (1972).
11. Blombäck, B., Blombäck, M., Edman, P., Hessel, B.: Biochim. Biophys. Acta 115, 371-396 (1966).
12. Dayhoff, M.O.: Atlas of Protein Sequence and Structure, National Biomedical Research Foundation, Washington (1972).
13. Henschen, A., Lottspeich, F., Kehl, M.: This volume.
14. Birken, S., Wilner, G.D., Canfield, R.E.: Thromb. Res. 7, 599-610 (1975).

CHARACTERISATION OF PIG FIBRINOGEN

John Lucas and Agnes Henschen
Max-Planck-Institut für Biochemie
D-8033 Martinsried bei München, Germany

Introduction

Since the completion of the human fibrinogen sequence by Henschen and co-workers in 1979 (1), fibrinogen sequences of a few other mammalian species have been investigated by our group and compared to that of human fibrinogen. This report deals with pig fibrinogen.

Results and Discussion

Chain separation of S-carboxymethylated pig fibrinogen.
Pig fibrinogen was provided by Sigma. After reduction of the disulfide bonds with mercaptoethanol and carboxymethylation of the resulting free sulfhydryl groups, the preparation was loaded on a column containing carboxymethyl-cellulose in 8 M urea and the individual Aα-, Bβ- and γ-chains separated by applying a constantly increasing pH-gradient. The progress of the chain separation of pig S-carboxymethylfibrinogen by CM-cellulose chromatography is shown in fig. 1.

When this chain separation is compared with the corresponding one in species such as human (2) and ox (3), it is apparant that for pig the yield of intact Aα-chain is much higher and contains only trace amounts of Aα-chain degradation products. This is in contrast to rat fibrinogen, where Aα-chain degradation is so extreme, that the Aα-chain can barely be seen in

Fibrinogen - Recent Biochemical and Medical Aspects

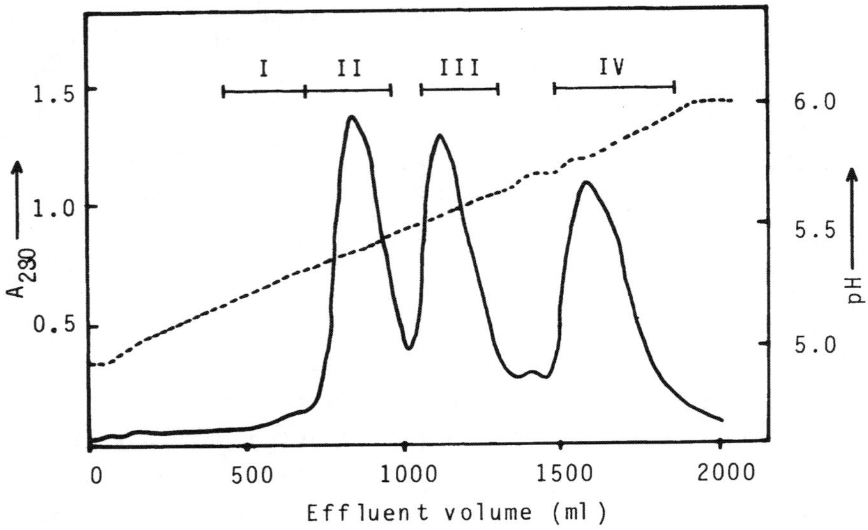

Fig. 1. Pig-Scm-Fibrinogen. CM-cellulose chromatography in 8 M urea. The numerals I-IV in figure 1 represent pooled fractions, from which samples were taken for SDS-gel electrophoresis. A picture of this gel is shown in figure 2.

the CM-cellulose chromatography and SDS-gel electrophoresis pattern (4). Normally, Aα-chain degradation products appear in the CM-cellulose chromatography between the peaks corresponding to the Aα- and Bβ-chains, resulting in low yields of intact Aα-chain. In SDS-gel electrophoresis, Aα-chain degradation products appear as bands occuring between the Aα- and Bβ-chain bands. As can be seen in the SDS-gel electrophoretic pattern from pig Aα-chain in fig. 2, pool IV, these bands are extremely weak, indicating that very little degradation has taken place.

It is still unclear as to what factors are involved in the marked differences in stability between the various mammalian species. For example, the observed differences in susceptibility to proteolysis could be the result of slight structural differences between the fibrinogens. A more likely explanation

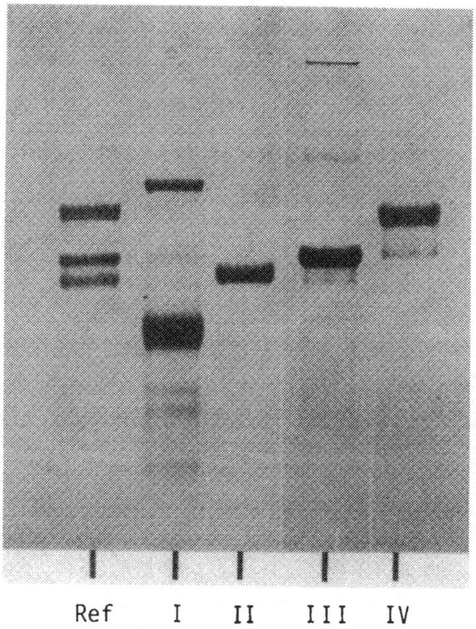

Fig. 2. Gel electrophoresis in SDS of S-carboxymethylated pig fibrinogen from pooled CM-cellulose column chromatography fractions. Ref: starting material before column chromatography. I: Pool I from column chromatography (fig. 1). The lower band consists of γ_y, a variant of the γ-chain, which differs in its size and charge (2). The distant upper band was not identified. II: Pool II. Pure γ-chain. III: Pool III. Pure Bβ-chain IV Pool IV. Aα-chain with very slight Bβ-chain contamination.

however may be the differing plasmin activity among the different species.

Finally, it might be suggested that because pig fibrinogen is more intact with respect to Aα-chain degradation, it would be better suited for physico-chemical investigations than fibrinogen from other species, where degradation is present to a greater extent.

Interspecial comparison of α-, β- and γ-chain amino acid sequence between man, pig and ox.

```
M      VV         D        E
P   GPRLGERHQSACKESDWPFCSDTD      pos.17-40
O       V  G       TG      E

M       L  K         AP    G
P   GHRPYDKRREEAPSLRPVAPPISAGGY   pos.15-41
O           K         P    G
```

Fig. 3. N-terminal amino acid sequence of α- and β- fibrin chains for Man (M), Pig (P) and Ox (O). Only sequences of Man and Ox different to those of Pig are shown. The position numbers refer to human fibrinogen.

```
M        D
P   YVATRENCCILDERFGSYCPTTCGIADFLS    pos.1-30
O        D                       N

M      AS L    S  Y            V L E  A    Q
P   KYENLITTHDSTIRFLQEIYNSNNQKISNIKQKVIQLEAK   pos.95-134
O      T  S  E                 V LRD  V    N

M   Q            QQ
P   ESGLYFIKPLKAPKKFLVY    pos. 163-181
O                QQ

M    I  F  LT
P   KLIPYNRIA    pos. 385-393
O    I  L  L
```

Fig. 4. Man (M), Pig (P), Ox (O) γ-chain amino acid sequences for positions shown.

The amino acid sequence of the N-terminal regions of the α- and β-fibrin chains for the species man, pig and ox in fig. 3. contain a number of positions, where the amino acids for all three species are non-identical. However, in the majority of these cases, the amino acids being considered have the same charge and character. An example of an exeption is found in the pig α-chain at position 39, where threonine is replacing the glutamic acid of man and ox.

In the γ-chain N-terminal sequence shown in fig. 4, only positions 6 and 30 have differing amino acids among the three species. This N-terminal region therefore posesses a marked conservatism in amino acid sequence in comparison with the internal regions of the γ-chain which are also shown in fig. 4. Since the fibrinopeptides are cleaved from the N-terminal side of fibrinogen, it would be expected that this region and in particular the fibrin side of the thrombin cleavage side would remain conserved. However, the observation that the N-terminal region of the γ-chain, which is linked to the Aα- and Bβ-chains in a "disulfide knot", posesses a much stronger structural conservatism than either the α- or β-chain N-termini has so far no obvious explanation.

Genetic analysis of the γ-chain amino acid sequences of man, pig and ox. From the 98 sequenced positions in the γ-chain segments shown in fig. 4, 21 of the positions contained for the three species being compared different amino acids. On the basis of the genetic code, the least number of nucleotides which have to be changed in the DNA in order to replace one amino acid for another may be calculated. When converting the amino acid sequence of one species into the corresponding sequence of another species, the number of nucleotides over the entire length of protein or peptide being considered are summed. This sum is a hypothetical measure of the mutational distance between the two corresponding but slightly different proteins (5).

	MAN	PIG
OX	15	19
MAN		21

Fig. 5. Mutational distances between Man/Pig/Ox calculated from γ-chain sequences given in fig. 4.

Using this method, a preliminary genetic analysis using the γ-chain sequences of man, pig and ox was done. The results shown in fig. 5 for the mutational distances calculated between the species imply that ox would be on the evolutionary scale closer related to man than to pig. However taxonomical evidence and sequence studies of hemoglobins and cytochrome c (6) support the theory that the primate line radiated from the other mammalian lines seperately. Reasons for the discrepancy in our findings may lie in the inability of detecting back-mutations, which could be determined only by analyzing a much larger variety of interrelated species. Also the evolutionary assumptions based on the preliminary genetic studies here may be unreliable due to lack of sufficient sequence information, since up to now only 25 % of the γ-chain sequences from pig has been determined.

References

1. Henschen, A., Lottspeich, F., Hessel, B.: Hoppe-Seyler's Z. Physiol. Chem. 360, 1951 (1979).
2. Henschen, A., Edman, P.: Biochim. Biophys. Acta 263, 351 (1972).
3. Kehl, M., Henschen, A.: unpublished results.
4. Kehl, M., Henschen, A.: this volume.
5. Fitch, W. M.: J. Mol. Biol. 26, 499 (1967).
6. Dayhoff, M. O.: Atlas of Protein Sequence and Structure Vol.4, Nat. Biomed. Res. Found., Md. (1969).

III. ABNORMAL FIBRINOGENS

ON DYSFIBRINOGENEMIA

Rudolf Marx and Wolfgang Schramm

Medizinische Klinik Innenstadt der Universität München,
Ziemssenstrasse 1, 8000 München 2

Introduction

We want to restrict our presentation to some clinical aspects of the relatively "young" clinical entity with an autosomal inheritance pattern: genetic dysfibrinogenemia.

Results

The first two cases of dysfibrinogenemia were reported from Parma (1958) and from Paris (1963) (table 1). Ménaché (10) described the first clear cut, genetic - familiar case PARIS I - according to (the nomenclature of) Beck (3) (1965). The genetic of this case was homozygote (no clot was formed).

Table 1

First Cases of Dysfibrinogenemia

1. 1958 Imperato and Dettori Fibrinogen Parma	 prolonged thrombin time (presumed heterozygote)
2. 1963 Menaché Fibrinogen Paris I	 no clot with thrombin (presumed homozygote)

Fibrinogen - Recent Biochemical and Medical Aspects

Table 2

Classification of 46 Cases of Dysfibrinogenemia (3 from Munich) According to Clinical Symptoms

Symptom	Frequency
1. bleeding	23
2. no symptoms	15
3. thrombosis	4
4. wound healing	2
5. bleeding/thrombosis	1
6. bleeding/wound healing	1

In general the most frequent symptom of dysfibrinogenemia (not confined to the homozygote) is a mild bleeding diathesis from mucosal membranes (table 2). In some cases cerebral hemorrhage and haemarthrosis are also reported (6). In a high percentage of cases reported in the literature no symptoms of bleeding or associated other disorders were found - at least up to the time of the publication of the respective papers. Thrombosis associated with dysfibrinogenemia is rare (e.g. dysfibrinogenemias Oslo (5), Paris II (11) and III (13), New York (1), Baltimore (4)). The pathogenesis of thrombosis is, as you know, a polyetiologic one. Therefore some of the episodes of thromboembolism associated with dysfibrinogenemia probably occurred in spite of dysfibrinogenemia, and were related to other factors. Only in very few families dysfibrinogenemia seems to be itself one factor of tendency for thrombosis (5,7).

Our 3 cases - dysfibrinogenemia Munich I (two sisters)(9) and Munich II (a young woman) - had also a mild bleeding tendency (hematuria; bleeding during pregnancy; skin hematoma; epistaxis) (table 3). In one case of dysfibrinogenemia Munich I and in the case Munich II the fibrinopeptide A generation assay of Schramm (12) revealed defects of the release of the fibrinopeptide A with the effect of a disturbed polymerization.

Table 3

Munich Dysfibrinogenemia

1.	FIBRINOGEN MUNICH I (1973) two sisters with mild bleeding diathesis (R. Marx, W. Schramm, G. Wanner, H. Murr, analysed by A. Henschen, 1979/1980)
2.	FIBRINOGEN MUNICH II (1980) a woman with (very) mild bleeding diathesis (R. Marx, G. Rindfleisch, F. Schüssler, H. Murr, M. Weinzierl, W. Schramm)

In the investigated case of dysfibrinogenemia Munich I the fibrinopeptide A release is - within 30 min. - reduced to 1/3, in the case dysfibrinogenemia Munich II to 1/6 of the normal amount. As you know, dysfibrinogenemia Munich I was studied by Henschen and Southan, and a molecular defect was found in position 19 of the alpha chain (8).

The hemostaseological parameters of the case Munich II are given in table 4. Case Munich II was first thought to be a case of primary hyperfibrino(geno)lysis, probably caused by a loss of the fast acting alpha-2-antiplasmin, because the staphylococcal clumping activity of serum from native venous blood was found to be very high on several occasions. A non-coagulable part of the heterozygote dysfibrinogen remains probably in serum. After 5 minutes of venous stasis the staphylococcal clumping activity in serum was 4 times higher than before. After treatment of the patient with 2 g of tranexamic acid for eight days the clumping activity in serum was much lower than before antifibrinolytic treatment. We concluded from this experiment that the dysfibrinogenemia Munich II had a fibrinogen more sensitive to plasmin than normal fibrinogen. It seems interesting to explore whether under normal conditions

Table 4

Dysfibrinogenemia Munich II

I. Clinical symptoms - nose bleeding and a tendency to hematoma (blue spots)

II. Hemostaseological analysis

A) <u>Coagulation:</u>

Lee-White-Coagulation time			- mild prolongation (13 min)
Thrombelastogram	r:	7 min	- normal
	k1:	10 min	- pathological
	ma:	59	- pathological
thromboplastin time	:	45 sec	- prolonged
partial thromboplastin time	:	47,5 sec	- mild prolongation
thrombin time	:	31 sec	- prolonged
coagulase time	:	50 sec	- prolonged
reptilase time	:	104 sec	- prolonged
fibrinogen-heat	:	180 mg%	- normal
fibrinogen-immunologically	:		- normal
fibrinogen-Clauss	:	37 mg%	- pathological
factor XIII	:		- normal

B) <u>Fibrinolysis:</u>

euglobulin lysis time	- normal
euglobulin lysis time (after stasis 10 min)	- accelerated
TEG lysis	- no lysis
clot lysis time	- after 24 h no lysis
Nilsson plate assay	- no lysis after stasis

C) <u>Studies of serum (staphylococcal clumping test):</u>

1. venous blood without stasis	1: 16384
2. item with 10 min stasis	1: 65536
3. item 2) with addition of thrombin (24 h)	1: 65530
4. after 8 days, dayly 2 g tranexamic acid	1: 1024
5. item 4) blood taken into aprotinin and thrombin	1: 1024

and without stasis the serum of the patient contains FDP aside from dysfibrinogen.

At the symposium of the "Deutsche Arbeitsgemeinschaft für Blutgerinnungsforschung" in Heidelberg in 1979 we published that dextran accelerates the thrombin coagulation time of the plasma of dysfibrinogen Munich I (9). It was shown by other authors that dextran is included into fibrin. This renders fibrin more susceptible to lysis by plasmin. Therefore it seems worthwhile to study the problems of hyperfibrino(geno)-lysis in genetic dysfibrinogenmia patients (Confer. Fibrinogen Valencia (2)) more thouroughly and also from a therapeutical point of view. Perhaps in the future an artificial dysfibrinogenemia may be of therapeutical value.

Summary: Dysfibrinogenemia seems to be of clinical relevance.

References

1. Al-Mondhiry, H. A., Bilezikian, S.B., Nossel, H.L.: Blood 45, 607-619 (1975).

2. Aznar, J., Fernandez-Pavón, A., Vila, V., Orellana, F.: Thrombos. Diathes. haemorrh. 32, 564-577 (1974).

3. Beck, E.A., Charache, P., Jackson, D.: Blood 24, 853 (1964).

4. Beck, E.A., Charache, P., Jackson, D.: Nature 208, 143-145 (1965).

5. Egeberg, O.: Thrombos. Diathes. haemorrh. 17, 176-187 (1967).

6. Flute, P.T.: Brit. med. Bull. 33, 253-260 (1977).

7. Graham, J.B., Barret, D.A., Blombäck, B., Cann, H.M., Hardisty, R.M., Larrieu, M.J., Renwick, J.H.: Thrombos. Diathes. haemorrh. 30, 2-11 (1973).

8. Henschen, A., Southan, C.: in: Fibrinolyse, Thrombose, Hämostase, E. Deutsch, K. Lechner (eds.), F.K. Schattauer, Stuttgart, 1980, 290-293.

9. Marx, R., Schramm, W.: in: Fibrinogen, Fibrin and Fibrin Glue, Kl. Schimpf (ed.), F.K. Schattauer, Stuttgart, 1980, 195-197.

10. Ménaché, D.: in: Proc. IXth Congr. Europ. Soc. Haematology Lisbon, II/2, Karger, Basel, 1963, 1255.

11. Samama, M., Soria, J., Soria, C.: Nouv. Rev. franc. Hémat. 9, 817-832 (1969).

12. Schramm, W.: Habil.-Arbeit, Munich 1981.

13. Soria, J., Soria, C.: Nouv. Rev. franc. Hémat. 14, 574 (1974).

DETECTION OF FIBRINOGEN ABNORMALITY IN DYSFIBRINOGENEMIA

SPECIAL REPORT ON FIBRINOGEN METZ CHARACTERIZED BY AN AMINO ACID SUBSTITUTION LOCATED AT THE PEPTIDE BOND CLEAVED BY THROMBIN.

Jeannette Soria, Claudine Soria, Meyer Samama, Agnes Henschen, Chris Southan

- Service du Prof. Bernadou et Laboratoires Centraux de Biochimie et d'Hématologie (Prof. Fabiani, Prof. Samama), Hotel-Dieu, F-75004 Paris
- Laboratoires Centraux de Biochimie et d'Hématologie (Prof. Rousselet, Prof. Caen), Hôpital Lariboisière, F-75010 Paris
- Max-Planck-Institut für Biochemie, D-8033 Martinsried/ München

Summary

A case of homozygous dysfibrinogenemia, characterized by an unclottability of citrated plasma, even in the presence of very high doses of human or animal thrombin, but, in contrast to this, only showing a prolonged whole blood clotting time, was studied. This abnormal fibrinogen was found to contain an abnormal Aα chain from which no fibrinopeptide A could be released. This led us to suspect a mutation in the vicinity of the peptide bond cleaved by thrombin. It turned out that arginine in position 16 in the Aα chain is replaced by cysteine. The platelet fibrinogen of this patient is normal.

Fibrinogen - Recent Biochemical and Medical Aspects

Introduction

Many cases of dysfibrinogenemia, a qualitative abnormality of fibrinogen, have recently been described ; however, the molecular abnormality has only been demonstrated in very rare instances : the study of this abnormal fibrinogen should provide useful information on the relationship between the structure of the molecule and the different functions of fibrinogen.

Two types of dysfibrinogenemia have so far been described.
- Acquired dysfibrinogenemias,in severe hepatic diseases are characterized by a prolonged thrombin time and an abnormal fibrin monomers aggregation. The anomaly in fibrin formation is related to a great increase in sialic acid content, since the desialidation by neuraminidase treatment, is capable of correcting the anomaly (1-5).
- Congenital dysfibrinogenemias are now very well known. However, in only two instances (fibrinogen Detroit (6) and fibrinogen Munich (7), the anomaly of the fibrinogen molecule has been localized : in the two cases, a substitution of arginine in position 19 of the Aα chain by serine and asparagine, respectively, has been demonstrated.

We have tried to localize the anomaly in congenital dysfibrinogenemia, in order to analyse by amino acid sequences the part of the molecule which is abnormal. In our study, we were able to detect, in only 3 cases, which fragment of the molecule contained the anomaly, as can be seen in table I.

For fibrinogen Metz, as predicted by our analysis, the protein chemical study revealed a mutation located at the peptide bond cleaved by thrombin, since in the Aα chain, arginine in position 16 was replaced by cysteine (14).

Table I : Main characteristics of 3 different abnormal fibrinogens

Dysfibrinogenemia tested	Anomaly detected by the study of the molecule	Location of the anomaly
METZ (8-10)	Abnormal migration of the Aα chain in disc electroresis. Absence of fpA release	Aα chain, at the site of thrombin cleavage
PONTOISE (11, 12)	Two types of Bβ chains detected by SDS-PAGE, in whole fibrinogen and in D fragment	Abnormal Bβ chain longer than normal (prolongation of the C terminal part)
HAIFA (13)	No protective effect of Ca in fragment D degradation by plasmin	Abnormal chain (γ_D fragment)

The analysis of this abnormal fibrinogen Metz is reported in this paper.

Case report.

The abnormality, in a 30 year old female, was observed because she had a history of moderate bleeding tendency, such as bleeding at the puncture sites when needles used were too large. She had also suffered repeated abortions.

Material and Methods

- Blood collection :

Nine parts of blood collected by venipuncture were drawn into one part of 0.13 M trisodium citrate, in the presence of apro-

tinin (Iniprol, Choay Lab. France). After centrifugation for 10 min. at 4°C, at 2,000 g, the plasma was either used immediately or stored at - 20°C.
Non citrated plasma was obtained after venipuncture by immediate centrifugation of native blood (4°C, 10 min.) in plastic disposable tubes and used immediately.

- Routine coagulation tests :
Bleeding time, whole blood clotting time, thromboplastin time, concentration of factor V, factors VII + X and factor II, thrombin and reptilase clotting times, factor XIII level and platelet aggregation were performed as described by Caen et al (15). The fibrinogen content was measured by three different technics : gravimetric (15), chronometric (15) and immunologic (Mancini (16), using Behring partigen). Fibrinogen related antigen in serum was estimated according to Merskey (17).

- Isolation of fibrinogen :
It was performed from plasma according to the Blombäck procedure (18) and from platelets according to Solum (19).

- Analysis of fibrin formation :
1. Spectrophotometric study of coagulation in the presence of thrombin was performed, according to Bouvier and Gruedlinger (20) by measuring the variations in time of the optical density of diluted native plasma in the presence of thrombin.
2. Fibrinopeptide A release from thrombin - treated fibrinogen was determined by an immuno enzymological assay (10).
3. Fibrin stabilization was studied by SDS-urea polyacrylamide gel electrophoresis (SDS-PAGE) in order to detect the formation by transamidation cross links of dimers of γ chains and polymers of α chains, induced by activated factor XIII (21). This study was performed from the clot obtained from native uncitrated plasma.

- Study of the fibrinogen molecule :

1. Immunoelectrophoresis was carried out according to Scheidegger (22).
2. Disc electrophoresis (in 4.5 % polyacrylamide gel) of a mixture consisting of purified fibrinogen and the 3 dissociated and alkylated chains was performed according to Soria et al (9).
3. SDS polyacrylamide gel electrophoresis (SDS-PAGE) of the 3 dissociated chains of fibrinogen was performed according to Schwartz et al (21).
4. Analysis of fibrinogen degradation products was performed according to Haverkate and Timan (23). A fibrinogen preparation in a medium containing either 2 mM $CaCl_2$ or 5 mM EGTA was digested with human plasmin. The digests were reduced with mercaptoethanol and subjected to SDS-PAGE using 9 % gels.
5. Amino acid sequence of the Aα chain was carried after mercaptolysis and alkylation of the fibrinogen. The three chains were separated by ion exchange chromatography according to Henschen (24). The amino acid sequence of the Aα chain N-terminus was determined according to the Edman procedure in a sequenator (25).

Results

1) Coagulation tests.

Resultats of clotting tests are summarized in table II.

Table II : Results of coagulation tests

	Normal	Patient
Bleeding time (min)	2-4	4
Whole blood clotting time (min)	6-12	40 (weak clot)
Clot retraction	+++	±
Prothrombin time (sec)	12	∞
Factors V, VII+X, II (%)	100	100
Fibrinogen (mg/100 ml)		
- Gravimetric	200-400	0
- Chronometric	200-400	0
- Immunologic	200-400	300
Thrombin clotting time (sec)	19	∞
Fibrinogen related antigen in serum (µg/ml)	$\leqslant$ 15	15

The anomaly was discovered because of an infinite prothrombin time with normal factor II, V, VII+X levels. Thrombin and reptilase clotting times were also infinite. However, a clot was formed in the absence of anticoagulant (whole blood clotting time very prolonged). Fibrinogen estimation by the immunodiffusion technique and by the clotting procedure were discordant : normal by Mancini (300 mg/100 ml), but not detectable by the clotting procedure.

Bleeding time, platelet count and platelet aggregation were normal.

2) Analysis of the prolonged thrombin clotting time

No clots appeared after addition of thrombin of various origins (at concentrations up to 5,000 U/ml) to citrated plasma. Neither did reptilase induce clotting.

Furthermore, the patient plasma remained unclottable after addition of thrombin in 0.01, 0.025 and 0.05 M $CaCl_2$.

However, for native plasma (in absence of citrate), the thrombin clotting time was only prolonged.

A partial correction of the apparent thrombin clotting time of citrated plasma by protamine sulfate was noted as it can be seen in table III.

Table III : Partial correction of the thrombin time of citrated plasma by protamine sulfate

	Thrombin time (sec)	
	Normal	Patient
Plasma	16	∞
Plasma + protamine sulfate 5,000 µg/ml 1,000 µg/ml 500 µg/ml	 6 8 12	 75 90 300

The thrombin and reptilase clotting times of the patient's plasma were corrected by the addition of normal plasma.

In a citrated solution of purified fibrinogen, no clot appeared after thrombin addition (5,000 U/ml). However, as in plasma, a clot was obtained in an uncitrated fibrinogen solution, but the thrombin time was very prolonged (control 10 sec, patient 5 min), and the clot was very weak.

3) Analysis of the fibrin formation abnormality

- The spectrophotometric study of coagulation revealed no modification of optical density after thrombin addition in the patient's plasma (citrated and uncitrated) or in purified fibrinogen.

- Kinetic determination of fibrinopeptide A release.
In our conditions, the release of fpA from thrombin-treated normal fibrinogen was very fast : 70 % of the fpA was released in 2 min., and it reached a plateau between 10 and 30 min. FpA was not released from Metz fibrinogen after treatment by thrombin (2.5 U/ml) (figure 1).

Figure 1 : Release of fpA after thrombin addition to normal fibrinogen and to Metz fibrinogen

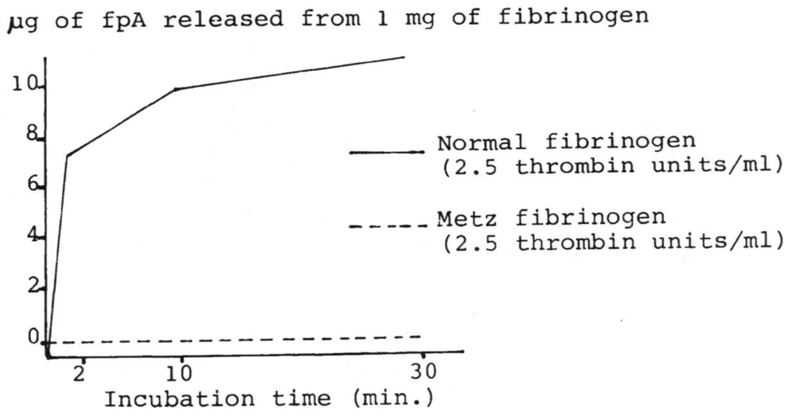

- Fibrin stabilization
SDS-PAGE revealed normal γ chain dimerization and normal α chain polymerization (figure 2)

Figure 2 : SDS-PAGE of reduced unstabilized and stabilized fibrin clots

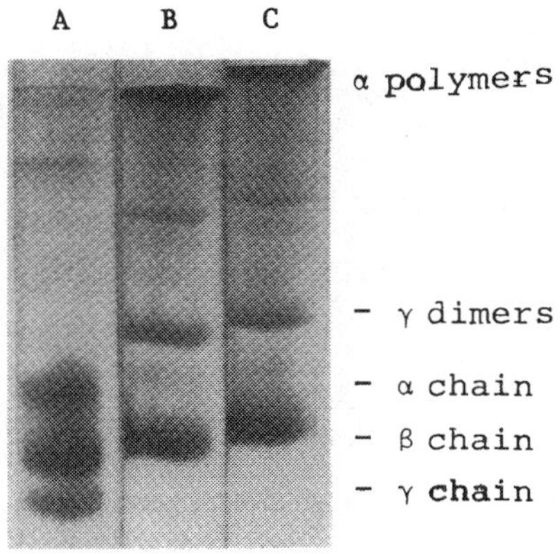

A : Normal unstabilized fibrin
B : Metz stabilized fibrin
C : Control stabilized fibrin

4) Study of the fibrinogen molecule

- The immuno-electrophoresis pattern was slightly abnormal (figure 3).

Figure 3 : Immuno-electrophoresis pattern

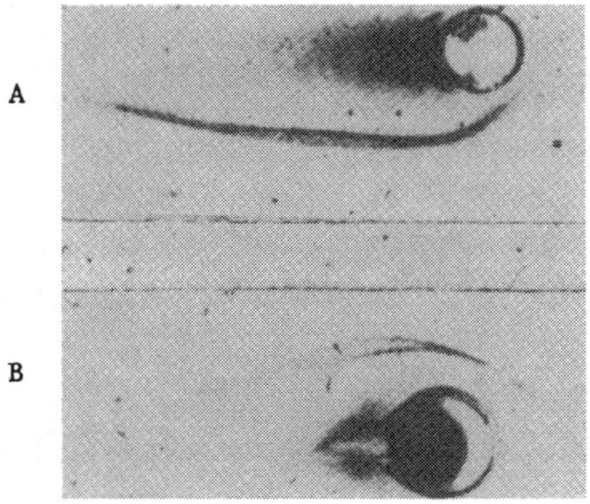

A : Metz fibrinogen
B : Control fibrinogen

- The SDS-PAGE patterns of the three dissociated chains of fibrinogen were identical for normal and abnormal isolated fibrinogens (figure 4).

Figure 4 : SDS-PAGE of reduced fibrinogen

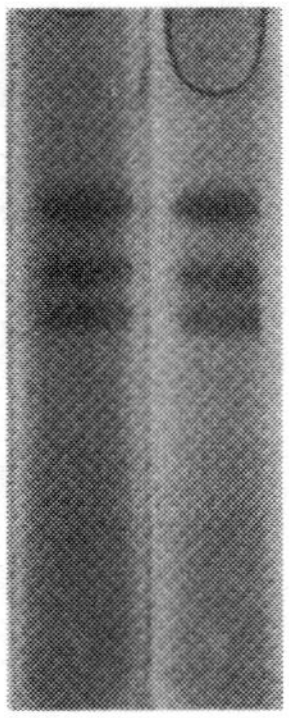

A : Metz fibrinogen

B : Control fibrinogen

- The SDS-PAGE of plasmin degradation products did not show any structural deviation in the fragment D moiety of fibrinogen Metz. Furthermore $CaCl_2$ had a protective effect upon further degradation of the Dγchain by plasmin, as has been shown previously by Haverkate et al. (12) (figure 5).

- Polyacrylamide disc electrophoresis in urea buffer of fibrinogen and mercaptolysed and alkylated chains of fibrinogen was performed :
  . for plasma fibrinogen, a more anodal migration was noted for the Metz Aα chain, than for the control (figure 6)
  . for platelet fibrinogen, no anomaly was detected : there were differences in the disc PAGE patterns between normal plasma and normal platelet's fibrinogens, but the migration of the Aα chain was similar. The abnormality in the electrophoretic mobility of the Aα chain of plasma fibrinogen is not found in the patient's platelet fibrinogen. (results not shown).

- The amino acid sequence of the Aα chain revealed a mutation characterized by the replacement in position 16 of artigine by cysteine (14).

<u>Figure 5</u> : SDS-PAGE of plasmin degradation products obtained in the presence or in the absence of $CaCl_2$

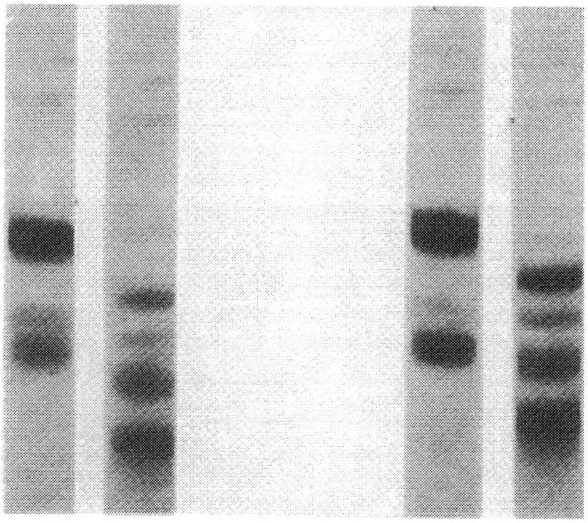

Plasmin treated fibrinogen - A : Normal + $CaCl_2$, B : Normal + EGTA, C : Metz fibrinogen + $CaCl_2$, D : Metz fibrinogen + EGTA

<u>Figure 6</u> : Polyacrylamide gel disc electrophoresis of fibrinogen and of mercaptolysed alkylated chains of fibrinogen

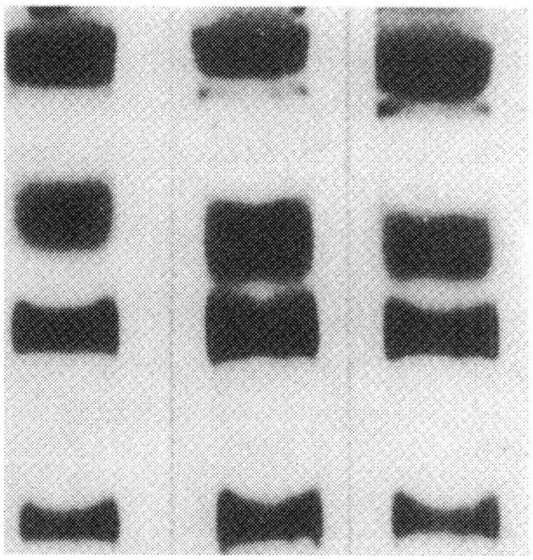

5) Family study

The parents of the propositus are cousins. The fibrin formation abnormality was found in both parents and in an aunt, but the intensity of the anomaly is more pronounced in the patient who is homozygous than in her parents who are heterozygous (figure 7).

Figure 7 : Familial study of Metz dysfibrinogenemia

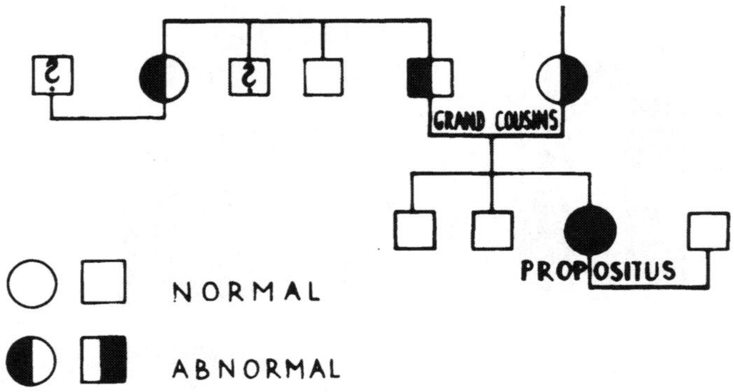

Discussion

This case of homozygous dysfibrinogenemia is characterized by the unclottability of citrated plasma even in the presence of very high doses of human or animal thrombin, and cannot be corrected by varying the calcium concentration. These results are contrasted by the prolonged whole blood clotting time and thrombin clotting time of native plasma obtained from blood collected in the absence of anticoagulant.

The abnormality of this fibrinogen is located in the $A\alpha$ chain which, on disc electrophoresis, migrated further towards the anode than the control. This anomaly is not related to a degradation of fibrinogen because of the normal pattern in SDS-PAGE on plasmin digestion.

The absence of fpA release, after thrombin addition to purified fibrinogen, associated with the Aα chain anodal mobility led us to suspect a mutation located on the peptide bond cleaved by thrombin. A further investigation confirmed this hypothesis : the arginine located in position 16 of the Aα chain was found to be replaced by cysteine. This mutation may be explained by a single substitution of the first base of the codon triplet for arginine (cytosine is replaced by uracil).

The amino acid substitution can explain the increase anodal mobility of the Aα chain.

The anomaly was not found in platelet fibrinogen. As the fibrinogen Metz anomaly reflects a genetic mutation, the absence of mutant Aα chain from platelet fibrinogen gives evidence that platelet and plasma fibrinogen pools are separate from each other (26). In related experiments, Jandrot-Perrus (27), comparing fibrinogens from plasma and platelets in patient having dysfibrinogenemia Paris I, has demonstrated also that the band corresponding to the γ-Paris I chain present in plasma fibrinogen is not present in platelet fibrinogen.

The discrepancy between the unclottability of citrated samples with thrombin and the clottability of whole blood has not been completely elucidated. Further experiments are needed to explain the mechanism of blood clotting in this patient.

References

1. SORIA J., COUPIER J., SAMAMA M., TIXIER P., BILSKI-PASQUIER G.: XII Congr. Internat. Soc. Haematol., Abstr. p. 180 (1968).

2. SORIA J., SORIA C., SAMAMA M., COUPIER J., GIRARD M.L., BOUSSER J., BILSKI-PASQUIER G.: Coagulation 3, 37 (1970).

3. FELTEN A. Von, STRAUB P.W., FRICK P.G.: New Engl. J. Med. 208, 8, 455 (1969).

4. MARTINEZ J., PALASCAZ J.E., KWASNIAK D. J. Clin. Invest.: 62, 2, 535 (1978).

5. SORIA J., SORIA C., RYCKEWAERT J.J., SAMAMA M., THOMSON J., POLLER L.: Thromb. Res. 19, 1/2, 29 (1980).

6. BLOMBÄCK B., BLOMBÄCK M., MAMMEN E.F., PRASAD A.S.: Nature 218, 134 (1968).

7. HENSCHEN A., LOTTSPEICH F., SOUTHAN C., TÖPFER-PETERSEN E.: In : Protides of Biological Fluids. Peeters H. ed. Pergamon Press 28, 51 (1980).

8. SORIA J., SORIA C., SAMAMA M., POIROT E., KLING C. Thromb.: Diath. Haemorrh. 27, 3, 619 (1972).

9. SORIA J., SORIA C., BOULARD C.: Biochimie 54, 415 (1972).

10. SORIA J., SORIA C., RYCKEWAERT J.J.: Thromb. Res. 20, 425 (1980).

11. SORIA J., SORIA C., GUYADER A.M., SABATER F.: XVII Congr. Internat. Soc. Haematol. Paris, Abstr. p. 940 (1978).

12. HAVERKATE F., TIMAN G., SORIA J., SORIA C., SAMAMA M.: Thromb. Res. 13, 689 (1978).

13. SORIA J., SORIA C., TAVORI S., SAMAMA M., RIMON A., TATARSKY I.: To be published.

14. HENSCHEN A., SOUTHAN C., SORIA J., SORIA C., SAMAMA M.: To be published.

15. CAEN J., LARRIEU M.J., SAMAMA M.: L'hémostase. Méthodes d'exploration et diagnostic pratique. Expansion Scientifique. Paris (1968).

16. MANCINI G., CARBONARA A.O., HERMANNS J.F.: Immunochemistry, 2, 235 (1965).

17. MERSKEY C., LALEZARI P., JOHNSON A.: Soc. Exp. Biol. (N.Y.) 131, 871 (1969).

18. BLOMBÄCK B., BLOMBÄCK M.: Arkiv Kemi 10, 415 (1956).

19. SOLUM N.O., LOPACIUK S.: Thromb. Diath. Haemorrh. 21, 419 (1969).

20. BOUVIER C., GRUEDLINGER J.: Schweiz Wschr. 93, 1441 (1963).

21. SCHWARTZ M.L., PIZZO S.V., HILL R.L., Mc KEE P.: J. Clin. Invest. 50, 1506 (1971).

22. SHEIDEGGER J.J.: Int. Arch. Allergy Appl. Immun. 7, 103 (1955).

23. HAVERKATE F., TIMAN G.: Thromb. Res. 10, 803 (1977).

24. HENSCHEN A., EDMAN P.: Biochim. Biophys. Acta 263, 351(1972).

25. EDMAN P., HENSCHEN A.: In : Protein sequence determination (Needleman S.B. ed.) 2nd edn, p. 232, Springer, Berlin (1975).

26. SORIA J., SORIA C., SAMAMA M., POIROT E., KLING C.: Path. Biol. 24 Suppl. 15 (1976).

27. JANDROT-PERRUS M., MOSESSON M.W., DENNINGER M.H., MENACHE D. Blood 54, 5, 1109 (1979).

BINDING OF CITRATE TO NORMAL FIBRINOGEN AND TO METZ FIBRINOGEN

Philippe Ladure, Yves Ricordel, Jeannette Soria, Claudine Soria, Meyer Samama.
- Service du Prof. Bernadou et Laboratoires Centraux de Biochimie et d'Hematologie (Prof. Fabiani, Prof. Samama), Hotel Dieu, F 75004 Paris
- Laboratoires Centraux d'Hématologie et de Biochimie (Prof. Caen, Prof. Rousselet), Hopital Lariboisière, F 75010 Paris
- Hopital des Armées du val de Grâce (Service de Biochimie-Département des Radio éléments), F 75005 Paris

Summary

The binding of citrate has been measured in normal and in Metz fibrinogens, using a technique of isotope exchange. Normal fibrinogen binds about 4 moles of citrate per mole of fibrinogen, but Metz fibrinogen binds about 15 moles of citrate per mole. These results allowed us to suggest that the excessive binding of citrate to Metz fibrinogen is responsible for the discrepancy between the prolonged whole blood clotting time and the absence of clotting of citrated plasma after thrombin addition, even in presence of large excess of calcium.

Introduction

The dysfibrinogenemia from Metz is a homozygous case. The fibrinogen of the patient is characterized by a mutation in the Aα chain located at the bond cleaved by thrombin, since the arginine located in position 16 of the Aα chain was found to be replaced by cysteine (1). Despite this anomaly, the whole blood of the patient clots, contrasting with an unclottability of citrated plasma even in the presence of very high doses of human or animal thrombin. The aim of this study is to explain the discrepancy between the prolonged whole blood clotting time, and the absence of clotting of citrated plasma after thrombin addition. In this work, we have studied the binding of labeled citrate to normal fibrinogen and to Metz fibrinogen.

Fibrinogen - Recent Biochemical and Medical Aspects

Material and Methods

Material used for the study of labeled citrate binding.

- Citric acid labeled on carboxylic functions with ^{14}C.Specific activity: 30 mc $nmol^{-1}$.This product was provided in sealed glass vials by the "Commissariat à l'Energie Atomique",Departement Biologie,Gif sur Yvette,F. Radioactive citric acid was used as trisodium citrate in a solution alkalinized with 0.01 N NaOH.
- Scintillation fluid :Counting solution A.C.S.Ref.196290,Amersham.
- Millipore ultrafiltration membranes : Worthington ultrafilter 4000 dalton
- Beta liquid scintillation counter SR 4000,Kontron.

Methods

-Whole blood clotting time was performed according to Lee and White (2).
- Plasma was obtained vy centrifugation at 2000 g for 20 minutes of blood collected in 0.13 M trisodium citrate (9vol/1vol.).
- Fibrinogen was isolated by the method of Blombäck and Bombäck (3)
- Thrombin time (4) was performed using thrombin of various origins (equine from Roussel,bovine from Roche and human thrombin was kindly given by Dr Steinbuch,C.N.T.S..F.)

Thrombin was either diluted in saline ,or in calcium chloride solution (0.01 - 0.025 and 0.05 M).Thrombin clotting time was also performed on citrated plasma diluted 1:2 with several dilutions of protamine sulfate (Lilly).

- Immunoelectrophoresis was performed by the technique of Scheidegger (5) using antifibrinogen from Behring.
- Fibrinogen electrofocusing was performed in a 4.5 % polyacrylamide gel using an equal part of ampholines pH 4-6 and ampholines pH 5-8.Each gel was loaded with 50 µg of purified fibrinogen in the presence or in the absence of trisodium citrate.The fibrinogen samples were obtained by adding either 1 volume of 0.13 M trisodium citrate or 1 volume of 0.13 M NaCl to 5 volumes of fibrinogen solution at 2.5 mg/ml (6).

Electrofocusing was also performed with citrated fibrinogen previously incubated with protamine sulfate (200 µg/ml,final concentration).

- Study of citrate binding to fibrinogen was carried out using a technique of isotope exchange between fibrinogen-bound citrate and free labeled citrate.5 nmoles of normal or Metz fibrinogen were diluted in 2ml 0.15 M NaCl. These solutions were incubated at 4°C for various lenghts of time from 0 to 2 hours with 0.5 ml of 0.13 M trisodium citrate (or 75,000 nmoles) containing a tracing dose of 16 nmoles trisodium ^{14}C citrate (in order to check if free citrate is removed from fibrinogen). Each sample was then ultrafiltered at 4°C under slight vacuum using ultrafiltration millipores capsules,until 2.4 ml ultrafiltrate were obtained.3washes were then performed,followed by 3 ultrafiltrations using 5 ml of 0.15 M NaCl.The radioactivity of the last ultrafiltrate was nearly zero (against back-ground). Fibrinogen remaining on the membrane was then diluted in 2.5 ml 0.15 M NaCl.

The citrate bound to fibrinogen was then labeled by isotope exchange procedure : fibrinogen solution was incubated for 3 hours at room temperature with 50 μl of a solution containing 33 nmoles of ^{14}C citrate (1 μcurie). Then,the solution was concentrated 5 times by ultrafiltration.Labeled citrate bound to fibrinogen and free labeled citrate were then separated using a precipitation technique with half-saturated ammonium sulfate. After centrifugation,the pellet was collected and then washed several times with 2 ml of half-saturated ammonium sulfate until the radioactivity in the supernatant was identical to the background (five washes were usually necessary).The pellet was then dissolved in 2 ml 0.15 M NaCl.The concentration of fibrinogen was determined using the specific absorption at 282 nm and β radioactivity was measured using 200 μl aliquots mixed with 10 ml scintillation fluid.The results are expressed as moles of citrate bound per mole of fibrinogen.

Results

In absence of citrate,whole blood clots : whole blood clotting time (40 minutes).

On the contrary,citrated samples do not clot after addition of thrombin of various origins up to 5,000 U/ml.The patient's citrated plasma remained unclottable after addition of thrombin dissolved in calcium chloride

(0.01 M,0.025 M and 0.05 M).
However,a partial correction of the thrombin clotting time of citrated plasma was noted after addition of protamine sulfate (patient : 90 sec / control : 8 sec,using a citrated plasma diluted 1 : 2 with 1,000 µg/ml protamine sulfate).

-Immunoelectrophoretic pattern : it appeared that for normal fibrinogen, no difference was noted between citrated and uncitrated sample.On the contrary,for Metz fibrinogen,the addition of citrate to fibrinogen induced a migration towards the anode (fig.1).

- Fibrinogen electrofocusing : the addition of sodium citrate to fibrinogen solution induces a modification of the fibrinogen isoelectric point which is very slight for normal fibrinogen and a little more evident for Metz fibrinogen.(fig.2).

- Binding of ^{14}C citrate to fibrinogen : After 60 minutes incubation of fibrinogen with citrate,one mole of normal fibrinogen binds 3.81 ± 0.71 moles of citrate (mean of 14 experiments),but one mole of Metz fibrinogen binds 14.80 ± 1.80 moles of citrate (mean of 5 experiments).The kinetic variations of citrate binding to normal and Metz fibrinogens are shown in table 1.

Table 1: Kinetics of citrate binding to normal fibrinogen and Metz fibrinogen

Incubation time (in minutes) of fibrinogen with citrate	Moles of citrate bound per mole of fibrinogen	
	Normal	Metz
0	4.1	8.9
30	4.1	11.5
60	4.2	15.0
120	3.9	14.5

Fig 1 : Immunoelectrophoretic pattern of Metz fibrinogen in the absence or presence of citrate.

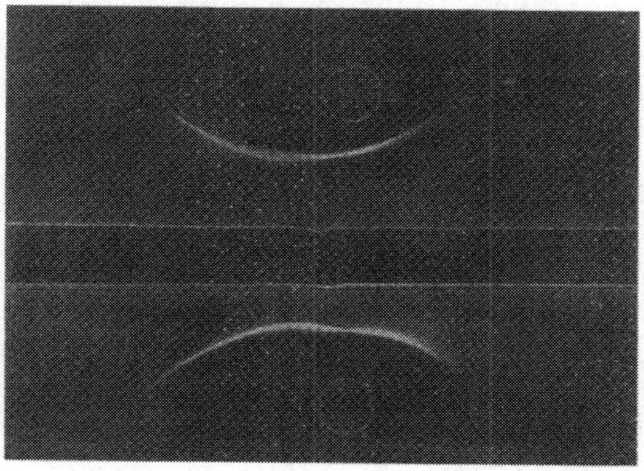

Uncitrated fibrinogen

Citrated fibrinogen

Fig.2 : Fibrinogen electrofocusing of non citrated and citrated normal and Metz fibrinogens

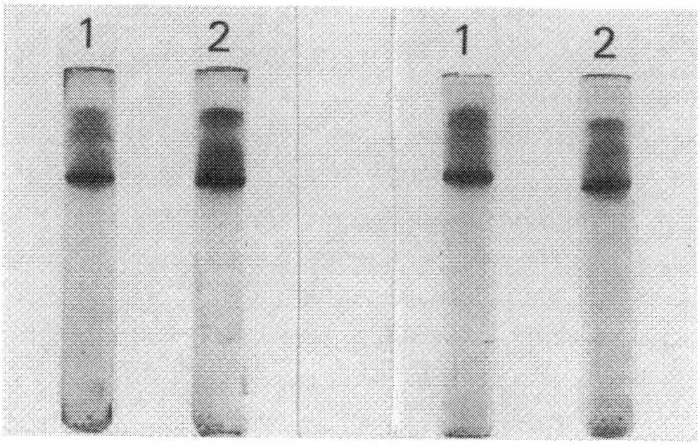

A : Normal fibrinogen B : Metz fibrinogen

(1 : uncitrated - 2 : citrated)

Discussion

From our results,it appeared that Metz fibrinogen in whole blood or native plasma (in absence of citrate)was clottable by thrombin.But the thrombin time was prolonged.In citrated plasma,fibrinogen was unclottable even with very high concentrations of thrombin of various origins.Calcium chloride added to thrombin did not induce any correction.A partial correction was only induced by addition of protamine sulfate to plasma.Therefore,it has been suggested that the unclottability of fibrinogen Metz induced by citrate might be related to a modification of the charge,partially corrected by addition of protamine sulfate.From our results,its appeared that Metz fibrinogen binds more citrate than normal fibrinogen,since immunoelectrophoretic and isoelectrofocusing patterns have shown an increase in negative charge in Metz fibrinogen induced by citrate addition.This modification of charge was not so obvious with control fibrinogen.Thus,immunoelectrophoresis and electrofocusing account for an increased binding of citrate on Metz fibrinogen.

A confirmation of this hypothesis was brought about by the experiment performed using labeled citrate.The experiment showed that Metz fibrinogen binds 4 times more citrate than normal fibrinogen.The binding reaction was carried out with a large excess of citrate,to bring the condition as close as possible to those of blood sampling.Such an excess of citrate was necessary because the binding of a single tracing dose of labeled citrate revealed binding values less than those obtained in excess citrate.Using a large excess of citrate,the number of citrate molecules involved (75,000) is very high when compared to the few molecules of bound citrate. This made it necessary to use a 2 step method : first,the citrate solution was incubated with fibrinogen,the excess being discarded,secondly,bound molecules were labeled with isotope exchange with ^{14}C citrate.

A direct method using labeled citrate in excess was not possible,because, it would have implied the manipulation of large quantities of radioactivity (in this condition,the precision and the reliability of the assay would be very poor).These results allowed us to suggest that the excessive binding of citrate to Metz fibrinogen prevent blood clotting.

References

1. Soria, J., Soria, C., Samama, M., Henschen, A., Southan, C.: In this volume.

2. Lee, R.I., White, P.D.: Am. I. Med. Sci., 145, 495 (1913).

3. Blombäck, B., Blombäck, M.: Arkiv. Kemi 10, 415 (1956).

4. Caen, J., Larrieu, M.J., Samama, M.: in: L'Hemostase-Methodes d'exploration et de diagnostic pratique. L'Expansion Scientifique. Paris (1975).

5. Scheiddeger, J.J.: Int. Arch. Allerg. 7, 103 (1955).

6. Wrigley, C.W.: Methods Enzymol. 22, 559 (1971).

THE SEARCH FOR MOLECULAR DEFECTS IN ABNORMAL FIBRINOGENS

Chris Southan, Agnes Henschen and Friedrich Lottspeich
Max-Planck-Institut für Biochemie,
D-8033 Martinsried, GFR

Introduction

The abnormal fibrinogens referred to in this article are those classified as the congenital dysfibrinogenaemias. For an introduction to this subject readers are referred to recent reviews (1,2) and to several of the articles appearing elsewhere in this volume. Congenital dysfibrinogenaemia or genetically abnormal fibrinogen is diagnosed in most cases by a prolonged thrombin time showing an autosomal dominant pattern of hereditary transmission. To date at least 46 cases have been reported, involving over 150 individuals. Consideration of the available data indicates that in many cases the genetic abnormality may be expressed by some or all of the fibrinogen molecules having an altered amino acid sequence. Out of the possible genetic rearrangements that can give rise to such sequence alterations by far the most common are the so-called point mutations by which a single, stable base exchange in the DNA leads to an amino acid substitution. For human haemoglobin, over 120 cases of this type of mutation have been characterised (3).

We have been concentrating our experimental attention on those abnormal fibrinogens where available data indicate the possible presence of an amino acid substitution in the vicinity of the thrombin cleavage site in the Aα-chain. Such an exchange has already been discovered in the case of fibrinogen Detroit (4) where in position 19 of the Aα-chain arginine has been substituted by serine.

Fibrinogen - Recent Biochemical and Medical Aspects

The following table gives an outline of the experimental strategy we have been using.

1) Consideration of clinical and in-vitro data. As far as this point is concerned we are fortunate in not having to involve ourselves with clinical investigations because several research groups have been very obliging in sending abnormal fibrinogens to us, either in the form of a patient's plasma or as purified fibrinogen material. Data for the individual cases of particular interest are: the kinetics and quantitation of thrombin-induced clot formation, indications of the presence of slow or non-clottable fractions of fibrin-(ogen), and the kinetics and stoichiometry of fibrinopeptide release. Such information, if available, gives a strong indication of whether the structural abnormality could be located in the N-Terminal region of the Aα-chain and if this is present in the heterozygous or homozygous form. The homozygous form, as would be expected, is rarer and seems to be confirmed so far for only a few cases, such as Detroit (4) and Metz. The heterozygous condition is experimentally more difficult because both normal and abnormal sequences will be present.

2) Isolation and purification of the abnormal fibrinogen. Considering the large number of purification techniques available for fibrinogen it would seem superfluous to develop a new one. However, on the basis of some theoretical and technical considerations we have developed a new method for the isolation of fibrinogen-like proteins from small amounts of plasma. This will be described later when referring to the isolation of fibrinogen Munich.

3) S-Carboxymethylation. This is a standard procedure for cyst(e)ine containing proteins and is used to prevent oxidation or cross-linking problems during subsequent experimental manipulations.

4) N-Terminal sequence analysis. A sequenator run of 3-4 steps gives information about the amount and purity of the preparations and the extent of Aα-chain N-terminal heterogeneity.

5) Chain separation. This is a routine procedure performed on a CM-cellulose ion exchanger using a pH gradient in 8 M urea (5). Such a separation gives a good yield of pure Aα-, Bβ- and γ-chains and can be carried out with between 10 mg and 10 g of S-carboxymethylated fibrinogen material.

6) PAGE-SDS profile of chain separation. This gives a qualitative and semi-quantitative analysis of the three major components, minor components and fragments. Overloades gels can detect minor impurities in the various fractions or pools. Such information in combination with N-terminal determination is useful for analysing the chain separation, especially the area between the Bβ- and Aα-chains where mixtures of fragments derived from both chains can occur.

7) Direct sequencing of Aα-chain material. After a chain separation, it is possible to prepare Aα-chain material for a direct sequencing run of about 30 residues into the N-terminal portion of the chain. This allows the'search for substitutions in those parts of the sequence directly involved in specific binding and cleavage by thrombin. In three cases reported here such runs have enabled identification of an amino acid exchange in the vicinity of the thrombin site.

8) Thrombin digest of Aα-chain material followed by N-terminal analysis. Such a test can give an indication of the relative amounts of cleavable and noncleavable components of Aα material in mixtures of normal and substituted chains derived from heterozygotes.

9) Fragmentation strategies for the separation of normal from substituted Aα-chain segments. It has already been mentioned that most abnormal fibrinogens can be expected to be heterozygotes. A complete phenotypic characterisation must include an experimental determination of the proportion of abnormal components present, i.e., quantitation of the heterozygous state. For many abnormal haemoglobins such determinations have been done (6) and the accumulated data has yielded important information regarding the genetic organisation and population variance of the haemoglobin molecule. Amino acid substitutions involving a difference in charge have allowed straightforward electrophoretic or isoelectric focussing techniques to be used for a reliable quantitation of the heterozygous state for many abnormal haemoglobins. Comparable experiments for fibrinogen would obviously be technically more difficult, but some preliminary experiments will be described below.

Fibrinogen Munich I

Abnormal fibrinogens can be found anywhere from Vancouver to Tokyo to Buenos Aires, but we decided to start our search here in Munich. Marx and Schramm from the University clinic in Munich have discovered two cases of dysfibrinogenemia now designated Munich I and Munich II. The clinical information has been reported elsewhere (7), but to summarise Munich I has a thrombin time of over 300 seconds, reduced fibrinopeptide A release, and a fraction of unclottable fibrin(ogen) seems to be present. Thus this patient, a woman, who shares the defect with her sister, appeared to be a good candidate for a heterozygous Aα-chain abnormality.

The new fibrinogen purification method, which was used for the Munich I plasma, was developed principally to reduce the possible selectivity of the classical precipitation and affinity techniques while at the same time giving good yields with small amounts of plasma (i.e. less than 50 ml) (8).

The first step consists of a gel filtration of the plasma through a Sephacryl column. The elution profiles are shown in Figure 1 for 5 ml (lower curve) and 30 ml (upper curve) of plasma. Figure 2 shows the SDS-PAGE patterns of fraction from this chromatography. The three bands characteristic of the three fibrinogen chains can be clearly identified. The fibrinogen containing section of the chromatography is marked by the broken lines in Figure 1.

Other gel filtration systems may give better resolution of the high molecular weight components of plasma, but the conditions were chosen for the combined advantages of rapid separation

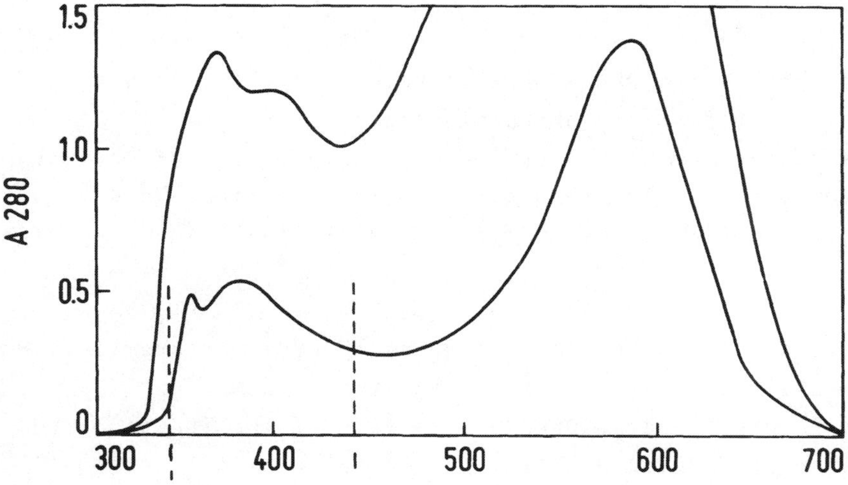

Fig. 1. Sephacryl S-300 gel filtration chromatography of human blood plasma.

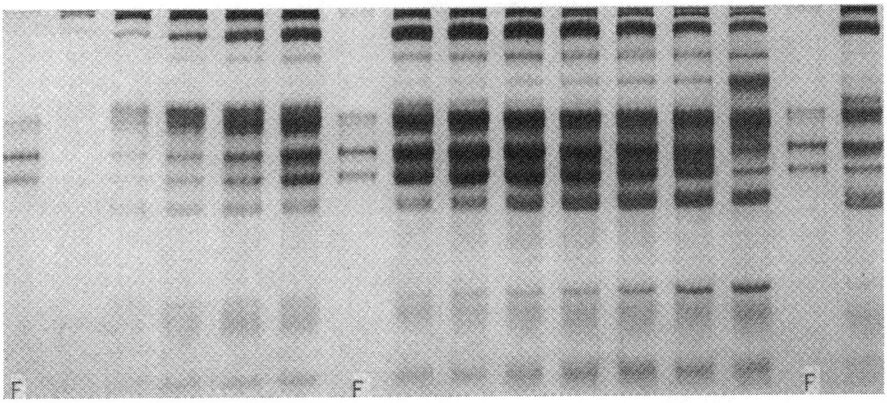

Fig. 2. SDS-PAGE (10% slab gel) of the fibrinogen containing section from the Sephacryl S-300 gel filtration chromatography of plasma. F denotes pure fibrinogen.

(less than three hours), high yield (85% based on clottability), and elution of the fibrinogen containing material in the same buffer as required for the subsequent purification step.

The second purification step, an ion-exchange chromatography, was performed with a DEAE-Sephacel column, the fibrinogen being eluted by a concave Tris-phosphate gradient (Figure 3). The peak, designated as pool I on the elution profile, was reduced in volume by ultrafiltration and then freeze-dried. This material consists of fibrinogen of sufficient purity for protein chemical analysis.

The purification can be followed directly by S-carboxymethylation and chain separation. This separation for Munich I is shown in Figure 4 and a SDS-PAGE analysis of the fractionated material shown in Figure 5. The pooled material of sections I to VII indicated in Figure 4 correspond to the first seven samples of the slab gel in Figure 5. Purified material from the γ-, Bβ- and Aα-chains can be seen in Lanes 1, 3 and 7 respecti-

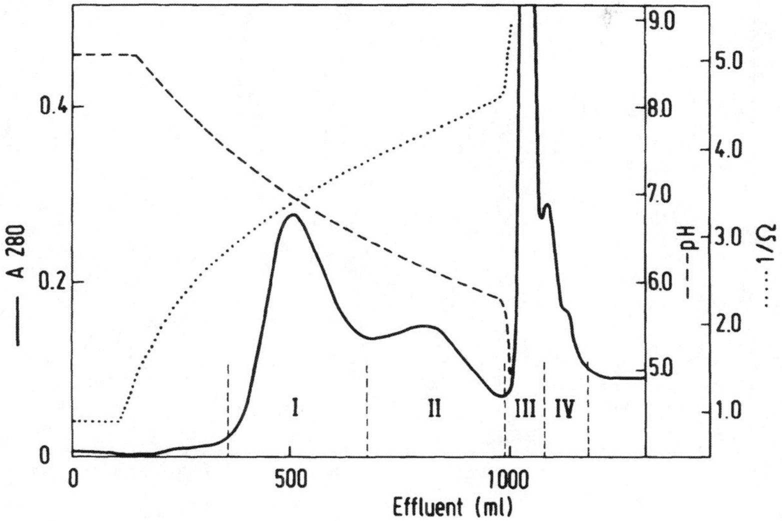

Fig. 3. DEAE-Sephacel ion exchange chromatography of fibrinogen-containing pool.

vely (left to right). Lane 8 is S-carboxymethylated fibrinogen Munich I and Lane 9 a standard fibrinogen preparation.

Pure Aα-chain material was subjected to sequenator analysis and showed in position 19 significant amount of asparagine as well as the expected arginine, thus making fibrinogen Munich the second fibrinogen (after fibrinogen Detroit) to show an amino acid exchange at position Aα 19. A short thrombin digestion of Aα material gave three sets of N-terminals Ala-Asp, Gly-Pro and a small amount of Val-Val. The result may be explained as due to the heterozygocity in the Aα-chain population, the abnormal Aα-chains being split at a much slower rate by thrombin.

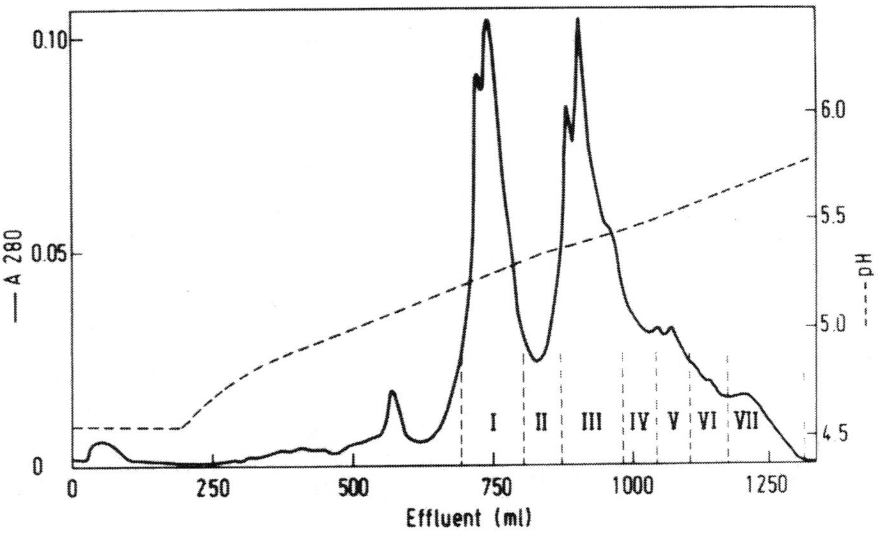

Fig. 4. CM-cellulose chromatography of Scm-fibrinogen Munich I, separation of the three component chains.

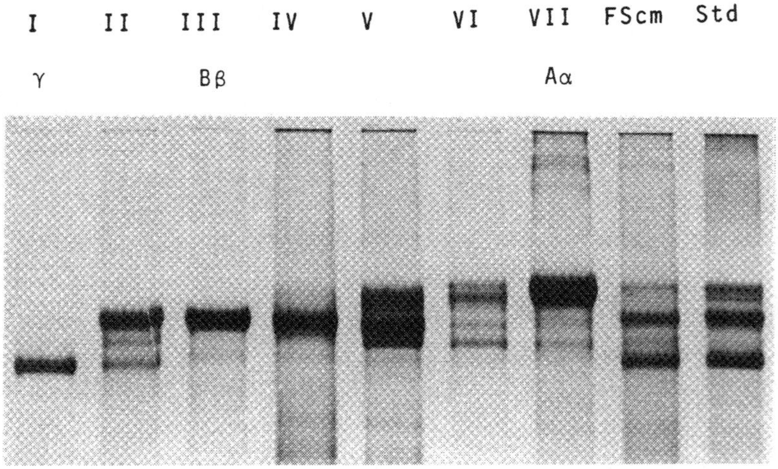

Fig. 5. SDS-PAGE 10% slab gel of pools from the CM-cellulose chromatography of Scm-fibrinogen Munich I showing the component chains.

Fibrinogen Metz

The second abnormal fibrinogen which will be reported on here is a homozygous case. The clinical and biochemical aspects of this case are reported in a preceding article in this volume and elsewhere (9). Fibrinogen Metz is characterised by an infinitely long clotting time and shows no fibrinopeptide A release at all with thrombin treatment. The fibrinogen was purified by Soria and Soria. After S-carboxymethylation of the material, a chain separation was carried out as shown in Figure 6.

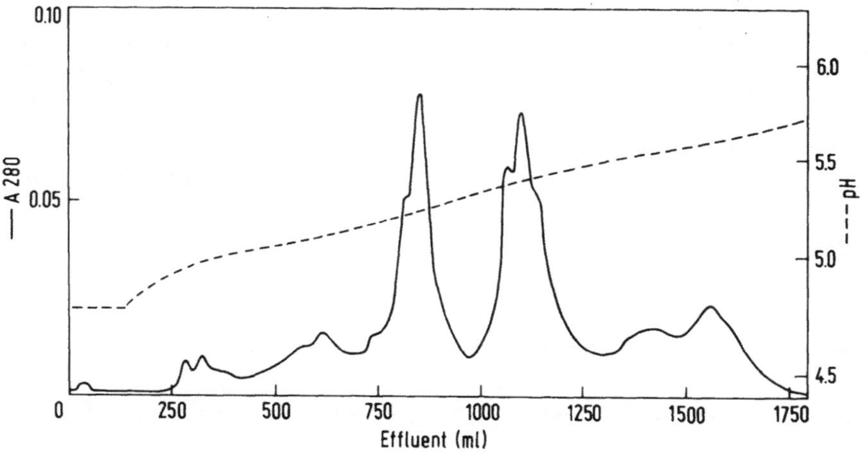

Fig. 6. CM-cellulose chromatography of Scm-fibrinogen Metz.

Sequencing of the Aα material showed the expected arginine at position 16 to be completely substituted by S-carboxymethyl cysteine, thus explaining the inability of thrombin to cleave fibrinopeptide A from the fibrinogen molecule.

Fibrinogen Zurich

The third abnormal fibrinogen to be reported on here comes from Zurich, and was isolated by Hofmann. Zurich I (10) has a prolonged thrombin time and from the stoichiometry and kinetics of fibrinopeptide release appears to be a heterozygote. When subjected to a protein chemical analysis similar to that already outlined for fibrinogen Metz, the presence of both S-carboxymethyl cysteine and arginine in position 16 of the Aα-chain was detected, thus making Zurich I a heterozygous case of the substitution occuring in Metz. Thrombin digestion of the same material also gave three N-terminals indicating a mixture of normal and substituted Aα-chains. A preliminary investigation by performing gel electrophoresis of native fibrinogens under non-reducing conditions indicates that the extra cysteine residues present in fibrinogens Metz and Zurich does not participate in inter-molecular cross-linking, i.e. significant amounts of molecular species with molecular weights above 340 000 were not observed.

Implication of Negative Results

The three cases described here represent positive results. However, negative results may be as important as positive ones, but are often more difficult to establish. For this reason, the possibility of negative results needs some qualification. Consideration of the sequencing detection system suggests a general confidence limit of about 20%, i.e. if an amino acid exchange was present in the sequenced material at a level of greater than 20% it should have been detected. However, the effects of low gene dosage, reduced genetic expression, or selectivity of purification procedures might give rise to heterozygous material in which the proportion of abnormal chains was less than 20%. The question arises if it would be possible to extend the search throughout the whole molecule even though very large amounts of material would be needed for fragmentation, fingerprinting, peptide purification, amino acid analysis, and sequencing to check through each of the 1482 amino acids. In practice, extensive protein chemical analyses of abnormal fibrinogens are not feasible without preliminary biochemical characterisation of the molecule to localise the site of the defect at least down to the level of the particular chain or chain segment.

Quantitation of the Heterozygous State

Attempts have been made to separate normal from substituted Aα-chains in a quantitative way in order to determine the ratio of the two components present in the heterozygous cases already described (Munich I and Zurich I). To perform such determinations on the total Aα-chains is difficult, considering the heterogenaity of Aα material due to the presence of C-terminally degraded chains. However, isolating and fractionating the N-terminal cyanogen bromide fragment, Aα 1-51 seemed to be feasible. Inspection of the abnormal sequences (Figure 9) suggested that, especially in the case of the Arg → Cys substitution in position 16, the charge difference between the S-carboxymethylated forms could allow a clear separation of normal and substituted components at the level of this fragment. Isolation of the fragments from normal Aα material is shown in Figures 7 and 8. A cyanogen bromide cleavage followed by a gel filtration and an ion-exchange chromatography were performed in a similar way to that already published (11).

On Sephadex chromatography the fragment, CB 1-51, is found in the pool indicated on Figure 7, together with some fragments of similar size. An ion-exchange chromatography, using a pH gradient, gives a quantitative recovery of pure N-terminal fragment (Figure 8). Preliminary experiments with fibrinogen Metz indicate that the cysteine substituted fragment elutes with the starting buffer under the conditions shown in Figure 8. Similar

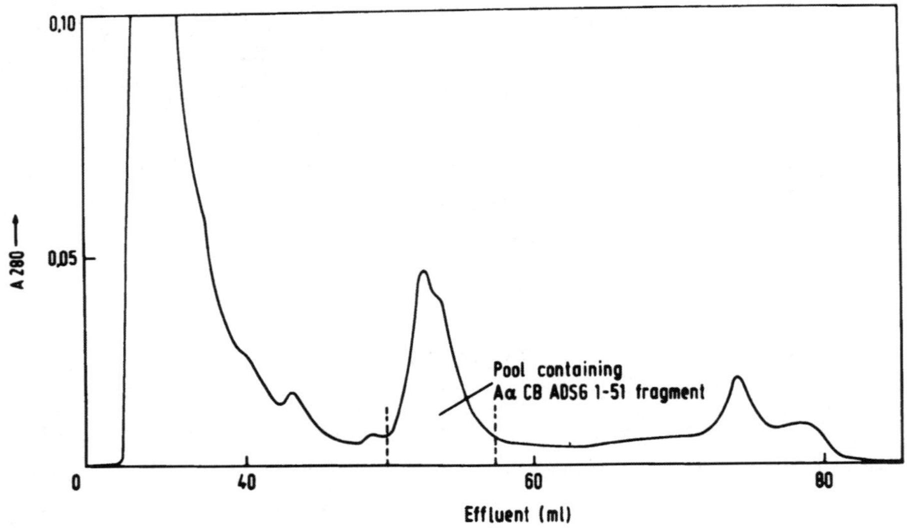

Fig. 7. Sephadex G-50 chromatography of Aα-chain cyanogen bromide cleavage products.

separations using fibrinogen Zurich I material should permit a quantitative separation of normal from abnormal fragment.

Conclusion

The three Aα-chain variants of human fibrinogen described here bring the total now known to four. A complete characterisation of these abnormal molecules requires, in addition to identification of the amino acid substitutions, quantitation of the heterozygous state and determinations of any epigenetic modifications such as alterations in carbohydrate composition.

A collection of genetic variants of fibrinogen, fully characterized at the molecular level, will obviously be of great importance for investigating the structure-function relationships of the fibrinogen molecule both at the in vitro and in

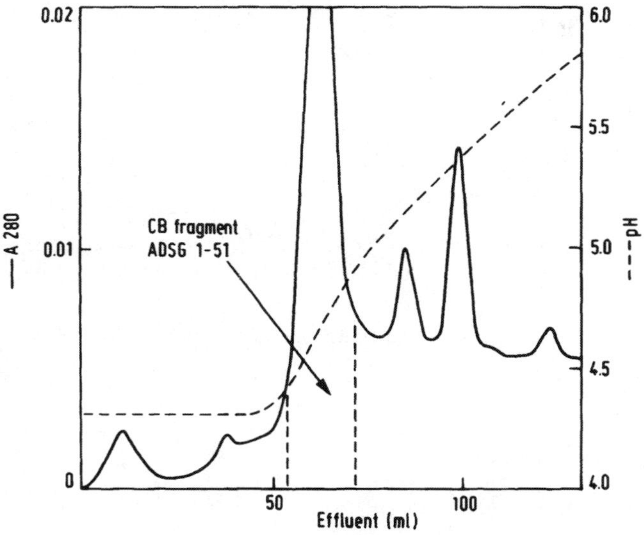

Fig. 8. CM-cellulose chromatography of N-terminal CB fragment containing pool derived from Aα-chain.

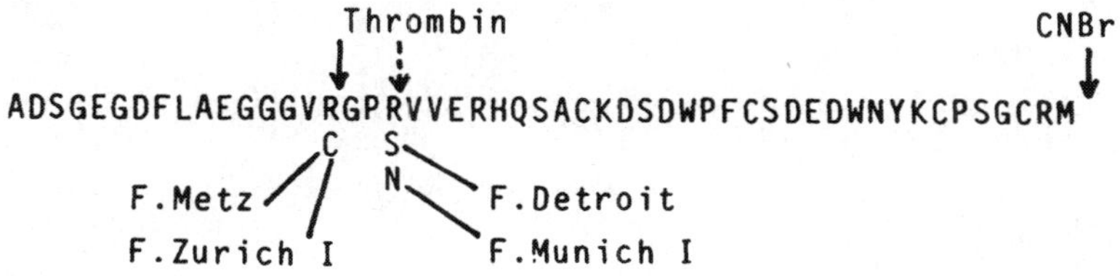

Fig. 9. Amino acid exchanges found in the abnormal fibrinogens. The first 51 residues of the fibrinogen Aα-chain, i.e. the CB fragment ADSG, are shown.

vivo level. They could also give significant insights into the genetic organisation, synthesis, assembly and catabolism of fibrinogen and its components.

The authors would like to thank the Deutsche Forschungs-Gemeinschaft for financial support during the course of this work.

References

1. Samama, M., Soria, J., Soria, C.: Recent Advances in Blood Coagulation, Churchill Livingstone, Edinburgh, 2, 313-335 (1977).
2. Flute, P.: British Medical Bulletin, 33, 253-259 (1977).
3. Huisman, T.H.J, Jonxis, J.H.P.: The Hemoglobinopathies, Clinical and Biochemical Analysis, Dekker, NY, 6, (1977).
4. Blombäck, M., Blombäck, B., Mammen, E., Prasad, A.: Nature 218, 134-137 (1968).
5. Henschen, A., Edman, P.: Biochim. Biophys. Acta, 263, 351-367 (1972).
6. Bain, P.: Proc. Natl. Acad. Sci., USA, 73, 3633-3636 (1976).
7. Marx, R., Schramm, W.: in Fibrinogen, Fibrin and Fibrin Glue, Schimpf, K., ed., Schattauer, Stuttgart, pp. 195-196 (1980).
8. Henschen, A., Southan, C.: Fibrinolyse, Thrombose, Hämostase, Deutsch, E., Lechner, K., eds., Schattauer, Stuttgart, pp. 296-298 (1980).
9. Soria, J., Soria, C., Poirot, E., Kling, C.: Thromb. Diath. Haemorrh. 27, 619-633 (1972).
10. Hofmann, V., Gati, W.P., Straub, P.W.: Thromb. Haem., 41, 709-715 (1974).
11. Lottspeich, F., Henschen, A.: Hoppe-Seyler's Z. Physiol. Chem., 359, 1451-1455 (1978).
12. Soria, J., Soria, C., Samama, M., Henschen A., Southan, C.: this volume.

FIBRINOGEN BERN I: A HEREDITARY FIBRINOGEN VARIANT WITH DEFECTIVE CONFORMATIONAL STABILIZATION BY CALCIUM IONS

Christoph Rupp, Cemal Kuyas*, André Häberli*, Miha Furlan, Beat A. Perret, Eugene A. Beck

Central Hematology Laboratory and *Thrombosis Laboratory, Inselspital, CH-3010 Bern, Switzerland

Summary

Inherited dysfibrinogenemia (fibrinogen Bern I) was detected in a young woman. The coagulation defect was characterized by a prolonged thrombin clotting time, decreased clottable fibrinogen and an only slightly subnormal fibrinogen concentration when measured by an immunological method. Fibrinopeptide A release was normal, but the aggregation of fibrin monomers was markedly delayed at low calcium concentrations. In contrast, fibrin polymerization was fully normalized at high calcium concentrations (5 mM). Plasmic degradation in the presence of calcium resulted in at least two terminal D fragments, compared with only one D derived from normal fibrinogen. This indicates a defective protection of fragment $D_{Bern\ I}$ against further plasmic digestion by calcium ions. In addition, a γ-chain lacking one net negative charge was demonstrated by isoelectric focusing of reduced fibrinogen Bern I. Our results indicate that the structural defect of fibrinogen Bern I is situated in the C-terminal part of the molecule, more precisely in the γ-remnant of the plasmic fragment D.

Fibrinogen - Recent Biochemical and Medical Aspects

Introduction

The number of described inherited fibrinogen variants is rapidly increasing. Up to 1980, 83 cases of congenital dysfibrinogenemia have been described (for review see 1). Defects include different functions of the fibrinogen molecule: fibrinopeptide release, fibrin monomer aggregation, crosslinking by factor XIII and fibrinogen proteolysis by plasmin. The molecular defect, however, an alteration in the primary structure of the protein has only in a small minority of fibrinogen variants been elucidated (2, 3).
Calcium ions play an important role in the functions of fibrinogen (4). By binding to the fibrinogen molecule, they stabilize a conformation most favourable for the aggregation of fibrin monomers. Calcium is also essential for fibrin crosslinking by factor XIII. Two of the three calcium high affinity binding sites (5) are located in each of the two plasmic fibrinogen fragments D (4, 6). Calcium protects this digestion fragment D against further degradation by plasmin (7). A partial correction of prolonged thrombin clotting times by the addition of calcium has been reported for many abnormal fibrinogens. The effect of calcium ions on the function of fibrinogen variants has been studied in more detail with six French cases of dysfibrinogenemia (8).
The present work deals with the fibrinogen variant Bern I, first described in 1978 (9). Dysfunction of this fibrinogen variant is strongly dependent on calcium concentration.

Case Report

The proposita, a young woman born in 1954, was first preoperatively studied in 1978. She had no history of unusual bleeding. The laboratory results, at first, lead to the diagnosis of

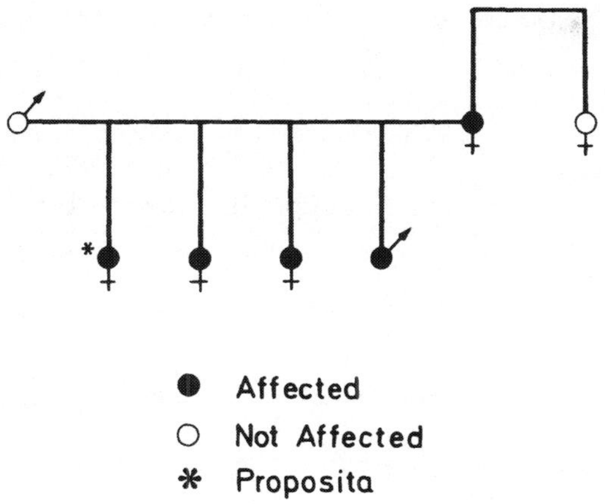

Fig. 1: Pedigree of the fibrinogen variant, fibrinogen Bern I

"hypofibrinogenemia". Subsequent studies of her plasma, as well as studies of family members, showed the typical pattern of inherited hypodysfibrinogenemia (9, 10). Fig. 1 shows the family pedigree of fibrinogen Bern I. It is consistent with an autosomal dominant inheritance pattern. Neither the mother nor the four affected children revealed any hemorrhagic or thrombotic episodes.

Materials and Methods

Coagulation studies. Thrombin clotting times were measured with a Greiner Micro Coagulometer by adding 20 µl of thrombin (Thrombin Roche, 7.5 NIH U/ml) to 40 µl of plasma. Fibrinogen in plasma was measured by three different methods: rate of clot formation according to Clauss (11), heat precipitation (56°C) according to Schulz (12) and antigen determination by immunoelectrophoresis according to Laurell (13).

Fibrinogen purification. Fibrinogen was prepared from citrated plasma by a new procedure, being a combination of modified methods (14, 15). In Table 1, a flow sheet of the steps is presented.

Table 1: Flow sheet of the applied purification method

Plasma

1. affinity chromatography on L-lysine-Sepharose 4B (16)
2. precipitation at 4 °C with either 26 % saturated ammonium sulfate or 10 % (v/v) ethanol (17,18)
3. gel filtration on Sephacryl S-300 (15)
4. affinity chromatography on gelatin-Sepharose 4B (19)
5. ion exchange chromatography on DEAE-Sephacel (14)

↓

purified fibrinogen

Chromatography gels were purchased from Pharmacia Fine Chemicals, Uppsala. All chemicals were reagent grade. The buffers contained 0.05 M triethanolamine (TEA), 0.1 M sodium chloride pH 7.4 for step 1; 0.04 M TEA, 0.01 M trisodium citrate, 100 KI U/ml Trasylol (Bayer, Leverkusen) pH 8.6 for step 3; 0.04 M TEA, pH 8.6 for step 4. For step 5, i.e., elution from DEAE-Sephacel, a linear gradient was employed, using 0.04 M TEA, pH 8.6 as the starting buffer and 0.15 M TEA, 0.1 M NaCl, pH 7.0 as the end buffer. Triethanolamine was chosen because it gave no interference with amino group determinations and with calcium ions. The elution of proteins was monitored on a LKB Uvicord UV-analyzer. Fibrinogen concentrations in purified solutions were determined spectrophotometrically ($A_{280}^{1\,\%} = 15.1$) with a Beckman model 26 spectrophotometer.

SDS polyacrylamidegel electrophoresis. The procedure was performed according to Weber and Osborn (20), using either 0.1 M Tris buffer, pH 7.4, 0.2 % SDS or 0.1 M sodium phosphate buffer, pH 6.8-7.0, 0.1 % SDS. The gel concentrations were as indicated in the legends to the figures. Reduction of disulfide bonds was achieved with 1 % 2-mercaptoethanol for 10 minutes at 95°C. Gels were stained with Coomassie brilliant blue and destained with acetic acid/methanol.

Isoelectric focusing. The method was developed by Kuyas (21). The highly crosslinked gels (7 %, 25 % crosslinked with DATD (N,N'-diallyltartardiamide)) contained 7 M urea and 1 % Triton X-100. Pharmalyte (Pharmacia, Uppsala) with a pH-interval 4-9 was used as a carrier ampholyte.

Sialic acid determination. The sialic acid content of fibrinogen was determined according to Aminoff (22) by the thiobarbituric acid method.

Polymerization studies. Fibrinogen solutions (0.5 mg/ml) were incubated for 10 minutes with buffer or calcium chloride solutions at 37°C. Turbidity at 350 nm was followed at constant temperature (37°C) in a spectrophotometer after addition of thrombin (0.5 NIH U/ml final concentration). The thrombin had first been purified on Sephadex G-10 (Pharmacia) to remove calcium. To obtain calcium free fibrinogen solutions at physiological conditions, they were first incubated with 5 mM EDTA and then dialyzed against 0.05 M TEA, 0.1 M NaCl buffer, pH 7.4 in pretreated dialysis tubes (23). To remove calcium ions from the buffer, it was purified by ion exchange chromatography (Chelex 100, Bio Rad, California).

Fibrinopeptide A release. Thrombin (0.5 NIH U/ml final concentration) was added to the fibrinogen solution (0.5 mg/ml). The mixture was incubated at 37°C for 0, 0.5 and 2 minutes and the samples were then measured by a radioimmunoassay according to Hofmann and Straub (24).

Plasmic degradation. Digestion of fibrinogen was carried out at 37°C for 3 hours in 0.05 M TEA, 0.1 M NaCl buffer, pH 7.4. Final concentrations: 2.4 mg/ml fibrinogen, 0.24 CU/ml plasminogen (KABI, Stockhom), 41 U/ml streptokinase (Streptase, Behring, Marburg), 5 mM $CaCl_2$. Plasmin was removed from the digestion mixture by affinity chromatography on L-lysine-Sepharose 4B (16).

Results

Coagulation tests showed the typical pattern for dysfibrinogenemia: prolonged thrombin clotting times and very low amounts of clottable fibrinogen (Table 2). Bleeding time, platelet count, other clotting factors and antithrombin III were within normal limits.

Table 2: Coagulation data for plasma samples of the proposita, her brother and her mother

	proposita	brother	mother	normal
Thrombin time (sec)	26	37	56	13-16
Fibrinogen (mg/ml)				
- Clauss (11)	0.3	0.2	0.4	1.5-3
- Schulz (12)	1.4	0.9	1.5	1.5-3
- Laurell (13)	1.3	1.0	1.4	1.5-3

Because the plasma of the proposita's mother gave the strongest prolongation of the thrombin time, further studies were performed using her plasma as starting material.

Fibrinogen purification. The first step of the purification

procedure, i.e., affinity chromatography on lysine-Sepharose, served to remove plasminogen from the plasma. The second step, a precipitation in the cold was essential to reduce the sample volume and to get rid of the α_2-macroglobulin which could not be removed without this step. Ethanol (10 %) was found to be superior to ammonium sulfate (26 % saturation) for it gave a better yield with less contaminants. Gel filtration was effective in separating fibrinogen from coprecipitating contaminants. The remaining contamination was mainly fibronectin which could be removed by affinity chromatography on gelatin-Sepharose. After the final ion exchange chromatography on DEAE-Sephacel (Fig. 2), a highly purified and intact fibrinogen was obtained. The use of triethanolamine as buffer system had quite a drastic effect on the fibrinogen elution from DEAE-Sephacel. A single peak resulted which was homogeneous when analyzed by SDS gel electrophoresis. This was in contrast to the two peak

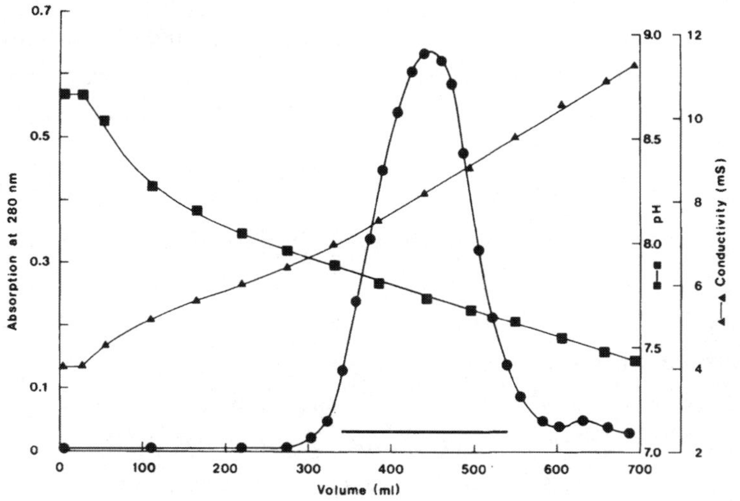

Fig. 2: DEAE-Sephacel chromatography of crude fibrinogen solution on a 1.5 x 25 cm column. Absorbance at 280 nm ●—●. Gradient was characterized by pH ■—■ and conductivity ▲—▲. The horizontal bar corresponds to the pooled fibrinogen fractions.

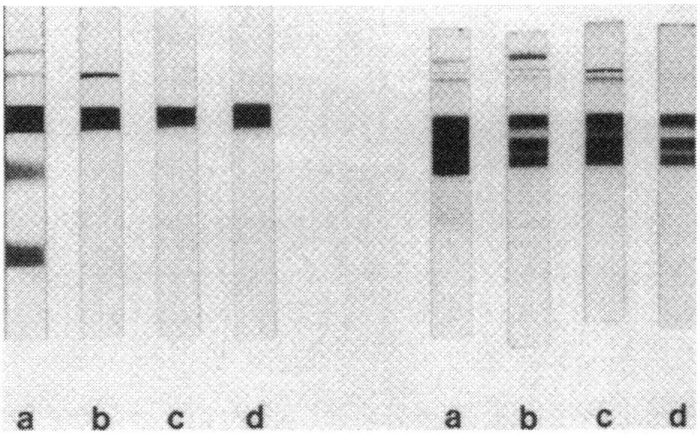

Fig. 3: SDS polyacrylamide gels of normal fibrinogen during purification. Left: unreduced samples (3 %/3 % gels); right: reduced samples (5 %/5 % gels); all gels were run in tris buffer. a, fibrinogen solution after ethanol precipitation; b, after gel filtration on Sephacryl S-300; c, following gelatin-Sepharose chromatography; d, after DEAE-Sephacel chromatography.

elution with the tris/phosphate buffer system (14). No difference in the elution of normal fibrinogen and fibrinogen Bern I was detected. The final yield of the fibrinogen was about 60 %. Fig. 3 shows SDS gels from various stages of the purification procedure.

Physicochemical characterization. Compared with normal fibrinogen, SDS gel electrophoresis of fibrinogen Bern I showed no difference in the size of the intact molecule and of the reduced polypeptide chains (Fig. 4).
By isoelectric focusing each of the three reduced chains was resolved into several lines. Whereas α- and β-chains of fibrinogen Bern I revealed no detectable abnormality in the pattern, an additional γ-line, lacking one net negative charge, was seen with the dysfibrinogen (Fig. 5). The γ-chain carrying this charge abnormality was named $\gamma_{Bern\ I}$.

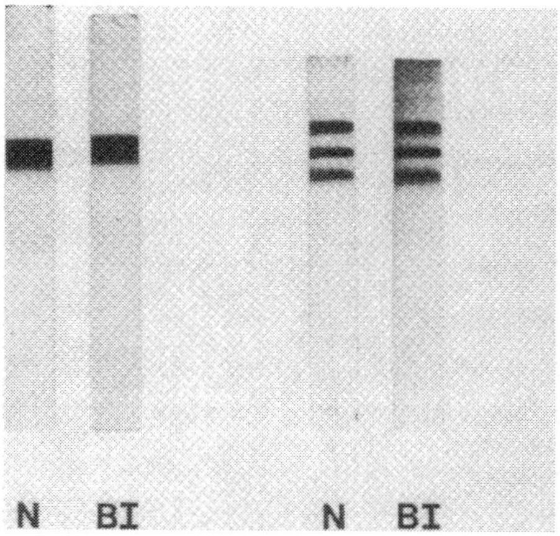

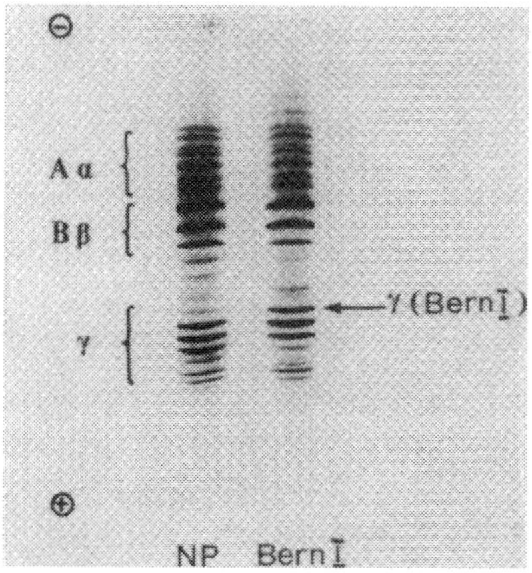

Fig. 4: SDS polyacrylamide gels of normal fibrinogen and fibrinogen Bern I. Left: unreduced samples (3 %/ 3 % gels, tris buffer); right: reduced samples (7 %/4 % gels, phosphate buffer).

Fig. 5: Isoelectric focusing gels with reduced samples of normal fibrinogen and of fibrinogen Bern I. pH range 4-9.

A densitometric evaluation of the isoelectric focusing gels was performed. Five γ-chain lines were evaluated and a median, considering the hight of every peak, was calculated. For normal fibrinogen this median lied between the lines 3 and 4 (3.14), whereas it was located between the lines 2 and 3 (2.26) for fibrinogen Bern I. This indicates a cathodal shift of the γ-chain pattern by 0.88 line intervals for fibrinogen Bern I, corresponding to 0.88 net charge units. If we assume that the $\gamma_{Bern\ I}$ chain is missing one net negative charge, this allows to conclude that 80 - 100 % of the fibrinogen is abnormal.
Sialic acid determination exhibited a significant increase in the sialic acid content of fibrinogen Bern I. Triplicate determinations gave 7.9 mol sialic acid per mol fibrinogen Bern I, compared with a value of 6.3 mol for normal fibrinogen.

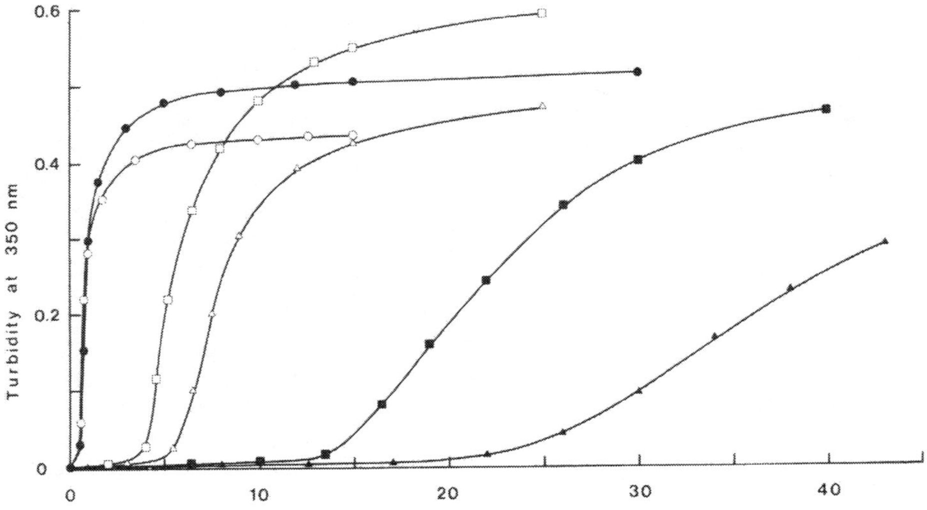

Fig. 6: Aggregation curves with normal fibrinogen (open symbols) and fibrinogen Bern I (closed symbols). ○/● , $5 \cdot 10^{-3}$M Ca^{2+}; □/■ , 10^{-5}M Ca^{2+}; △/▲ , < 10^{-6}M Ca^{2+}. Ordinate shows the increase in turbidity at 350 nm, abscissa the time after addition of thrombin

Functional characterization: polymerization. Thrombin induced aggregation curves with normal and Bern I fibrinogen (Fig. 6) were identical at $5 \cdot 10^{-3}$M Ca^{2+}, where the molecule is saturated with calcium ions. At 10^{-5}M Ca^{2+}, where only the high affinity calcium binding sites of normal fibrinogen are occupied (5), the aggregation of fibrin, derived from fibrinogen Bern I, was markedly delayed. The lag phase was prolonged and the rate decreased, whereas the final turbidity was normal. This effect was even more accentuated at a calcium concentration of less than 10^{-6}M, where fibrinogen has no bound calcium ions.

Fibrinopeptide A release. The release of fibrinopeptide A was measured after treatment with thrombin under the same conditions as for the aggregation experiments. Both the rate and the extent of the release from fibrinogen Bern I were normal. This indicates that the delayed aggregation is caused by impaired fibrin monomer aggregation and not by defective thrombin cleavage.

Plasmic degradation. Normal fibrinogen and fibrinogen Bern I were digested for 3 hours with plasmin in the presence of 5 mM Ca^{2+}. Fig. 7 shows the characterization of the degradation products by SDS gel electrophoresis. Normal fibrinogen digest revealed the expected pattern of D_{cate} fragment and fragment E. (D_{cate} is the term used by Haverkate (7) for the calcium terminal fragment D). In contrast, digest of fibrinogen Bern I contained a smaller D fragment (D'). (D' seems to have identical electrophoretic mobility as Haverkate's fragment D_{EGTA}). After reduction the corresponding smaller D'γ-chain appeared with fibrinogen Bern I. Prolonged digestion (24 hours) of either normal or abnormal fibrinogen did not significantly al-

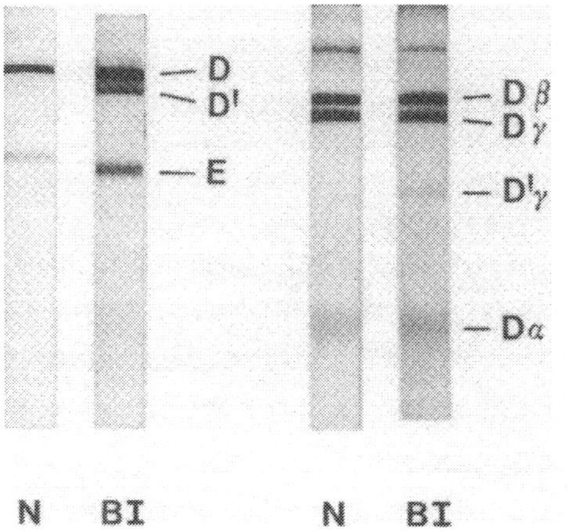

Fig. 7: SDS polyacrylamide gels of plasmic digests of normal fibrinogen and fibrinogen Bern I. Left: unreduced samples (7 %/4 % gels); right: reduced samples (9 %/4 % gels), all in phosphate buffer.

ter the electrophoretic patterns. The chain remnants of the fragment E migrated either with the Dα-chain or too fast to be detected on the gels. If we compare the staining intensities of fragments D and D' with those of the reduced remnants Dγ and D'γ (Fig. 7), we observe that the relative staining intensity of D'γ is much weaker than that of Dγ. This difference can only partly be explained by the weight loss during degradation of Dγ to D'γ (about 34 %). Thus, we conclude that the staining intensities of Dγ and D'γ are not representative for their mass ratio.

Further degradation experiments were performed at a calcium concentration of 1 mM. This concentration was sufficient to protect normal D_{cate}. In contrast, D_{cate} from fibrinogen Bern I was almost completely degraded within 3 hours. After 24 hours, only the D' fragment and the D'γ-remnant were found on electrophoretic gels of unreduced and reduced digests, respectively. The degradation of the γ-chain from 38 000 MW to 25 000 MW (7) in fragment $D_{Bern\ I}$ indicates a lacking protection by calcium ions against further plasmic digestion of this degradation fragment.

Discussion

According to our results, fibrin monomers, derived from fibrinogen Bern I, exhibit abnormal polymerization at low calcium concentrations. At 5 mM Ca^{2+}, however, the aggregation reaction was fully normalized. Thus, clot formation could well be normal at physiological calcium concentrations in the circulating blood, which would explain the lack of clinical symptoms.

SDS gel electrophoresis revealed a normal size of the polypeptide chains of fibrinogen Bern I. However, isoelectric focusing demonstrated a less negatively charged γ-chain, called $\gamma_{Bern\ I}$. An increased sialic acid content was found (7.9 instead of 6.3

residues) for fibrinogen Bern I. Increased sialic acid has been observed in several other dysfibrinogens (1). The sialic acid attached to the carbohydrate in the N-terminus of the γ-chain thus cannot be the reason for the decreased negative charge of the $\gamma_{Bern\ I}$-chain. Quantitative evaluation (isoelectric focusing patterns and plasmic degradation experiments at 1 mM Ca^{2+}) lead us ot the conclusion that almost all fibrinogen of the proposita's mother is abnormal.

Fibrinogen Bern I is one of the few fibrinogen variants which yielded abnormal plasmic derivatives. This phenomenon has been described in the case of fibrinogen Paris I (longer Dγ-chain (25, 26)), fibrinogen Tokyo (abnormal tryptic peptides of fragment D (27)) and fibrinogen Lille (abnormal size and charge of fragments E (28)). However, only with fibrinogen Pontoise the digestion experiments were performed at controlled calcium concentrations (Dβ-chain with higher MW (8)). In the presence of 5 mM calcium fibrinogen Bern I yielded a faster migrating D fragment (D') besides D_{cate}. Reduced gels showed a smaller Dγ-remnant which, with normal fibrinogen, was only seen after removal of calcium ions (7). Thus calcium does not seem to be able to protect the $D_{Bern\ I}$ fragment against further degradation by plasmin. Preliminary experiments with purified D_{cate} fragments showed again a charge shift of the $D_{Bern\ I}$, carrying less negative charge than normal D. Thus, the charge abnormality, located in the γ-remnant of fragment $D_{Bern\ I}$, could be responsible for deficient calcium binding. This assumption is in good agreement with the recent model of Doolittle (29), where the calcium binding site of the D fragment lies in a hole, formed by acidic amino acids of the β- and the γ-chain; calcium binding influences also directly the nearby situated polymerization binding sites for the fibrin N-termini of α- and β-chains.

Acknowledgements

This work was supported by Research Grants No. 3.936-078 and No. 3.915-080 from the Swiss National Foundation. We wish to thank Mr. Reto Sievi for the great practical help in this study.

References

1. Beck, E.A.: In Textbook of Thrombosis and Haemostasis (ed. Colman, R.W., Hirsh, J., Marder, V.J., Salzman, E.W.) J.B. Lippincott, Philadelphia, in press (1981).
2. Blombäck, M., Blombäck, B., Mammen, E.F., Prasad, A.S.: Nature 218, 134-137 (1968).
3. Henschen, A., Lottspeich, F., Southan, C., Töpfer-Petersen, E.: Protides biol. Fluids 28 (ed. H. Peeters), 51-56 (1980).
4. Purves, L.R., Lindsey, G.G., Franks, J.J.: S. Afr. J. Sci. 74, 202-209 (1978).
5. Marguerie, G., Chagniel G., Suscillon, M.: Biochim. biophys. Acta 490, 94-103 (1977).
6. Van Ruijven-Vermeer, I.A.M., Nieuwenhuizen, W., Nooijen, W.J.: FEBS Lett. 93, 177-180 (1978).
7. Haverkate, F., Timan, G.: Thromb. Res. 10, 803-812 (1977).
8. Haverkate, F., Timan, G., Soria, J., Soria, C., Samama, M.M.: Thromb. Res. 13, 689-692 (1978).
9. De Fliedner, V., Perret, B.A., Furlan, M., Beck, E.A.: Blut 37, 157 Abstr. (1978).
10. Beck, E.A.: Clin. Haematol. 8, 169-181 (1979).
11. Clauss, A.: Acta haemat. 17, 237-246 (1957).
12. Schulz, F.H.: Acta hepat. 3, 306-310 (1955).
13. Laurell, C.B.: Analyt. Biochem. 15, 45-52 (1966).
14. Mosesson, M.W., Finlayson, J.S.: J. Lab. clin. Med. 62, 663-674 (1963).
15. Henschen, A., Southan, C.: In Fibrinolyse, Thrombose, Haemostase (ed. Deutsch, E., Lechner, K.), Schattauer, Stuttgart, 290-293 (1980).

16. Castellino, F.J., Sodetz, J.M.: Meth. Enzym. XLV, 273-286 (1976).

17. Finlayson, J.S.: In Fibrinogen (ed. Laki, K.), M. Dekker, New York, 39-59 (1968).

18. Blombäck, B., Blombäck, M.: Ark. Kemi 10, 415-443 (1956).

19. Vuento, M., Vaheri, A.: Biochem. J. 183, 331-337 (1979).

20. Weber, K., Osborn, M.: J. biol. Chem. 244, 4406-4412 (1969).

21. Kuyas, C., Haeberli, A., Straub, P.W.: Eur. J. Biochem. (submitted).

22. Aminoff, D.: Biochem. J. 81, 384-392 (1961).

23. Brewer, J.M.: In Experimental techniques in biochemistry, Prentice Hall Series, 1-9 (1974).

24. Hofmann, V., Straub, P.W.: Thromb. Res. 11, 171-181 (1977).

25. Budzynski, A.Z., Marder, V.J.: J. Lab. clin. Med. 88, 817-825 (1976).

26. Mosesson, M.W., Denninger, M.H., Ménaché, D.: Thromb. Res. 9, 115-121 (1976).

27. Samori, T.: J. Tokyo Med. Coll. 36, 583-594 (1978).

28. Denninger, M.H., Finlayson, J.S., Reamer, L.A., Parquet-Gernez, A., Goudemand, M., Ménaché, D.: Thromb. Res. 13, 453-466 (1978).

29. Doolittle, R.F., Laudano, A.P.: Protides biol. Fluids 28, (ed. H. Peeters), 311-316 (1980).

CARBOHYDRATE IN ABNORMAL FIBRINOGENS

David A. Lane, Antony K. Allen, John Markwick,
Elisabeth Thompson and Ian Mackie

Departments of Haematology and Biochemistry, Charing Cross Hospital Medical School and Central Middlesex Hospital, London, U.K.

Introduction

Normal human fibrinogen contains one oligosaccharide on the γ-chain linked to Asp52 (1) and at least one oligosaccharide on the β-chain, again attached to an asparagine residue (2). The function of this carbohydrate remains uncertain. In contrast to many other plasma glycoproteins, exposure of galactose by removal of sialic acid does not appreciably influence catabolism (3). Some abnormal fibrinogens contain altered amounts of carbohydrate and evidence has been advanced for its involvement in the functional defects they exhibit (4, 5). In this article the role of carbohydrate in the function of abnormal fibrinogens has been re-examined by studying five abnormal fibrinogens that have defects differing in type or extent. One of these fibrinogens was isolated from a patient whose plasma concentration was one-half that of normal and was therefore a case of hypodysfibrinogenaemia. In the course of an examination of the clotting defect exhibited by this patient the catabolism of her purified fibrinogen was studied. Because of the possible connection between carbohydrate and catabolism it was felt it was appropriate to study the catabolism of the four other abnormal fibrinogen and the results of this study are also included here.

Materials and Methods

All haematological and coagulation tests were performed as

Fibrinogen - Recent Biochemical and Medical Aspects

previously described (6, 7). Carbohydrate composition was studied by methanolysis followed by GLC for neutral sugars and sialic acid (8), and by amino acid analysis techniques for amino sugars (9). Catabolism of normal and abnormal fibrinogens was studied in New Zealand White rabbits using a method similar to that adopted elsewhere (10), with some modifications (12). Adult abnormal fibrinogens were purified from individual patient plasma A.J. (6), R.M. (7), T.O. (11) and T.H. (12). Fetal fibrinogen was purified from a pool of five plasma samples prepared from cords of normal full-term deliveries (12).

Results

One of the fibrinogens studied, T.H., was isolated from a female patient with the same initials, who presented with a clotting defect associated with repeated thrombophlebitis. Her plasma fibrinogen concentration was approximately one-half the normal concentration (1 mg/ml) and the thrombin clotting time of her purified fibrinogen was prolonged compared to that of normal. This was, therefore, a case of hypodysfibrinogenaemia. Some of the results of a haematological investigation of the patient are contained in Table I. Experiments conducted with purified

	Control	T.H.
Prothrombin time (sec)	12	14
PTTK (sec)	33	44
Thrombin time (sec)	15	30-180
Reptilase time (sec)	11	45
Fibrinogen conc:		
clottable (mg/ml)	2.2	1.0
immunological (mg/ml)	2.2-5.5	1.3
FDP (μg/ml)	6.0	6.0
Thrombin time of purified fibrinogen (sec)	12.6	30-180

Table I: Haematological investigation of patient T.H.

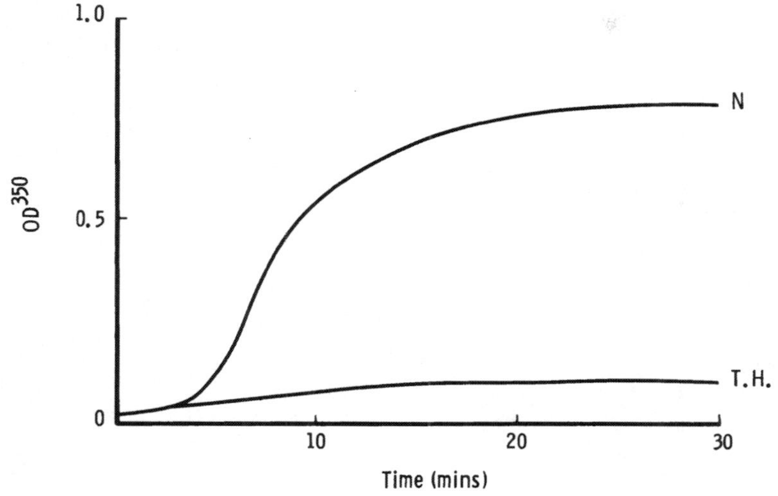

Fig. 1: Polymerisation of fibrin monomers prepared from normal (N) and patient (T.H.) fibrinogens

fibrinogen and fibrins established that the major defect was one of impaired polymerisation of fibrin monomers (Fig. 1), while the release of fibrinopeptide A was normal (Fig. 2). Additionally, a series of electrophoretic experiments in polyacrylamide gels did not demonstrate any abnormality associated with polypeptide chains or factor XIIIa crosslinking. Unfortunately, we were unable to establish a family history of the defect, partly because of past medical mismanagement of the patient, which rendered her ill-disposed to further investigations.

Fibrinogen	FpA	FpB	Polymerisation	Plasma conc.
London	N	Ab?	Ab	N
Manchester	Ab	Ab	Ab	N
Oslo II	N	N	Ab	N
T.H.	N	-	Ab	Ab
Fetal	N	N	Ab	N

Table II: Functional properties of abnormal fibrinogens. (abbreviations: N, normal; Ab, abnormal; -, not performed).

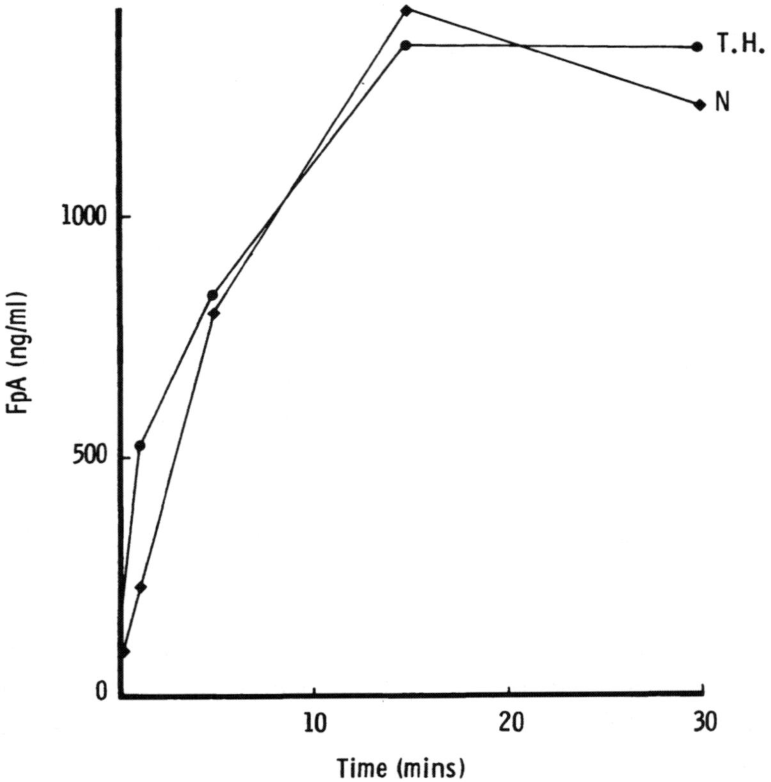

Fig. 2: Release of fibrinopeptide A (FpA) by thrombin from solutions of normal (N) and patient (T.H.) fibrinogens.

In Table II are listed the five abnormal fibrinogens that were studied, together with some of their properties. The defect of fibrinogen London is predominantly one of impaired polymerisation (7), but preliminary results suggest that there may be a slightly impaired FpB release (Hessel and Blomback, private communication). Fibrinogen Manchester has impaired FpA (6) and FpB (Hessel and Blomback, private communication) release, as well as a polymerisation defect. Fibrinogens Oslo II and that from T.H. do not polymerise as well as normal fibrin (11, 12) and, as already mentioned, the polymerisation defect of T.H. is associated with a reduced plasma fibrinogen concentration. When the

	Man	Gal	GlcNAc	GalNAc	S.A.
Normal	22	18	19	5	11
London	19	14	20	-	11
Manchester	20	16	-	-	-
Oslo II	22	17	20	11	12
T.H.	25	22	24	4	12
Fetal	15	21	13	20	20

Table III: Carbohydrate analysis of normal and abnormal fibrinogens (mol/mol). (-, not performed).

fetal fibrinogen was reconstituted at the same concentration of normal fibrinogen (2 mg/ml), its thrombin clotting time was appreciably prolonged. Experiments conducted with fibrin monomers have indicated that this is caused by impaired polymerisation.

Carbohydrate analysis of normal fibrinogen in mols of sugar/mol of fibrinogen, Table III, revealed a higher content of mannose (Man) than galactose (Gal), 19 mols of N-acetylglucosamine (Glc NAc), 5 mols of N-acetylgalactosamine (GalNAc) and 11 mols of sialic acid (S.A.). The adult abnormal fibrinogens showed a similar number of Man and Gal residues; always the number of mols of Man were higher than the number of Gal. The content of amino sugars was also similar to normal fibrinogen, with the possible exception of that of Oslo II, which had an elevated number of GalNAc residues. Not all the amino sugars and sialic acid residues are listed in the table as precise quantitative data is not yet available for them. However, preliminary results suggest that the numbers of these residues in the adult abnormal fibrinogen do not differ appreciably from those of normal fibrinogen. The carbohydrate composition of fetal fibrinogen appears substantially different from that of the adult fibrinogens. Firstly, the Man to Gal ratio was inverted, there being less Man than Gal residues. More important, the number of GlcNAc residues was reduced and there were 2-4 times greater GalNAc

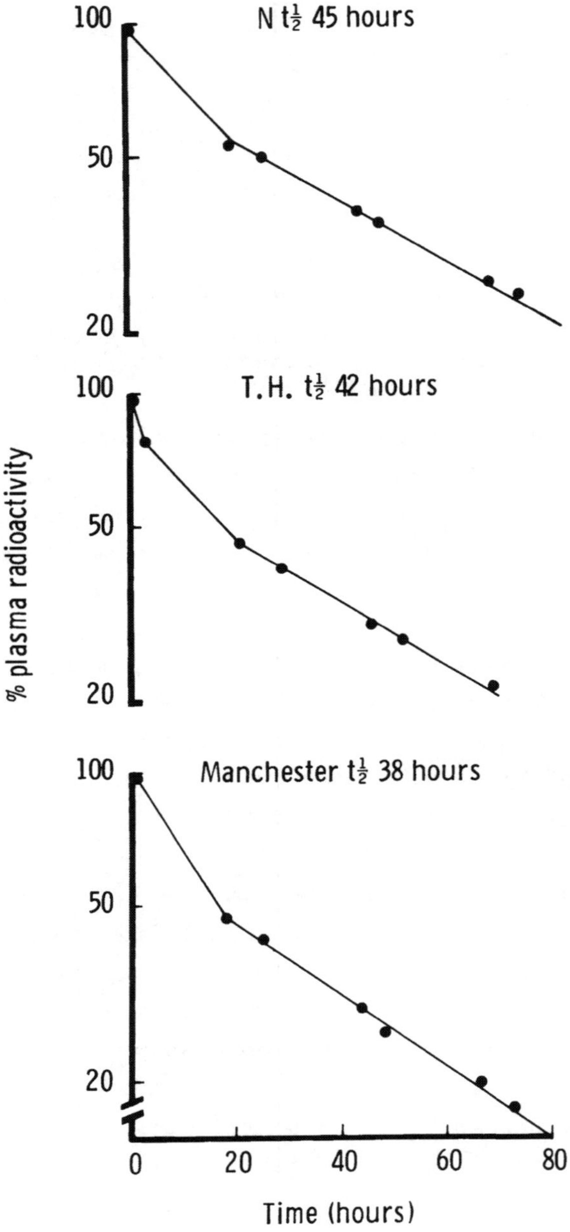

Fig. 3: Representative turnover experiments of I^{125}-labelled normal (N) and abnormal fibrinogens (T.H., Manchester)

	$t\frac{1}{2}$ (hrs)
Normal	mean 50.2 (n=5, range 42-60)
London	39
Manchester	38
Oslo II	44
T.H.	42
Fetal	mean 47.7 (n=5, range 39.5-70)

Table IV: Catabolism of normal and abnormal fibrinogen

residues together with an increased sialic acid content.

When I^{125}-labelled human fibrinogens were injected into rabbits they were initially eliminated rapidly, then exponentially with a $t\frac{1}{2}$ of approximately 2 days. Representative experiments are illustrated in Fig. 3 and all the results of $t\frac{1}{2}$ determinations are summarised in Table IV. Normal fibrinogen had a mean $t\frac{1}{2}$ of 50.2 hrs and fetal fibrinogen did not differ significantly from this, $t\frac{1}{2}$ 47.7 hrs. While the fibrinogen from patient T.H. and fibrinogen Oslo II had $t\frac{1}{2}$'s within the normal range, the half lives of fibrinogens London and Manchester were marginally shorter.

Discussion

Oligosaccharide chains have been located on the γ- and β-chains of fibrinogen, in both cases linked to an asparagine residue (1,2). Our carbohydrate analyses indicate that most of the carbohydrate on normal fibrinogen is linked to asparagine, as only small amounts of GalNAc are present. The increase in GalNAc observed with fibrinogen Oslo II requires further investigation, but aside from this it does seem that the adult abnormal fibrinogens are basically similar in their carbohydrate composition to normal fibrinogen. This makes it likely that altered carbohydrate is not responsible for their abnormal functional behaviour. It should be noted that our results for sialic acid content of adult fibrinogens are higher than those quoted by others. This is probably due to different methodology as it is widely known that sialic acid determinations made with GLC are higher than those made with the thiobarbituric acid method.

The elevated number of sialic acid residues obtained with fetal fibrinogen is in agreement with other work that has further suggested a role for this sugar in the polymerisation defect exhibited by this variant (4). The demonstration of an associated large increase in GalNAc with a reduced GlcNAc content implies that some of the oligosaccharides present on fetal fibrinogen may be attached by an O-glycosidic linkage to threonine or serine residues, rather than asparagine. However, this should be regarded as speculative as we have not completely eliminated the possibility that tryptophan degradation products may be interfering with the GalNAc peak obtained from the amino acid analyser.

The turnover studies were carried out in rabbits rather than humans because of the risk of hepatitis from patient plasma preparations. A previous study has demonstrated the essentially similar catabolic properties of rabbit and human fibrinogen (10). Although the dysfibrinogenaemia observed with patient T.H. was associated with a reduced plasma concentration, hypercatabolism of the abnormal fibrinogen was not responsible because the $t_{\frac{1}{2}}$ was normal. This distinguishes the presently described case from three other cases of hypodysfibrinogenaemia, fibrinogens Bethesda II, Bethesda III and Philadelphia, all of which are hypercatabolised (15). The present case can also be distinguished from the above mentioned variants and another case of hypodysfibrinogenaemia, fibrinogen Leuven, by a comparison of clotting test results, electrophoretic mobilities and the conditions for optimal polymerisation of fibrin monomers (15,16).

Despite its increased sialic acid content, fetal fibrinogen was catabolised with the same half-life as normal fibrinogen. This supports the proposition that exposure of galactosyl residues by cleavage of terminal sialic acid residues is not an important mechanism involved in the catabolism of normal fibrinogen (10) Fibrinogen is therefore similar to transferrin in this respect but differs from other plasma glycoproteins such as ceruloplasmin and orosomucoid which are removed by such treatment. While a minority of human fibrinogen is eliminated from the circulation following limited thrombin or plasmin (-like) action on the NH_2 (13) and COOH (14) terminal regions of the α-chain, the mechanism by which the majority is catabolised remains obscure. In the present work abnormal fibrinogens with a range of different functional abnormalities were shown to be catabolised at approximately normal rates. That is, the full integrity of the regions of the fibrinogen molecule involved in the polymerisation or peptide releasing functions were not essential for an adequate catabolism. Presumably, therefore, the region of fibrinogen involved in the regulation of normal catabolism is not involved in the dysfunction of the presently studied fibrinogens.

Acknowledgements

Drs. Kierulf, Brosstad and Godal kindly supplied fibrinogen Oslo II for this study. Mrs. J. Bottomley and Dr. B. Cuddigan kindly provided plasma from patients A.J. and R.M., respectively.

References

1. Blombäck, B., Gröndahl, N.J., Hessel, B., Iwanaga, S., Wallén, P.: J. Biol. Chem. 248, 5806 (1973).

2. Töpfer-Petersen, E., Lottspeich, F., Henschen, A.: Hoppe-Seylers Z. Physiol. Chem. 357, 1509 (1976).

3. Morrell, A.G., Gregoriadis, G., Scheinberg, I.H.: J. Biol. Chem. 246, 1461 (1971).

4. Gralnick, H.R., Givelber, H., Abrams, E.: N. Engl. J. Med. 299, 221 (1978).

5. Soria, J., Soria, C., Bertrand, C., Samama, M.: Biochem. Biophys. Res. Commun. 82, 442 (1978).

6. Lane, D.A., VanRoss, M., Kakkar, V.V., Bottomley, J., Dhir, K., Holt, L.P.J., MacIver, J.E.: Br. J. Haematol. 46, 89 (1980).

7. Lane, D.A., Cuddigan, B., VanRoss, M., Kakkar, V.V.: Br. J. Haematol. 44, 483 (1980).

8. Chambers, R.E., Clamp, J.R.: Biochem. J. 125, 1009 (1971).

9. Allen, A.K., Neuberger, A.: FEBS Lett. 60, 76 (1975).

10. Martinez, J., Palascak, J., Peters, C.: J. Lab. Clin. Med. 89, 367 (1977).

11. Godal, H.C., Brosstad, F., Kierulf, P.: Scand. J. Haematol. 20, 57 (1978).

12. Lane, D.A., Allen, A.K., Markwick, J., Thompson, E., Owen, J. To be published (1981).

13. Nossel, H.L., Yudelman, I., Canfield, R.E., Butler, V.P., Spanondis, K., Wilner, G.D., Qureshi, G.: J. Clin. Invest. 54, 43 (1974).

14. Mosesson, M.W.: Semin. Thromb. Hemostas. 1, 63 (1974).

15. Gralnick, H., Coller, B.S., Frantantoni, J.C., Martinez, J.: Blood 53, 28 (1979).

16. Verhaeghe, R., Verstraete, M., Vermylen, J., Vermylen, C.: Br. J. Haematol. 26, 421 (1974).

OBSERVATION ON A NEW VARIANT OF HEREDITARY DYSFIBRINOGENEMIA (HOMBURG I) WITH DELAYED FIBRIN POLYMERIZATION AND WITHOUT CLINICAL ABNORMALITY

Chieko Miyashita, Karl-Heinz Nienhaus, Ernst Wenzel, Michael Köhler, Peter Hellstern,
Abteilung für Klinische Haemostaseologie und Transfusionsmedizin, Universitätskliniken, Fb 4/3, D-665o Homburg/Saar, BRD

Introduction

In congenital deficiences of fibrinogen synthesis different degrees of bleeding tendencies were described, e.g. Baltimore: 1965 (1), Detroit: 1969 (2), Cleaveland II: 1974 (3) and Gießen II: 1975 (4). On the other hand some asymptomatical courses of congenital dysfibrinogenemia with distinct pathological findings in laboratory analysis were also observed like Zürich I: 1965 (5) or Cleaveland I: 1968 (6).

Results

A 28-year-old sportsman was further examined because of a pathologically prolonged prothrombin time and low concentration of clottable fibrinogen (table 1). In spite of a low value of 47% of quick-test the hepatoquick-time was not conspicuous and the determination of each clotting factors II - XIII had normal values too. The concentration of clottable fibrinogen amounted to 58 mg% and the distinctly pathologically prolonged thrombin time (44 sec.), reptilase time (58 sec.) and thrombin-coagulase time (75 sec.) suggest a severe disturbance of fibrin polymerization. Platelet function test, platelet morphology as well as platelet count were found to be normal (table 2). A thromboelastogram showed a little prolonged k and r + k time and other parameters were normal. However, when measurement was done a second time, a distinct phenomenon of tearing fibrin was registered (fig. 1).

Fibrinogen - Recent Biochemical and Medical Aspects

Table 1. Various plasma coagulation times and concentration of clotting factors in patient compared with normal values.

	PAT."H"	NORMAL PL.
PTT	43 SEC.	< 45 SEC.
PROTHR. T. HEPATOQUICK	47 % 75 %	> 75 % > 65 %
PLASM.THR.T. REPTILASE T. THR.COAG. T.	44 SEC. 58 SEC. 75 SEC.	< 23 SEC. < 18 SEC. < 25 SEC.
FIBR.STAB.T.	NORMAL	
A) F.II-XIII	NORMAL	
B) F.I	58 MG/100	125-(-280)

Table 2. Laboratory parameters found in normal ranges in patient.

BLEED.TIME DUKE	2'45
BLEED.TIME STAND.	3'00
THR.ADH.(HELLEM II)	84 % (> 65*)
COLLAG.INDUC.AGGR.	M_A INSIGNIFICANT
ADP-INDUCED AGGR.	REDUCED
M_ϵ : TEG	104 (>78*)
EXTENSION	UP TO NORMAL RANGE
VOL.DISTR.CURVE	INSIGN.SMALL THROMB.

* NORMAL VALUES

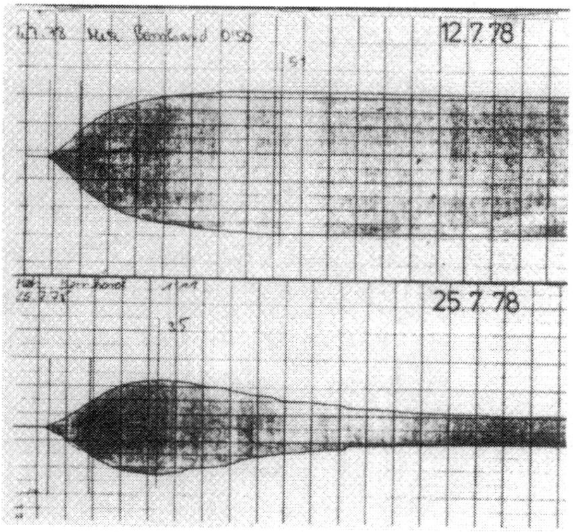

	N-Value	Pat. „H"	
		12.7.78	25.7.78
r	<5'00	4'50	9'11
k	<3'30	5'30	7'30
r+k	<8'50	10'20	16'41
$\frac{am}{r+k}$	>78	104	54

Figure 1. Thromboelastogram in patient, with a normal (above) and an abnormal (below) figure and some evaluated parameters compared with normal values.

In contrast with the low concentration of clottable fibrinogen measured by means of Clauss-method, fibrinogen concentration was normal when measured immunologically (nephlometric or according to Mancini-method using anti-fibrinogen serum of rabbit(Behring Werke) and by means of heat precipitation technique (table 3). After the incubation of patient's plasma with high concentrated thrombin-calcium (5o NIH Units/ml) at 37° C for 6o minutes, no more than 3o% of fibrinogen was removed. The rest (7o% of fibrinogen) was investigated immunoelectrophoretically according to Laurell-method with rabbit antiserum against human fibrinogen respectively against human anti-E-determinant of fibrinogen. They showed immunologically normal values (figure 2).

Table 3. Concentration of fibrinogen measured immunologically as well as by heat precipitation method and of clottable fibrinogen (Clauss-method) in patient and in normal plasma.

COMP."FGN.IMMUNOL."(ANTI-FGN-SER.FR.RABBIT)
TO CLOTT. AND HEAT PRECIP. FGN.

MG/100	IMMUNOLOG.		CLOTT.	HEAT-PREC.
	NEPHEL.	MANC.	FIBRINOGEN	
POOL PLASMA	162	270	178	280
"H"-PLASMA	111	222	58	250

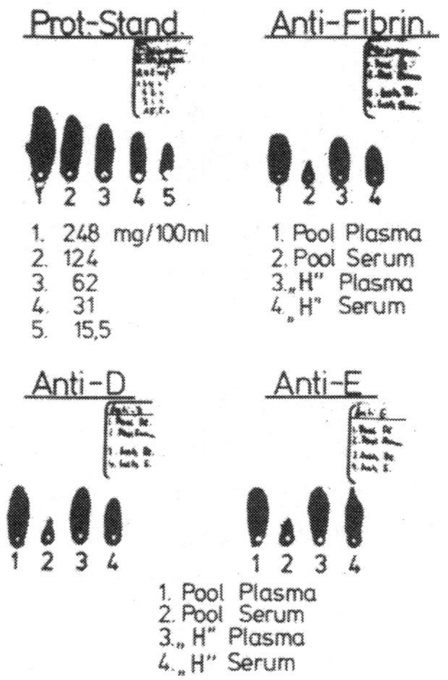

Figure 2. Immunoelectropherogram (Laurell-method) of normal and patient's plasma against anti-fibrinogen, anti-D as well as anti-E. Left above: Presentation of standard curve with antifibrinogen-serum.

Table 4. Coagulation status before and after addition of fibrinogen to normal and patient's plasma.

	Ø			+ FIBRINOGEN*		
	FGN*	PROTHR.T.▲		FGN*	PROTHR.T.▲	
POOL-PL.	175	97	12.6	277	100	12.3
"H"-PL.	58	36	21.0	167	66	15.2
	MG/100	%	SEC.	MG/100	%	SEC.

* KABI HUMAN FGN. (95 % CLOTTABLE)

▲ BOEHRINGER MANNH. THROM. (MONKEY BRAIN)

After adding purified human fibrinogen to the patient's plasma in vitro the prothrombin time normalized to the same extent as concentration of clottable fibrinogen increased (table 4).

From an examination of the relatives of the patient (n = 8), in the plasma of patient's father and two brothers no pathological parameters were found. However, particularly remarked disturbances of fibrin polymerization were recognized in his mother, a sister and a brother as well as in the patient himself. In other members of his family this function was comparably pathological (table 5). The ratio of concentration of clottable fibrinogen to that of immunologically measured fibrinogen amounted to 1/1o to 1/5. Moreover, two children of the second sister with hemorrhagic tendency suffered from spontaneous appearing blue spots and considerable complication of bleeding after operation like teeth extraction, appendectomie or Billroth II. Bleeding time and morphology of blood platelet were normal in all members of the family except in one sister, whose bleeding time was one time measured prolonged (7 minutes).

The data of conventional coagulation analysis at our patient are similar to those of dysfibrinogenemia "Amsterdam" (7), Zürich I" (5) or Zürich II" (8). Particularly these are resemble to "Amsterdam" in which inhibition of fibrin polymerization are determined only in presence of a_2-plasmafraction of patient's plasma. While in our case these defects normalized after addition of fibrinogen to the patient's plasma. Further biochemical qualitative investigation should be done for the further discussion.

Table 5. Haemorrhagic tendencies and laboratory parameters in coagulation status in members of patient's family.

Name		Haemor.	TC	Laboratory Examinat. R sec.	TZ	PTZ %	Clott. Fibrinogen mg%	Prec. mg%
H.,Al.	Ø	Ø	22,4	19,6	15,8	96	335	465
H.,He.	Ø	Ø	20,2	16,1	15,4	100	240	300
H-J,Ch.	+(+)	Ø	69,2	65,8	38,8	54	57	370
H.S,An.	+	+	66,9	52,3	36,1	57	62	370
H.,Otto	Ø ?	Ø	19,8	16,1	16,8	96	159	300
H.,Mag.	++	Ø	131,0	121,5	46,9	47	51	370
H.,Hans	+	+	77,9	58,5	35,0	55	58	
H.,K.H.	+	Ø	65,6	49,0	35,2	51	61	500
H.,Ber.	+(+)	Ø	87,9	67,8	39,8	47	43	240

References

1. Beck, E.A., Charache, P., Jackson, D.P.: Nature 2o8 , 143 (1965).
2. Mammen, E.F., Prasad, A.S., Barnhard, M.I., Au, Chi C.: J. Clin. Invest. 48, 235 (1969).
3. Crum, E.D., Schainoff, J.R., Graham, R.C., Ratnoff, O.D.: J. Clin. Invest. 53 , 13o8 (1974).
4. Krause, W.H., Huth, K., Heene, D.L., Lasch, H.G. Klin. Wschr. 53, 781 (1975).
5. Felten, von, A., Duckert, F., Frick, P.: Schweizerische Medizinische Wochenschrift 95, 1453 (1965).
6. Forman, W.B., Ratnoff, O.D., Boyer, M.H.: Lab. Clin.Med. 72, 455 (1968)
7. Janssen, C.L., Vreeken, J.: Brit. J. Hemat. 2o, 287 (1971).
8. Felten, von, A., Frick, P.G., Straub, P.W.: Brit. J. Hemat. 16, 353 (1969).

IV. FIBRINOPEPTIDE DETERMINATION

FPA-RELEASE IN NORMAL AND ABNORMAL FIBRINOGEN

Wolfgang Schramm and Rudolf Marx
Medizinische Klinik Innenstadt der Universität München
D - 8000 München , Germany

Introduction

In the following we present a short and preliminary comment about fpa-release from normal and abnormal fibrinogen.

Clinical data from our three patients with dysfibrinogenemia were already demonstrated and the impaired release of fibrinopeptide A was mentioned (see Marx, this volume).

The major defect, observed in some abnormal fibrinogens (Baltimore, Bethesda, Detroit, Giessen, Metz) was an impaired release of fibrinopeptide A after incubation with thrombin. The release of fpa occured more slowly than normally and the total amount of this peptide released was less than the expected amount (1, 2, 3).
One of the commonly used methods is described by Conn and Davis in which the guanidino-group of arginine is measured fluorometrically after the addition of aqueous ninhydrin at alkaline pH (4). Arginine is the COOH-terminal amino acid in both fibrinopeptides A and B; thus, its appearance in the supernatant fluid after clotting is a measure of released fibrinopeptides.
Other methods are e. g. bidimensional paper-electrophoresis, and the radioimmunologic estimation of fibrinopeptide A (5). We are working on the radioimmunologic determination of fpa. The interpretation of single fpa-levels from patient plasma samples proved to be questionable due to the pronounced

---

Herrn Professor Dr. E. Buchborn zum 60. Geburtstag gewidmet.

Fibrinogen - Recent Biochemical and Medical Aspects

influence on the results both of the sampling technique and of the short half-life of fpa (6, 7).

In the treatment of hemophiliacs with circulating inhibitors we found, in agreement with Manucci and Sixma, elevated fpa-levels in plasma after the substitution of activated factors like FEIBA (8). The usual parameters for controlling therapy, PTT and r+k value in the thrombelastography, did not correlate well with the replacement therapy and its clinical effectiveness. These difficulties in the laboratory control of FEIBA effectiveness in different hemorrhagic disorders led us to apply additional methods for adequate control (8). Thrombin arises continuously during coagulation and therefore we made an approach to detect fpa as an product of thrombin originating during the coagulation.

Results

For this purpose we used a commercially available test kit from IMCO. However, we modified the test procedure by replacing the time consuming dialysis method with adsorption extracting using DEAE A-25 Sephadex ion exchanger and the charcoal technique with a double antibody method. Data processing was performed by a computer program using spline approximation.
We examined the in vitro generation and release of fpa which served as a parameter for coagulability in a sample of native blood. The kinetics of the in vitro fpa-generation were assumed to be an improved parameter of the coagulability of the respective plasma sample.
Following blood sampling using a standardised procedure, fpa-concentrations were determined at defined intervalls (0, 4, 10, 20 and 30 mins). The results were graphically illustrated as "fpa-generation-curves". The normal range as determined from 16 normal individuals is shown in figure 1.

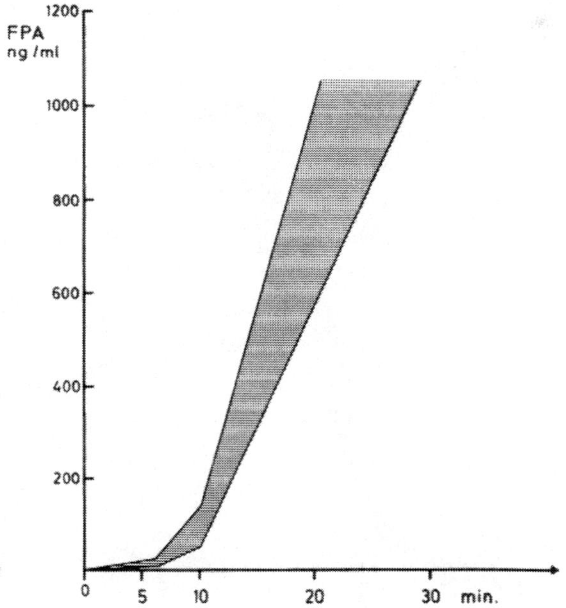

Fig. 1. FPA-generation (normal range n=16)

In addition figure 2 shows the curves obtained from a hemophiliac with circulating inhibitors proceeding and following substitution with factor VIII and FEIBA.

While in non-substituted patients only poor fpa-generation is demonstrable under the conditions mentioned above, the resulting flat curve is raised in the direction of normal range following substitution with FEIBA. We were able to find a comparable behaviour of the curves in another patients following replacement therapy.

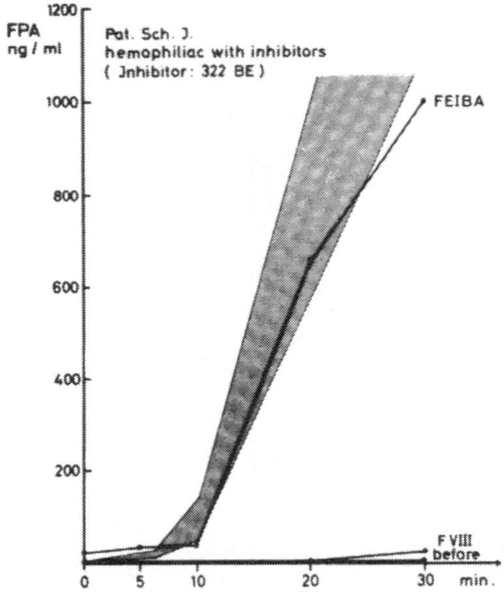

Fig. 2. FPA-generation before and after subst. of:
Faktor VIII: 273 E/kg KG
FEIBA: 202 E/kg KG

Following the explanation of our testsystem we describe the "fpa-release or -generation" in patients with dysfibrinogenemias or abnormal fibrinogen. With the method mentioned before we examined 2 of our patients with abnormal fibrinogen Munich I and II and 4 additional patients with dysfibrinogenemia. The plasma samples of the latter patients were obtained from Prof. Scharrer from Frankfurt.

The curves of these patients with abnormal fibrinogen are recorded on the figure 3 and the impaired fpa-generation is demonstrated. We found only a slight difference between the individual patients with dysfibrinogenemia.

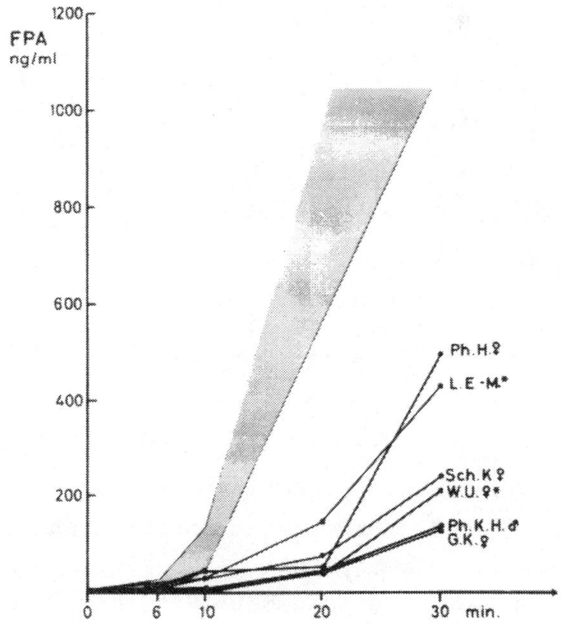

Fig. 3. FPA-generation in abnormal fibrinogen (n=6)

The severity of the "coagulopathy" as measured by the fpa-release or generation is similar to the one of patients with mild or slight hemophilia or patients during anticoagulant-therapy.

In conclusion, the method presented allows a limited statement about the coagulopathy. We were not yet able to obtain further details on the pathogenesis of this disease, despite of the always impaired fpa-generation in these patients.
An exchange of an aminoacid, as observed by Dr. Henschen fibrinogen Munich, may be a possible reason for the impaired fpa-release.

References:

1. Beck, E. A., Shainoff, J. R., Vogel, A. and Jackson, D.P.: J. Clin. Invest. 50, 1874 - 1884 (1971).

2. Gralnick, H. R., Givelber, H. M., Shainoff, J. R. and Finlayson, J. S.: J. Clin. Invest. 50, 1819 - 1830 (1971).

3. Crum, E. D., Shainoff, J. R., Graham, R. C. and Ratnoff, O.W.: J. Clin. Invest. 53, 1308 - 1319 (1974).

4. Conn, E. B. and Davis, R. B.: Nature (Lond.) 183, 1053 (1959).

5. Nossel, H. L. et al.: Proc. Nap. Acad. Sci. 68, 2350 (1971).

6. Schramm, W., Drost, W., Schmidt, M., Borlinghaus, P. and Marx, R.: Deutsche Arbeitsgemeinschaft für Blutgerinnungsforschung, Heidelberg, Febr. 1979. in: Fibrinogen, Fibrin und Fibrinkleber, Herausgeber: K. Schimpf, F. K. Schattauer-Verlag, 391 - 397 (1979).

7. Schramm, W.: 1. Kongreß für Thrombose und Blutgerinnung der Deutschen Arbeitsgemeinschaft für Blutgerinnungsforschung in Wien. Herausgeber: E. Deutsch u. K. Lechner, F. K. Schattauer-Verlag, 242 - 245 (1980).

8. Schramm, W.: 2nd workshop on factor VIII-inhibitor. Further experience with FEIBA and leading disorder, Vienna, 19. 3. 1979, in press.

SIMULTANEOUS CLINICAL EVALUATION OF RADIOIMMUNOASSAYS FOR FIBRINOPEPTIDES AND β-THROMBOGLOBULIN

David A. Lane, Helen Ireland and Sylvia Wolff

Department of Haematology, Charing Cross Hospital Medical School, Hammersmith, London, W6 8RF, U.K.

Introduction

The measurement of the *in vivo* concentration of serine proteases of the blood coagulation system, such as thrombin and plasmin, would be useful in the diagnosis and treatment of thrombosis and disseminated intravascular coagulation. Because these proteases are rapidly inactivated by plasma inhibitors their direct measurement in plasma is impractical. An alternative approach is to quantitate stable derivatives that are produced as a consequence of protease action on their natural substrates. Fibrinopeptide A (FpA) is the first peptide cleaved from fibrinogen by thrombin when fibrin is formed. Nossel and co-workers (1) have developed a specific radioimmunoassay to FpA and a number of studies have shown that an elevated FpA in patients' plasma can reflect intravascular thrombin action. Plasmin is thought to be the major enzyme of the fibrinolytic system responsible for fibrin dissolution. Nossel and co-workers (2) have also developed a radioimmunoassay to fibrinogen fragment Bβ1-42, a fragment that is cleaved from fibrinogen or fibrin I early in their digestion by plasmin. The measurement in plasma samples of both FpA and Bβ1-42 therefore enables a simultaneous assessment of the presence of the major clot producing and clot dissolving proteases of plasma.

A third component of the haemostasis system, the platelet, also participates in normal coagulation and in thrombosis. A number

Fibrinogen - Recent Biochemical and Medical Aspects

of markers of the *in vivo* platelet release reaction are presently under evaluation. β-Thromboglobulin (βTG) is a convenient protein with which to set up a radioimmunoassay and numerous studies with it have demonstrated enhanced platelet release in various disorders predisposing towards thrombosis (3). However, what has often remained uncertain is the relationship existing between platelet release and activation of coagulation and fibrinolysis.

We have recently performed a preliminary simultaneous clinical evaluation of assays to FpA, Bβ1-42 and βTG. It was hoped that a better perspective could be obtained on the nature of the imbalance of haemostasis that occurs in thrombotic disease. A summary of some of our results is presented here.

Materials and Methods

Blood samples were collected within 45 sec of trauma-free venepuncture and added to anticoagulants contained in tubes maintained in melting ice. Blood for FpA and Bβ1-42 was added (9 vol) to 1000 u/ml heparin and 1000 u/ml aprotinin (1 vol), also containing EDTA and ε-aminocaproic acid (4). Blood for βTG analysis was added to the anticoagulant recommended by Ludlam and Cash (5), namely a mixture of EDTA, theophylline and prostaglandin E_1.

Blood was then centrifuged to remove the cellular components and processed prior to assay. For FpA the method of Nossel *et al.* (1) was used, with some modifications (6). The concentration of Bβ1-42 was determined by establishing its FpB content, using a specific radioimmunoassay to that peptide (2, 6). The βTG assay was performed as described by Bolton *et al.* (7), with some modifications (6).

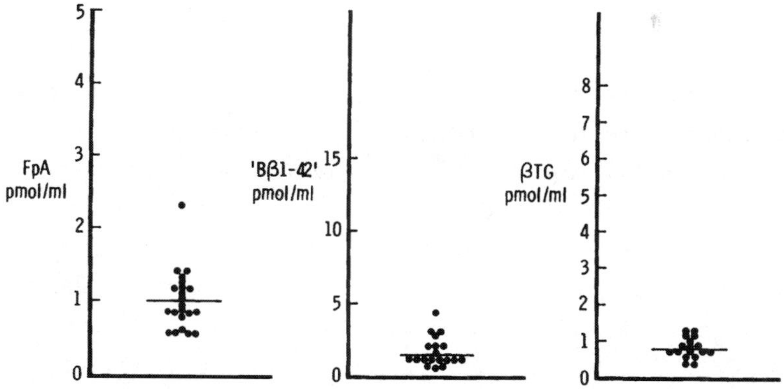

Figure 1: Plasma concentrations of FpA, Bβ1-42 and βTG in 20 normal volunteers. The solid line in this and subsequent figures indicates the position of the mean of the results. In subsequent figures a dashed horizontal line indicates the position of the upper limit of the 95% normal range at 1.5, 3.5 and 1.3 pmol/ml, respectively.

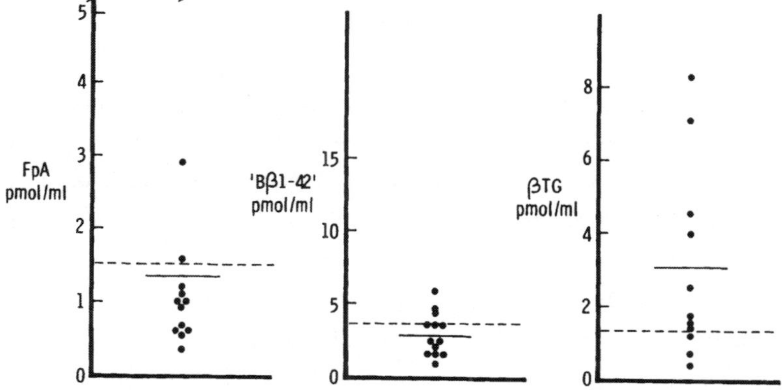

Figure 2: Plasma concentrations of FpA, Bβ1-42 and βTG in 13 patients with myeloproliferative disorders.

Results

Plasma was initially collected from normal laboratory personnel (n=20) and normal means and ranges established for FpA, Bβ1-42 and βTG. The individual results are illustrated in Fig. 1 and mean values were found to be 1.06, 1,59 and 0.80 pmol/ml, respectively. A 95% range was thought to be more representative of the distribution of the results than the complete range because a minority of individuals with no apparent thrombosis have elevated levels of one or more of the activation products. In subsequent figures this 95% range is represented by a dashed horizontal line at 1.5, 3.5 and 1.3 pmol/ml, respectively.

Thirteen patients with myeloproliferative disorders were studied, Fig. 2, and mean results for FpA, Bβ1-42 and βTG were 1.40, 2.90 and 3.04 pmol/ml, respectively. The plasma concentration of βTG is therefore raised much more than the FpA or Bβ1-42 levels. These results suggest that the acquired storage pool defect seen in these disorders (8) occurs, in general, by disseminated intravascular platelet release in the absence of thrombin and plasmin generation. However, a small number of patients do have evidence of elevated thrombin action and this may contribute further to the increased platelet release.

Twelve patients were examined, Fig. 3, who had had clinical evidence of thrombotic stroke, the infarct of which was confirmed by computer tomographical scanning. Mean plasma concentrations of FpA, Bβ1-42 and βTG were 2.12, 4.76 and 1.94 pmol/ml, respectively. These mean values are clearly outside the normal range and demonstrate a simultaneous activation of the three components of the haemostasis system. This is despite the fact that the stroke occurred in some of these patients greater than two years before examination. Alkjaersig *et al.* (9) have previously demonstrated activation of coagulation following stroke by the gel filtration technique. The presently used methods allow greater numbers of samples to be

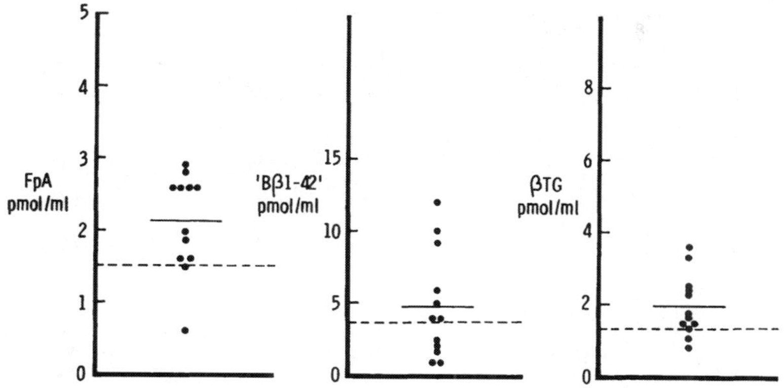

Figure 3: Plasma concentrations of FpA, Bβ1-42 and βTG in 12 patients who had had thrombotic stroke.

more readily analysed. We are currently investigating a larger group of stroke patients in order that the influence of age, therapeutic intervention and length of time after stroke on plasma levels of activation products can be ascertained (10).

We have also studied a number of patients with renal failure. In Table I are presented results from four patients who were studied before and after haemodialysis. Plasma FpA levels were sporadically elevated before dialysis. The reason for this occasional elevation is not clear at present but may be a consequence of indwelling shunts. The reduction of FpA by dialysis is not surprising in view of the low MW of the peptide. Plasma concentrations of βTG and Bβ1-42 were consistently elevated but were not reduced by dialysis. Recent work (11) has demonstrated that βTG, unlike platelet factor 4 (PF4), is actively catabolised by the kidney and the large plasma concentration seen during renal failure are probably due to impaired catabolism of the protein rather than increased activation of platelets. Comparable data is not yet available for Bβ1-42 but the plasma concentration of fibrinogen degradation product

		FpA	Bβ1-42	βTG
J.T.	Pre	1.87	31.3	4.11
	Mid	0.77	33.3	5.55
J.C.	Pre	39.5	90.6	8.16
	Mid	3.23	82.6	9.72
R.P.	Pre	5.5	76.6	4.72
	Mid	8.3	66.6	-
C.J.	Pre	1.46	82.6	3.61
	Mid	1.46	80.0	3.80

Table I: Plasma concentrations of FpA, Bβ1-42 and βTG before (pre) and during (mid) renal dialysis for chronic renal failure. Results are in pmol/ml.

fragment E is also known to be dependent upon active renal tissue (12). It seems reasonable to infer therefore that the very large plasma concentrations of Bβ1-42 are also produced because of impaired renal catabolism.

Patients (n=20) have also been studied prior to and following total hip replacement operations in which no prophylaxis was given to prevent deep venous thrombosis (DVT). Results of FpA, Bβ1-42 and βTG determinations are summarised in Fig. 4. Mean FpA and Bβ1-42 plasma concentrations were slightly elevated in the pre-operative samples, possibly reflecting tissue damage or some other underlying disease such as rheumatoid arthritis, or even prolonged bed rest. A further small rise in mean FpA was observed on post-operative day 1 and on post-operative day 6 the mean FpA was 2.25 pmol/ml. Bβ1-42 rose much more rapidly and by a greater amount to 12.4 pmol/ml on day 4. Plasma βTG concentrations were also minimally raised post operatively with the largest rise on post-operative day 1, but the mean values were not elevated outside the upper limit of the 95% normal range.

Three possible explanations for these elevated levels of activation products can be advanced. Firstly, the trauma of

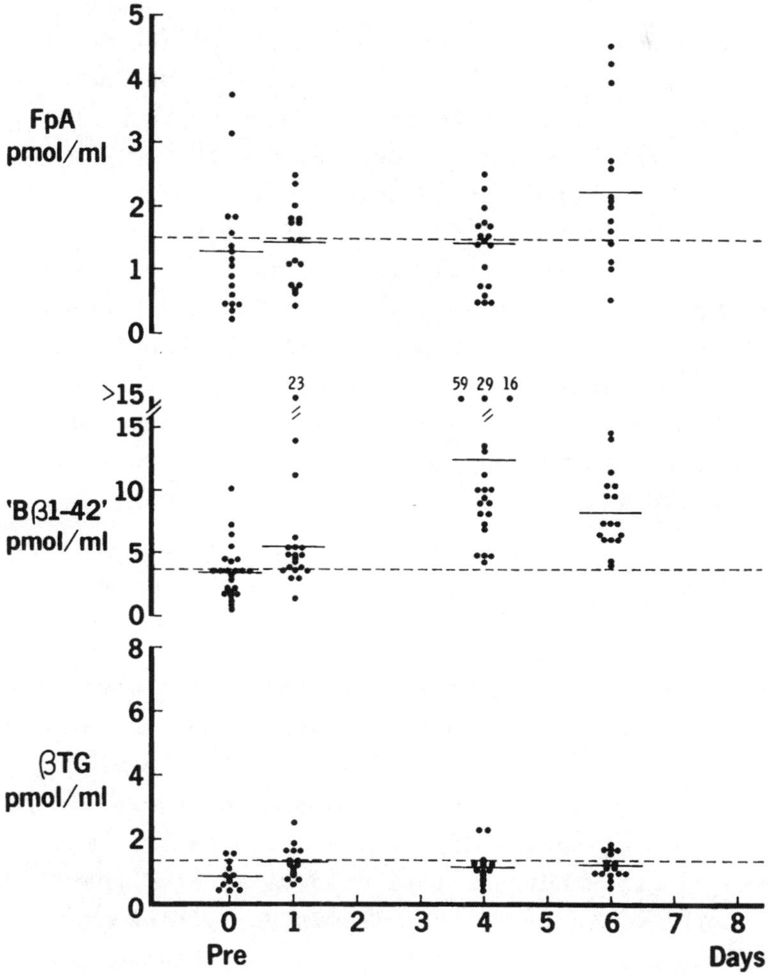

Figure 4: Plasma concentrations of FpA, Bβ1-42 and βTG in 20 patients undergoing total hip replacement operations. Blood samples were withdrawn pre-operatively and on days 1, 4 and 6 post-operatively.

the operation and associated tissue damage may cause a localised wound healing response that involves fibrinogen and platelets. Secondly, there is a high incidence of post operative intravascular fibrin deposition following total hip replacement and this process is associated with peptide cleavage. Finally, alterations in vascular volumes may cause transient changes in activation products. Currently we are engaged in a larger study of these patients so that each of these alternatives can be challenged (13). Preliminary results of this continuing study (in which 26 patients have been examined for DVT by phlebography and I^{125} fibrinogen scanning) are suggesting that patients who have developed DVT have different concentrations of activation products than those who have not developed thrombosis.

Discussion

It was Nossel *et al.* (2) who initiated the simultaneous measurement of FpA, Bβ1-42 and βTG to study alterations in haemostasis in thrombotic disease. They investigated the disseminated intravascular coagulation induced in women undergoing intra-uterine infusions of saline to terminate pregnancy. The present preliminary study has further demonstrated that a combined assessment of thrombin and plasmin activities and platelet releasing stimuli may be a useful approach with which an imbalance of haemostasis can be monitored. The assay methods are conceptionally superior to many existing techniques because of their sensitivity and specificity. However, in some clinical situations additional information may be required to distinguish between (a) impaired renal catabolism or clearance, (b) redistribution between intra- and extravascular spaces and (c) activation of coagulation and fibrinolytic systems and platelets.

Acknowledgements

We would like to thank Dr. H. Nossel, College of Physicians and Surgeons, New York, and Dr. D. Pepper, S.E. Scotland Blood Transfusion Service, Edinburgh, for gifts of reagents, without which this work could not have been carried out.

References

1. Nossel, H.L., Yudelman, I., Canfield, R.E., Butler, V.P., Spanondis, K., Wilner, G.D., Qureshi, G.D.: J. Clin. Invest. 54, 43 (1974).

2. Nossel, H.L., Wasser, J., Kaplan, K.L., LaGamma, K.S., Yudelman, I., Canfield, R.E.: J. Clin. Invest. 64, 1371 (1979).

3. Pepper, D.S., Ludlam, C.A.: In 'Platelets and Thrombosis', ed. Mills, D.C.B., Paretti, F.I. Academic Press (1977).

4. Peuscher, F.W., van Aken, W.G., Flier, O.T.N., Stoepman-van Dalen, E.A., Cremer-Goote, T.M., van Mourik, J.A.: Thrombos. Res. 18, 33 (1980).

5. Ludlam, C.A., Cash, J.D.: Br. J. Haematol. 33, 239 (1976).

6. Lane, D.A., Ireland, H., Wolff, S., Boots, M., Pegrum, G.D.: Submitted for publication, 1981.

7. Bolton, A.E., Ludlam, C.A., Moore, S., Pepper, D.S., Cash, J.D.: Br. J. Haematol. 32, 233 (1976).

8. Boughton, B.J., Allington, M.J., King, A.: Br. J. Haematol. 40, 125 (1978).

9. Alkjaersig, N., Laursen, B., Fletcher, A.: J. Clin. Invest. 51, 2a (1972).

10. Lane, D.A., Gawel, M., Wolff, S., Ireland, H., Clifford-Rose, F.: Thrombos. Haemostas., in press (1981).

11. Guzzo, J., Niewiarowski, S., Musial, J., Bastl, C., Grossman, R.A., Rao, K., Berman, I., Paul, D.: J. Lab. Clin. Med. 96, 102 (1980).

12. Ilo, A., Rutherford, W.E., Wochner, R.D., Spilberg, I., Sherman, L.A.: J. Lab. Clin. Med. 87, 934 (1976).

13. Lane, D.A., Allen-Mersh, T., Ireland, H., Wolff, S., Grant, R.: Thrombos. Haemostas., in press (1981).

ANALYSIS OF HUMAN FIBRINOPEPTIDES BY HIGH PERFORMANCE LIQUID CHROMATOGRAPHY : QUANTITATIVE DETERMINATION AND KINETICS OF RELEASE

Maria Kehl, Friedrich Lottspeich and Agnes Henschen

Max-Planck-Institut für Biochemie, D-8033 Martinsried bei München, Germany Federal Republic

Introduction

The mechanism of the reaction between fibrinogen and thrombin, whereby the fibrinopeptide A and B are released and fibrin is formed, has already been the subject of numerous studies. However, determination of the fibrinopeptides is still of considerable interest, as the release of these peptides may be correlated to the state of the blood coagulation system, both in vivo and in vitro. Thus, an increased fibrinopeptide blood plasma level can be used as an indicator of increased intravascular coagulation. Abnormality in the rate or amount of fibrinopeptide released on incubation of fibrinogen with thrombin is of great diagnostic value for the identification of a dysfibrinogenemia.

Many determination methods for human fibrinopeptides have been developed since Blombäck et al. in 1966 (1) succeeded in separating the peptides by precipitation of the B-peptide and ion-exchange chromatography of the A-peptides. However, all methods described have shown the disadvantage of being either very time- and labor-consuming or somewhat unspecific.

In this report a novel method, based on high performance liquid

Fibrinogen - Recent Biochemical and Medical Aspects

chromatography (HPLC), is outlined. The method has the merit of being simple, rapid, specific and applicable to all types of fibrinopeptides and their most common degradation products.

Procedure

Thrombin digestion. Human fibrinogen (Forschungs-Fibrinogen, Deutsche Kabi, München) was dissolved to 10-15 mg/ml in 0.15M ammonium acetate, pH 8.5 with ammonia. Bovine thrombin (Behringwerke, Marburg/Lahn) was added to a final concentration of 0.04 NIH units/ml. For kinetic studies samples were placed after various times (0.5 - 18 h, see Figure 1 below) in a boiling waterbath for 1 min., then cooled and the heatprecipitate removed by centrifugation. A part of the peptide-containing supernatant liquid was subjected to HPLC analysis.

HPLC conditions.

Column	Hibar 250-4, LiChrosorb RP-18 (5µ) Merck, Darmstadt
Solvent	A: 0.02 M ammonium acetate pH 6.0 with ortho-phosphoric acid/acetonitrile, 94:6 (v/v). B: 0.05 M ammonium acetate pH 6.0 with ortho-phosphoric acid/acetonitrile, 75: 25 (v/v).
Gradient	A straight concentration gradient from 0 to 100 % B in 66 min.
Flow	1.5 ml/min.
Temperature	Ambient.
Detection	210 nm at 0.4 AUFS.
Sample volume	250 µl.

Results

A reversed phase C-18 column was employed for the separation of all fibrinopeptides in the supernatant of thrombin-digested fibrinogen. To begin with, several isocratic elution systems were tested, but the separations which could be obtained were unsatisfactory. Thus, with an acetonitrile concentration of 6% no peptides were eluted from the column; with 8 or 8.5% all peptides, except the B-peptide, were eluted and separated, but were obtained as broad peaks; with 10% the AP-peptide was unretarded and remaining peptides eluted as broad peaks, and with 15% all peptides were completely unretarded.

When, instead, gradient elution was used all fibrinopeptid species could be separated and eluted as sharp peaks, the gradient between 6 and 25 % acetonitrile giving the optimal resolution, as demonstrated in Figure 1. The figure shows the superimposed HPLC elution profiles of the fibrinopeptides released during 0.5 to 18 hours of thrombin digestion. The elution order and structure of the peptides is given in Table 1, the peaks being identified by the analysis of their amino acid composition and /or sequence and also by comparison with the elution times of fibrinopeptides isolated according to the classical procedure (1).

Table 1. Human fibrinopeptide components separated by HPLC.

Peak number	Designation	Sequence and charges
1	AP	+ A D S(Ⓟ) G E G D F L A E G G G V R + −
2	A	+ A D S G E G D F L A E G G G V R + −
3	AY	+ D S G E G D F L A E G G G V R + −
4	B -Arg	Z G V N D N E E G F F S A −
5	B	Z G V N D N E E G F F S A R + −

Ⓟ denotes phosphate and Z pyroglutamic acid

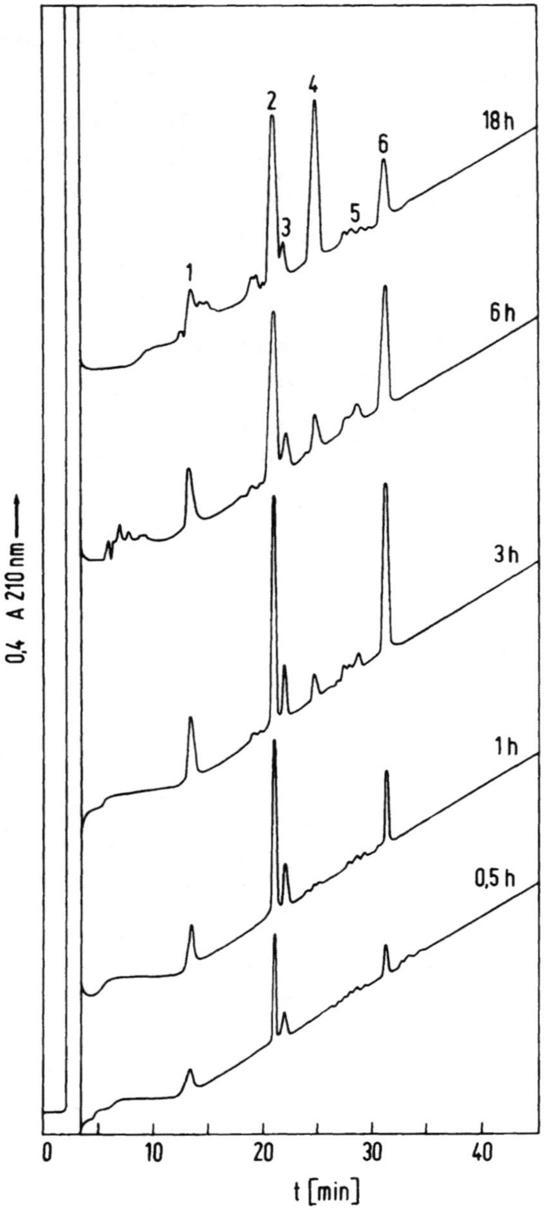

Figure 1. HPLC elutions profiles of human fibrinopeptide release, determined at various times.

The first peak, which appears soon after sample injection, is due to solvents ans salts in the sample. It also contains some free amino acids, i.e. glycine from the fibrinogen preparation and arginine released from the B-peptideby the action of a carboxypeptidase which is present in blood (2) and contaminates fibrinogen preparations. It seems noteworthy that the B-peptide, but not the A-peptides, is readily attacked by this enzyme.

The elution order of the fibrinopeptides is mainly in good agreement with that beeing predictable from the amino acid compositions using retention coefficients as calculated by Meek (3). Amino acids with aromatic or aliphatic side chains have been shown to give major positive contributions to the retention, those with acidic side chains major negative contributions and those with basic or neutral side chains only minor contributions. Thus, the group of A-peptides, which contain fewer aromatic and more acidic amino acids among their 16 or 15 residues, would indeed be anticipated to elute before the group of B-peptides, which contain more aromatic and fewer acidic amino acids among their 14 or 13 residues. Furthermore, the peptide AP with its additional negative charge, i.e. the phosphate group, is also expected to have the shortest retention time. The relative order of the peptides A and AY might be explained as due to an interference between the N-terminal amino group and the carboxyl group of the first aspartic acid residue in peptide AY, but not in peptide A, causing AY to behave as a less acidic peptide than A.

A similar explanation might be given for the B-peptides, where B-des-Arg elutes before the B-peptide. The B-peptide without the C-terminal arginine might have more free acidic side chains than the complete B-peptide, in which the C-terminal carboxyl group might be counteracted by the basic arginine side chain.

In Figures 2 and 3 the kinetics of release for all three types of A-peptide may be compared. In Figure 2 the peak areas of A-peptides are plotted as a function of time. Under the conditions used, the release of all A-peptides levels off at 6 h. When the ratios between each A-peptide and the total A-peptides are plotted against time, straight lines are always obtained, the relative amount of peptide A being 62 %, of AP 25 % and AY 13 % (Figure 3). This indicates that no type of A-peptide is preferentially released by the action of thrombin.

Figure 4 and 5 illustrate the kinetics of appearance or disappearance of the two forms of B-peptide. The plot of the B-peptide peak area against time in Figure 4 shows clearly that at least two interfering reactions are involved, i.e. the thrombin-catalysed release and the carboxypeptidase-catalysed degradation. The amount of the B-peptide species lacking arginine, however, increases steadily with time. Figure 5 gives the time course of the ratios between each B-peptide form and the total B-peptide.

In Figure 6, finally, the sum of all A-peptides and the sum of all B-peptides is plotted against incubation time. The well-established fact that the A-peptide is released at a higher initial rate than the B-peptide is confirmed also by the present results.

Conclusion

With the aid of high performance liquid chromatography the human fibrinopeptides could easily and quickly be separated. The conditions used are quite different from those earlier described for the separation of bovine fibrinopeptides by Martinelli and Scheraga (4).

The method outlined in this report offers for the first time the possibility to identify and quantitate in parallel all

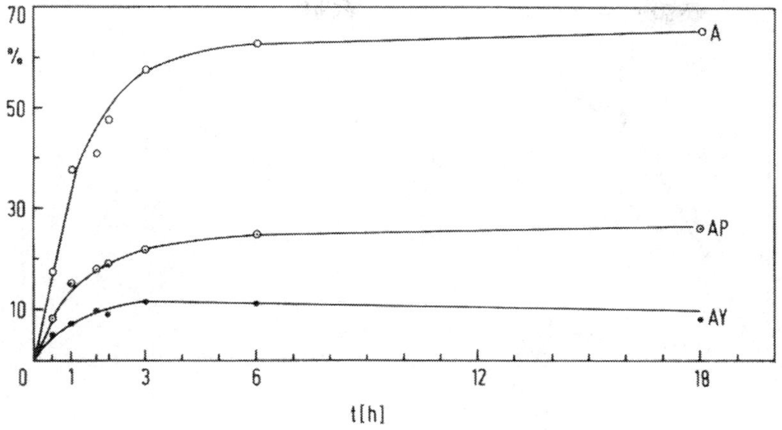

Figure 2. Human fibrinopeptide A release, peak areas of each A-peptide plotted as function of time.

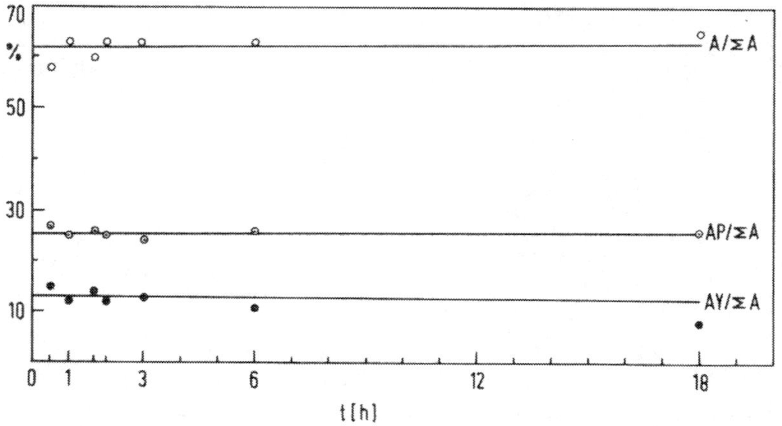

Figure 3. Ratio between each A-peptide and the sum of all A-peptides plotted as function of time.

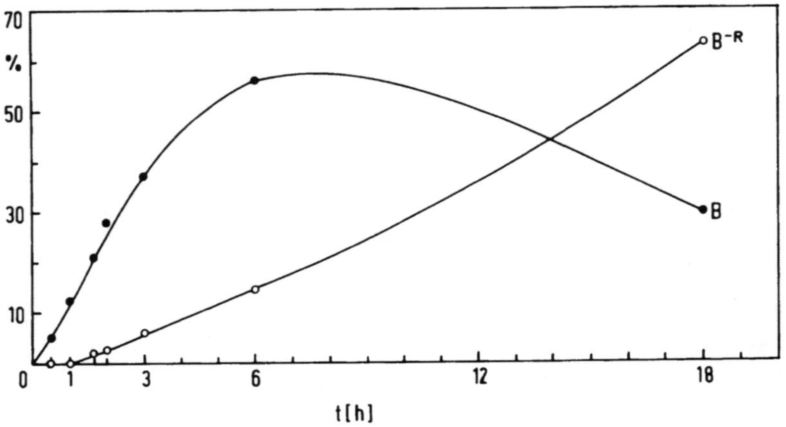

Figure 4. Human fibrinopeptide B release,peak areas of B- and B-des-Arg-peptide plotted as function of time.

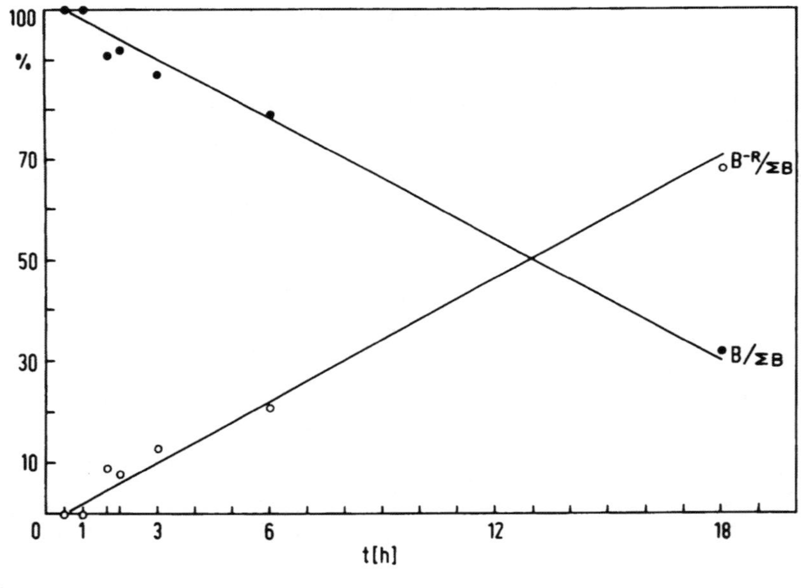

Figure 5. Ratio between each B-peptide and the sum of all B-peptides plotted as function of time.

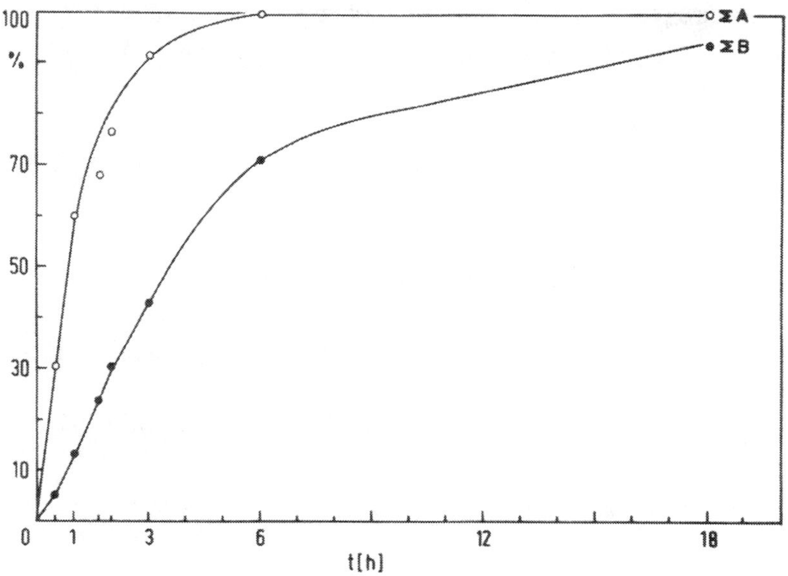

Figure 6. Time course of human fibrinopeptide release expressed as procentage of total fibrinopeptide A and B.

variants of the human fibrinopeptides. In comparison with the recently widely used radioimmunological fibrinopeptide assay (5,6) the procedure has the drawback of being much less sensitive. However, the HPLC method has the great advantage of identifying the whole peptide, not just an antigenic determinant, which might be part of many undefined peptides.Furthermore, all A- and B-peptides and their degradation products can be determined simultaneously. Thus, the HPLC assay is much more informative and less time-consuming.

An additional advantage of the HPLC fibrinopeptide determination is that it is ideally suited for microscale peptide preparation. Since the solvents involved are volatile, the separated components may conveniently be analysed for, e.g.,amino acid composition and sequence directly after lyophilisation. This is of special importance for the analysis of abnormal fibrinopeptides,derived from genetically abnormal fibrinogen.

This investigations was supported by a grant-in-aid from the Deutsche Forschungsgemeinschaft, which is gratefully acknowledged.

References

1. Blombäck,B., Blombäck,M. Edman,P., Hessel,B.: Biochim. Biophys. Acta 115, 371-396 (1966).
2. Teger-Nilsson,A.C. : Acta Chem. Scand. 22, 3171-3182 (1968).
3. Meek,J.E.: Proc. Natl. Acad. Sci. USA, 77, 1632-1636 (1980).
4. Martinelli,R.A., Scheraga,H.A.: Anal. Biochem. 96, 246-249 (1979).
5. Nossel,H.L., Younger, L.R., Wilner, G.D., Procupez,T., Canfield, R.E. and Butler,V.P.: Proc. Natl. Acad.Sci.USA, 68, 2350-2353 (1971).
6. Bilezikian,S.B., Nossel,H.L., Butler, Jr.,V.P., Canfield, R.E.: J. Clin. Invest. 56, 438-445 (1975).

V. FIBRINOGEN - FIBRIN COMPLEXES

SOLUBLE FIBRINOGEN-FIBRIN COMPLEXES IN PLASMA AT 37°C

Peter Kierulf and Frank Brosstad

Central Laboratory and Hematological Research Lab., Dept. IX, Ullevål Hospital, Oslo, Norway

Introduction

The existence of soluble fibrinogen-fibrin complexes in circulation in vivo has been at doubt. Recent evidence has suggested such complexes to exist in plasma at 20°C, but not at 37°C, implying such intermediates to be non-existent in vivo (1).

Material and Methods

Ca^{++}-boosted (2.5 mmol) prothrombin adsorbed plasma (=Ca^{++}-plasma) was prepared by adsorbing thrice with $Ba(SO_4)_2$ and dialyzing against Ca^{++}-containing Ringer-solution. Ca^{++}-plasma showed negligible thrombin generation as evidenced by Fibrinopeptide A (RIA-FPA) appearance, even when continuously chromatographed at 20°C for 48 hours (FPA < 15 ng/ml). Plasma containing 125-I fibrinogen and/or 131-I fibrin Des AA (or Des AABB) monomers (2) was subjected to either gel chromatography or to enzyme-fixation. Gel chromatography at 37°C was performed on Sepharose CL-6B, freeze-dried, reconstituted in Ca^{++}-plasma, and developed with Ca^{++}-plasma or Ca^{++}-plasma with EDTA. Enzyme-fixation was done by adding thrombin-activated (removable, insolubilized thrombin) Factor XIII (XIII^X). Subsequently XIII^X-acitivity was quenched with EDTA. Enzyme-fixed, radiolabelled complexes were either harvested by thrombin-clotting in the presence of EDTA or chromatographed on Sepharose CL-6B (0.15 M Tris-HCl pH 7.0, 0.15 M NaCl, 2.5 M urea, 10 mmol EDTA). Harvested clots were mercaptolysed and SDS-gel electrophoresed. Gel chromatographed specimens were radioprofiled and freeze-dried.

Results

In Ca^{++}-plasma both Des AA and Des AABB 131-I fibrin eluted in front of the 125-I fibrinogen, whereas in the presence of EDTA 131-I fibrin Des AA eluted completely within the 125-I fibrinogen peak. SDS-gel electrophoresed, mercaptolysed, clots, containing enzyme-fixed, radiolabelled, fibrinogen derivatives, demonstrated 125-I and 131-I activity in γ-γ-dimers and to lesser extent in α-oligomers. Gel chromatographed, enzyme-fixed specimens, demonstrated radiolabelled dimerized fibrinogen derivatives (125-I and 131-I activity) RIA-FPA of such dimers demonstrated approx. 2 moles FPA/mol dimer.

Conclusion

Soluble, covalently-linkable largely dimeric fibrinogen-fibrin complexes exist in Ca^{++}-plasma at $37^{o}C$.

References

1. Müller-Berghaus, G., Mahn, I., Krell, W.: Formation and dissociation of soluble fibrin complexes in plasma at $20^{o}C$ and at $37^{o}C$. Thrombos.Res., 14, 561-572 (1979).

2. Brosstad, F., Godal, H.C., Kierulf, P.: Preparation of fibrin monomers from human fibrinogen in urea, using soluble or insolubilized thrombin. Haemostasis, 6, 225-235 (1977).

SOLUBLE DES-AA FIBRIN CLEARANCE IN HEALTHY HUMAN SUBJECTS

Frank Brosstad,Tor Hamborg,Bjørn Holm,Hans C.Godal,Peter Kierulf[x]
Hematological Research Laboratory and The Central Laboratory[x]
Ullevål Hospital, Oslo,Norway

Introduction

Although fibrinaemia is a common feature of both surgical,obstetrical,infectious and medical conditions of some severity, it has not yet been decided wether such soluble des-AA fibrin per se further hazards the underlying disease or not. From the fact that des-AA fibrin solubility of plasma is limited(approx. 3% of the fibrinogen level), the possibility that soluble fibrin formation in excess of this might lead to thromboembolic complications does exist, implying that therapeutic measures are indicated whenever fibriniaemia is detected. However, in order to reach a definite conclusion as to a potentially harmful effect of fibrininaemia does exist or not, knowledge as to soluble fibrin formation rates as well as soluble fibrin clearance rates in the diseased and healthy man, will be necessary. However,such knowledge has hitherto been virtually lacking, and the present study gives some preliminary results as to the clearance rate of soluble des-AA fibrin in the healthy man.

Materials and Methods.

Labelled monomers(I^{125} or ^{131})were made by incubation of labeled fibrinogen in 2.5 M urea with insolubilized reptilase(1). Such monomers were solubilized upon addition to test-subject plasma,and subsequently injected.At intervals,citrated blood

Fibrinogen - Recent Biochemical and Medical Aspects

samples were tested for :I.Total Radioactivity(Tr).II.Radioprecipitability (Rp) with trichloroacetic acid III.Radioclottability (Rc) with plasmin-free thrombin(pH 6.3,1 H)IV.Radiodistribution after ultracentrifugation (Airfug,Beckmann) and V.Radiodistribution in SDS-PAGE gels after electrophoresis of harvested EDTA-clots, solubilized in urea-SDS.

Results and Discussion

Prior to injection, plasma demonstrated: Rp=100%,Rc=94%, single peak ultracentrifuge pattern and 95% of total radioactivity moving as fibrinogen on SDS-PAGE gels.Almost identical results were obtained in all 8 subjects tested. Thus, after 30 min, Tr had dropped to 50% of the Tr value at 5 min, being 10% and 2% after 420 min and 48 h,respectively. A parallell drop in Rp and Rc, scarcely noticeable at 10 min, reached a minimum of 30% at 360 min; the Airfug and SDS-PAGE patterns substantiating these findings. Due to non-linearity on semilogarithmic plots, a definite half-life could not be stated from Tr decay parameters alone, but metabolic data as outlined above(Rp,Rc,Airfug, SDS-PAGE gel patterns) suggest a half-life of des-AA fibrin of 30-60 min. These data correlate well with those obtained by Müller-Berghaus et al(personal communication) and those suggested from the work of Shainoff et al.
In five subjects,I^{131}-des-AA fibrin and I^{125}-fibrinogen were injected simultaneously, resulting in clearance rates characteristic of that of the individual molecular species.
The extremely rapid elimination rate of fibrin des-AA compared to that of fibrinogen suggests specific receptors on RES-cells to be involved. Studies are know in progress to assess the elimination rate of des-AA fibrin in fibrinaemic patients.

References.

Brosstad,F.,Kierulf,P,Gravem,K. and Godal,H.C.:Thromb.Res.13 715-723(1978).

COMPLEX FORMATION OF SOLUBLE FIBRIN POLYMERS WITH AGAROSE-COUPLED FIBRINOGEN AND FIBRIN WITH REGARD TO THEIR MOLECULAR WEIGHT

René von Hugo, Reimar Hafter, Henner Graeff
I. Universitäts-Frauenklinik München
D-8000 München, Germany

High molecular weight derivatives of fibrin can be observed in plasma of patients during the so-called prethrombotic state and in cases of intravascular coagulation (1). The fibrinogen-fibrin conversion by thrombin is the prerequisite for the occurrence of fibrin derivatives. It has been shown that fibrinogen in high concentrations can keep small amounts of fibrin in solution (2). In the earliest stage of fibrin generation soluble fibrin monomers interact with fibrinogen and form dimeric fibrin-fibrinogen complexes on nearly equimolar base (3). These complexes dissociate easily and it is not yet clear in which form they occur in the circulating blood.

When the polymerization process goes on, higher ordered polymers of fibrin can be observed. It was the aim of the present investigation to elucidate the interaction of these polymers with fibrinogen.

A fibrinogen solution containing various soluble molecular populations of fibrin was prepared by incubation of a 0.5 % fibrinogen solution in Tris citrate buffer with small amounts of thrombin. The thrombin action was inhibited after 5 minutes by addition of hirudin and the sample was allowed to stand for 85 minutes at room temperature (4).

The filtration of the preparation on 4 % agarose gel showed two main fractions which are eluted in front of the fibrinogen peak (Fig. 1). Fraction a, which was eluted at the void

Fibrinogen - Recent Biochemical and Medical Aspects

volume contains high molecular weight fibrin polymers showing α-chain degradation but no dimeric γ-chains as result of F XIII induced γ-chain crosslinking could be demonstrated. Fraction b, which was eluted in a shoulder prior to the fibrinogen peak showed intact α-, β- and γ-chains and is in accorrdance with the elution position and the electrophoretic mobility in 5 % PAA-gel of a fibrin dimer.

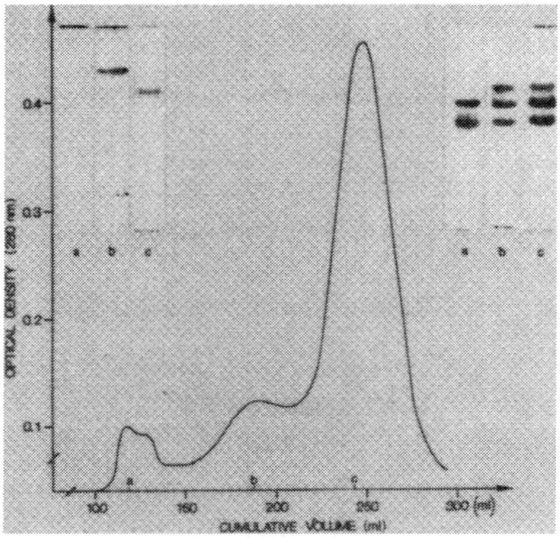

Fig. 1: Elution profile from 4 % agarose gel filtration of fibrinogen incubated with thrombin. PAA gel patterns of unreduced fractions a, b and c (left) and SDS PAA gel patterns of reduced fractions (right).

The fractions a and b were further analyzed by affinity chromatography on fibrinogen-agarose. Only 21 % of HMWFP are adsorbed to fibrinogen agarose whereas 34 % of dimeric fibrin bind to fibrinogen (Fig. 2). The difference is even more impressive if one considers that fraction b contains small amounts of fibrinogen since it is not completely separated from the fibrinogen peak. If the adsorption rate is corrected for the containing fibrinogen, as we did in one of our earlier studies (3), about 43 % of the dimeric fibrin is adsorbed to fibrinogen-agarose.

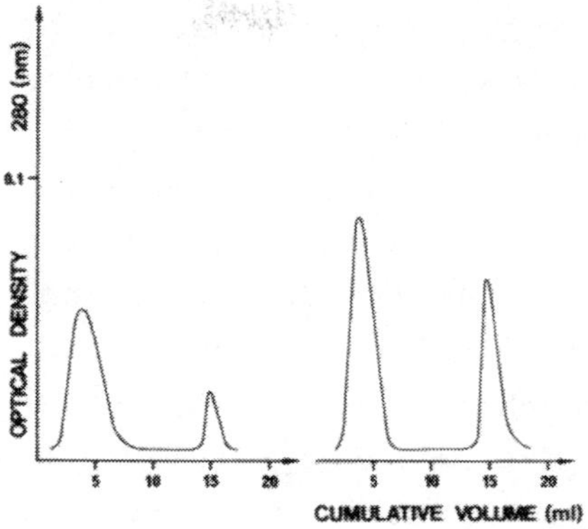

Fig. 2: Elution profiles from affinity chromatography of fractions a and b from Fig. 1 on fibrinogen-agarose. Protein which was not adsorbed was eluted at 4 ml. The adsorbed protein was eluted at 15 ml.

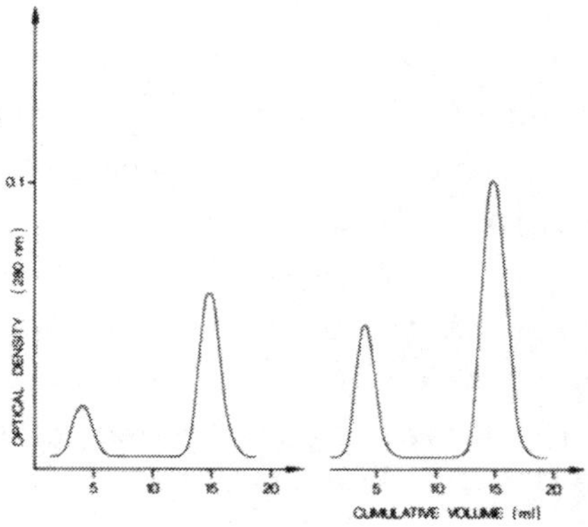

Fig. 3: Elution profiles from affinity chromatography of fractions a and b from Fig. 1 on fibrin-agarose. Protein which was not adsorbed was eluted at 4 ml. The adsorbed protein was eluted at 15 ml.

The observations presented indicate that high molecular weight fibrin has lower affinity to fibrinogen than fibrin dimers. This could mean that the efficacy of fibrinogen in mediating solubility of fibrin decreases with increasing fibrin polymerization. Furthermore progressively larger fibrin polymers become essentially irreversible with regard to fibrinogen under physiological conditions.
The interaction of fibrin polymers with agarose-coupled fibrin monomers might give an interpretation of this observation. High molecular weight fibrin polymers of fraction a and fibrin dimers of fraction b were both adsorbed to a higher rate on fibrin agarose (fraction a: 83 %; fraction b: 72 %) than on fibrinogen agarose (fraction a: 21 %; fraction b: 34 %) (Fig. 2). Complex formation of soluble fibrin results from the interaction of a thrombin activated binding site in the N-DSK region with a second binding site in the fragment D portion of the molecule. Following this concept fibrin monomers can bind fibrinogen on aequimolar ratio by the interaction of one activated binding site (fibrinmonomer) and one binding site in the fragment D region (fibrinogen). However in the binding of two fibrin monomers, the interaction of two activated binding sites results in a higher affinity between two fibrin monomers than between fibrinogen and fibrin.
The data presented restrict the efficacy of fibrinogen in solubilizing fibrin to the very early stage in the process wherein the thrombin activated fibrinogen molecules polymerize in solution.

1. Graeff, H., von Hugo, R., Hafter, R.: Thrombos. Res. 3, 465-476 (1973).

2. Shainoff, J. R., Page, J. H.: J. Exp. Med. 116, 687-707 (1962).

3. von Hugo, R., Hafter, R., Stemberger, A., Graeff, H.: Thrombos. Diathes. Haemorrh. 34, 216-222 (1975).

4. Ly, B., Kierulf, P., Jakobsen, E.: Thrombos. Res. 4, 509-522 (1974).

TEMPERATURE-DEPENDENT REVERSIBLE ASSOCIATION AND DISSOCIATION OF SOLUBLE FIBRIN IN PLASMA *

Gert Müller-Berghaus, Ingo Mahn, Wolfgang Krell and Johannes-Christoph Bernhard

Clinical Research Unit for Blood Coagulation and Thrombosis of the Max-Planck-Gesellschaft, Division of Haemostaseology of the Department of Medicine and Central Division of the Department of Nuclear Research, University of Giessen
Klinikstr. 36, D-6300 Giessen, Germany

Soluble fibrin can be isolated from citrated as well as EDTA plasma from patients with disseminated intravascular coagulation (1-7). The isolated fraction contains fibrin and degraded fibrin as well as fibrinogen (8,9). Since purified fibrin does not stay in solution in an isotonic buffer at a pH of 7.4 but needs fibrinogen or fibrinolytic degradation products to be kept in solution, complex formation between the two components has been claimed.

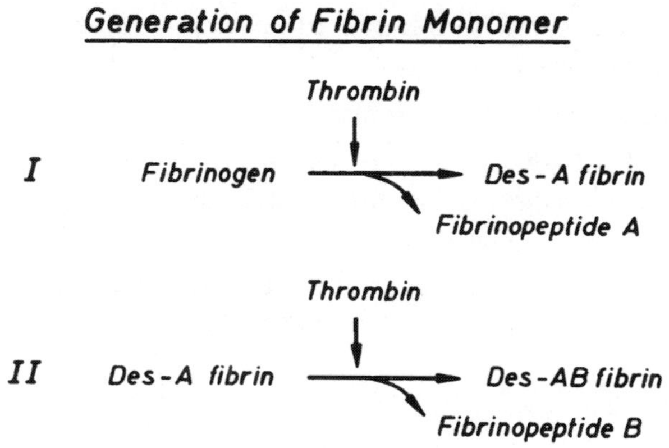

Fig. 1. Generation of des-A fibrin and des-AB fibrin.

* Supported by the Deutsche Forschungsgemeinschaft, Bonn.

When thrombin is generated in the blood either in vitro or in vivo, fibrinogen is converted to fibrin monomer. It has been shown that thrombin first splits off fibrinopeptide A generating des-A fibrin or type-I fibrin (Fig. 1). Only after a lag phase is the fibrinopeptide B removed (10). If both peptides have been split off, the resulting product is named des-AB fibrin or type-II fibrin. Several soluble fibrin monomers may aggregate to a clot, if enough fibrin molecules are present. If, however, the concentration of fibrin is low in comparison to the amount of fibrinogen, fibrin will remain in solution. This fibrin has been designated as soluble fibrin.

Gel Filtration of Soluble des-AB Fibrin through Buffer-Equilibrated Agarose Columns

One of several methods recommended for demonstrating soluble fibrin monomer complexes is the agarose gel filtration (11, 12). In order to differentiate between fibrin and fibrinogen within a plasma sample, fibrin as well as fibrinogen were labeled with different isotopes. Fibrin was prepared by clotting of ^{125}I-fibrinogen. The isolated clot was dissolved in urea and consecutively added to plasma containing ^{131}I-fibrinogen. This mixture was applied to sepharose-6B columns equilibrated with buffer containing Tris, NaCl, EDTA, EACA, aprotinin, pH 7.4. Details have previously been published (13). In comparison to other similar experiments we tried to use relatively physiologic conditions, as a ionic strength of less than 0.2, a pH of 7.4 and a temperature of 37°C.

When a mixture of labeled fibrin and labeled fibrinogen were filtered through a buffer-equilibrated agarose column at 20°C and at 37°C, fibrin could be separated from fibrinogen (Fig. 2). Surprisingly, not all the fibrin applied to the column

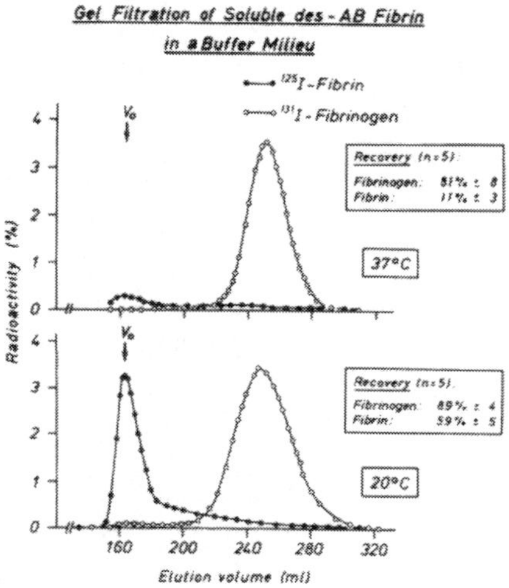

Fig. 2. Gel filtration of plasma containing purified ^{125}I-des-AB fibrin and ^{131}I-fibrinogen on sepharose-6B columns at 20°C and 37°C. The fractions were eluted with buffer, pH 7.4. The mean recovery rates ($\bar{x} \pm s$) of ^{125}I-fibrin and ^{131}I-fibrinogen are given from 5 experiments. The radioactivities are given in % of the total radioactivities applied to the columns. (From: Krell et al. (13)).

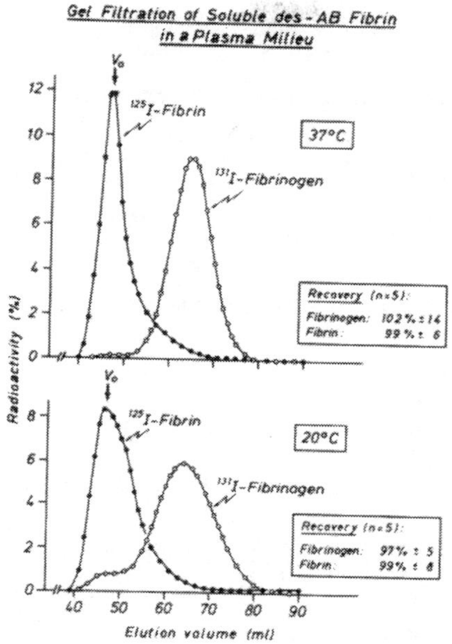

Fig. 3. Gel filtration of plasma containing ^{125}I-des-AB fibrin and ^{131}I-fibrinogen on sepharose CL-6B columns at 20°C and 37°C. The fractions were eluted with buffered plasma, pH 7.4. The mean recovery rates ($\bar{x} \pm s$) of ^{125}I-fibrin and ^{131}I-fibrinogen are given from 5 experiments. The radioactivities are given in % of the total radioactivities applied to the columns. (From: Mahn et al. (15)).

could be recovered in the effluent. At 20°C, a mean of 59%, and at 37°C, even only a mean of 11% of the fibrin applied to the columns could be eluted with the elution buffer. Obviously, fibrin precipitated within the column when it was separated from fibrinogen necessary to keep fibrin in solution. Furthermore, the recovery rates were influenced by the temperature used for gel filtration. Thus, the gel filtration data in a buffer medium do not seem to allow conclusions on the fibrin content in a plasma sample. These results indicate that fibrinogen and/or other plasma proteins are necessary for keeping fibrin in solution. Furthermore, these data question gel filtration as a suitable method for the quantification of fibrin in plasma. In the next series, we describe the gel-filtration procedure allowing the complete recovery of fibrin applied to an agarose column.

Agarose Gel Filtration of Soluble Fibrin through Plasma-Equilibrated Columns

In order to have a fibrin-solubilizing agent present during the entire gel-filtration procedure, plasma and, later on, fibrinogen were used as an admixture to the equilibration and elution medium (14,15,16). The elution medium contained Tris, EDTA, EACA, aprotinin and hirudin in addition to plasma. Under these conditions, we observed neither an activation of the fibrinolysis system nor of the coagulation system. When plasma samples containing labeled fibrin and differently labeled fibrinogen were passed through these columns, fibrin as well as fibrinogen could completely be recovered (Fig. 3). Similar recovery rates for fibrin and fibrinogen were obtained when plasma samples were applied to fibrinogen-equilibrated agarose columns (17). Thus, this procedure is suitable for studying fibrin-fibrinogen interactions on a quantitative basis.

Gel Filtration of des-AB Fibrin through Plasma-Equilibrated Agarose Columns at 37°C

If mixtures of ^{125}I-des-AB fibrin and ^{131}I-fibrinogen were filtered through sepharose CL-6B columns equilibrated with buffered urea, fibrin and fibrinogen were eluted together as a single peak at the same position indicating that des-AB fibrin was homogenous and monomeric. If des-AB fibrin was added to plasma and applied to a column equilibrated with buffered plasma, the previously monomeric des-AB fibrin formed high-molecular weight fibrin aggregates either in plasma or during gel filtration. Fig. 3 demonstrates that des-AB fibrin could be separated from fibrinogen at 20°C as well as at 37°C. Fibrin was eluted in the void volume, whereas fibrinogen was mainly recovered with a volume corresponding to monomeric fibrinogen. The high-molecular weight fibrin aggregates did not precipitate on the columns as the recovery rates of the isotopes were nearly 100% (Fig. 3).

At 37°C, about 99% of the ^{131}I-fibrinogen were eluted as a single peak corresponding to monomeric material, whereas only a mean of 1.2% of the labeled fibrinogen applied to the columns was recovered with des-AB fibrin. Thus, gel filtration on sepharose-plasma columns at 37°C did not show interactions between labeled fibrin and labeled fibrinogen. Such interactions would indicate the existence of stable fibrin-fibrinogen complexes in plasma at 37°C. In contrast, these experiments question the existence of stable fibrin-fibrinogen complexes in vivo and are in accordance with previous studies showing the in vivo dissociation of soluble fibrin-fibrinogen complexes prepared in vitro (18).

Gel Filtration of des-A Fibrin through Plasma-Equilibrated Agarose Columns at 37°C

Des-A fibrin was produced by clotting of ^{125}I-fibrinogen by thrombin in the presence of 2.5 M urea (19). After an incubation time of 2 hours, thrombin action was quenched by the addition of hirudin. Des-A ^{125}I-fibrin was added to plasma containing ^{131}I-fibrinogen, and these mixtures were applied to

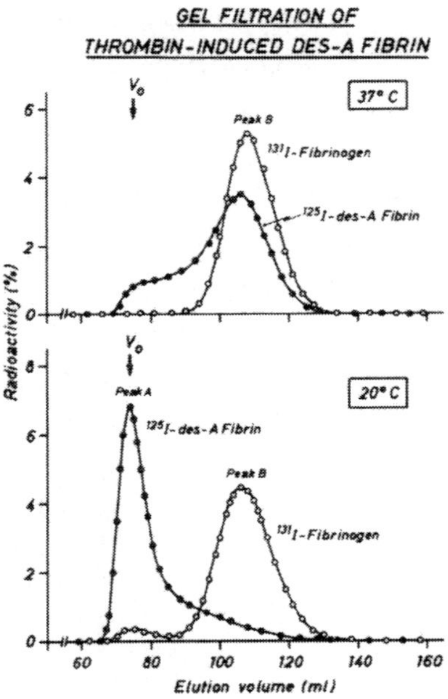

Fig. 4. Gel filtration of plasma containing ^{125}I-des-A fibrin and ^{131}I-fibrinogen on sepharose columns. The fractions were eluted with buffered plasma. The radioactivities are given in % of the total radioactivities applied to the columns. Two ml of peak A of the 20°C run and 2 ml of peak B of the 37°C run were rechromatographed (see Figs. 5 and 6).

a plasma-equilibrated sepharose CL-6B column. During gel filtration at 20°C, again as in the des-AB fibrin experiments, fibrin could be separated from fibrinogen. At 37°C, however, most of the fibrin was eluted together with fibrinogen (Fig. 4).

In order to substantiate the different elution behaviour at 20°C and 37°C, rechromatography experiments were performed. If peak A fractions eluted at 20°C (Fig. 4) were rechromatographed at 20°C and at 37°C, different elution patterns were observed depending on the temperature used for the rechromatography. At 20°C, most of the applied material was eluted with the void volume, whereas at 37°C, the main peak was found with the elution volume corresponding to that of fibrinogen (Fig. 5). Thus, the high-molecular weight fibrin aggregates isolated at 20°C dissociated when rechromatographed at 37°C.

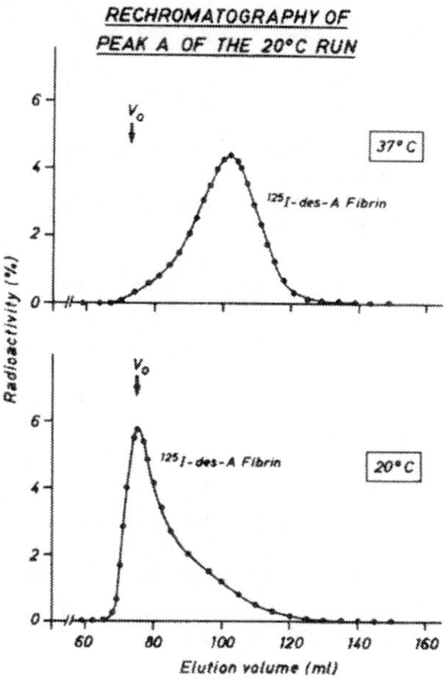

Fig. 5. Rechromatography of isolated fractions of peak A of the 20°C run (see Fig. 4). The fractions were divided into 2 equal portions and rechromatographed on plasma-equilibrated columns.

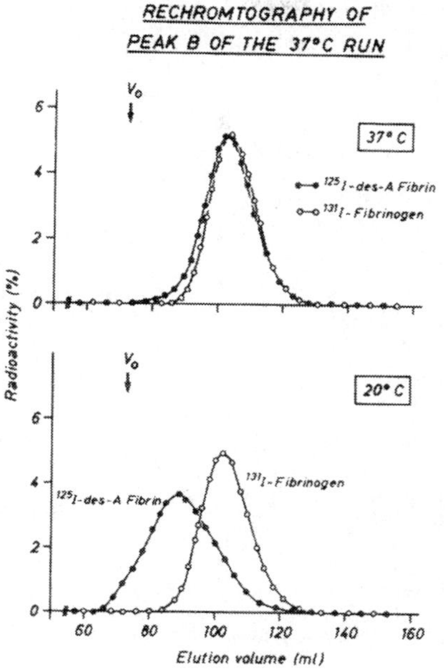

Fig. 6. Rechromatography of isolated fractions of peak B of the 37°C run (see Fig. 4). The fractions were divided into 2 equal portions and rechromatographed on plasma-equilibrated columns.

If the fractions of peak B eluted at 37°C and containing des-A fibrin as well as fibrinogen (Fig. 4) were rechromatographed at 37°C, fibrin as well as fibrinogen eluted at the same position as in the original experiment (Fig. 6). At 20°C, however, the originally monomeric fibrin formed high-molecular weight fibrin aggregates (Fig. 6). From these experiments it can be concluded that des-A fibrin in the absence of calcium ions is monomeric at 37°C but forms high-molecular weight aggregates at 20°C.

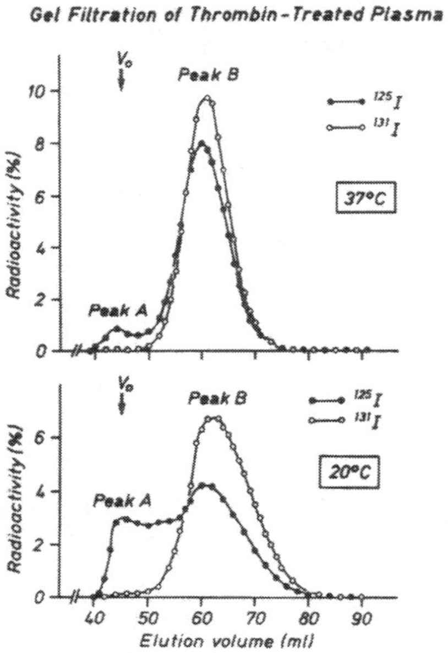

Fig. 7. Gel filtration of plasma each containing thrombin-treated ^{125}I-fibrinogen and untreated ^{131}I-fibrinogen. Fibrin-fibrinogen mixtures in plasma were prepared at room temperature, divided into two equal portions and chromatographed at 20°C and at 37°C. Equilibration and elution medium: buffered plasma. The radioactivities are given in % of the total radioactivities applied to the columns. Two ml of peak A of the 20°C run as well as 2 ml of peak B of the 37°C run were rechromatographed (see Figs. 8 and 9). (From: Müller-Berghaus et al. (14)).

Gel Filtration at 37°C of Soluble Fibrin Produced by the Addition of Thrombin to Plasma

In the next experiment, we intended to produce a more physiologic fibrin. By adding thrombin to plasma containing labeled fibrinogen, part of the plasma fibrinogen as well as of the labeled fibrinogen was converted to fibrin. After an incubation time of 20 min, hirudin was added to stop thrombin action and a second differently labeled fibrinogen was added to have an internal standard. Thereafter, the sample was split, and one half applied to a plasma-equilibrated column at 20°C and the second half to a plasma-equilibrated column at 37°C. At 20°C, ^{125}I representing thrombin-treated fibrinogen was eluted at two different positions (Fig. 7). Peak A eluted with the void volume and peak B at the position of fibrinogen. At 37°C, only a small amount of ^{125}I was eluted in the void volume, again indicating a dissociation of soluble fibrin.

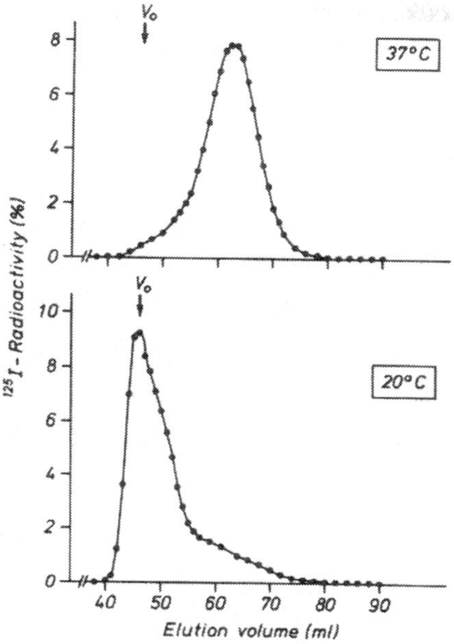

Fig. 8. Rechromatography of isolated fractions of peak A of the 20°C run (see Fig. 7). The fractions were divided into 2 equal portions and rechromatographed on plasma-equilibrated columns. (From: Müller-Berghaus et al. (14)).

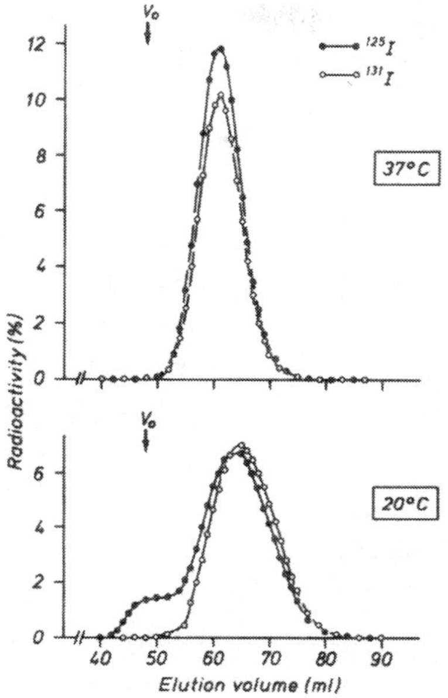

Fig. 9. Rechromatography of isolated fractions of peak B of the 37°C run (see Fig. 7). The fractions were divided into 2 equal portions and rechromatographed on plasma-equilibrated columns. (From: Müller-Berghaus et al. (14)).

In order to substantiate temperature-dependent reversible association and dissociation of soluble fibrin in plasma, peak A as well as peak B were rechromatographed at different temperatures. If peak A of the original 20°C experiment was rechromatographed at 20°C, fibrin was again eluted with the void volume, whereas at 37°C, the material was eluted at the position of fibrinogen (Fig. 8). This experiment demonstrates the dissociation of the original high-molecular weight material when rechromatographed at 37°C. If peak B was rechromatographed at 37°C, thrombin-treated fibrinogen (^{125}I) was eluted at the same position as untreated fibrinogen (^{131}I). At 20°C, however, a certain amount of the thrombin-treated fibrinogen, but not of the untreated fibrinogen, eluted with the void volume (Fig. 9).

These experiments indicate that soluble fibrin in plasma generated by the addition of thrombin to plasma behaves like des-A fibrin. This is in accordance with experiments showing that in plasma thrombin preferably removes the fibrinopeptides A from fibrinogen before a visible clot occurs (20,21).

Conclusions

The experiments presented demonstrate that at a physiologic ionic strength, a pH of 7.4 and a temperature of 37°C, but in the absence of calcium ions, fibrin is present as a variable molecular entity depending on the type of fibrin. Des-AB fibrin is a high-molecular weight aggregate with a molecular weight of more than 4×10^6 daltons. Such an oligomer is kept in solution in plasma as well as in a fibrinogen solution. Des-A fibrin in contrast to des-AB fibrin aggregates at room temperature. Obviously, soluble fibrin in plasma, produced by the addition of thrombin to plasma, is monomeric at 37°C and associates at room temperature. Fibrin-fibrinogen complexes could not be demonstrated by gel filtration at a physiologic ionic strength, a pH of 7.4 and a temperature of 37°C. Most likely, the interactions between fibrin and fibrinogen are so weak that they cannot be demonstrated by gel filtration. So-called complexes observed by other investigators using gel filtration may either not represent complexes, but rather high-molecular weight fibrin aggregates, or complexes may generate at higher ionic strength than physiologic strength, an alkaline pH or a temperature lower than 37°C.

References

1. Godal, H.C., Abildgaard, U.: Scand J. Haematol. 3, 342-350 (1966).
2. Fletcher, A.P., Alkjaersig, N., O'Brien, J., Tulevski, V. G.: Transact. Assoc. Am. Physicians 83, 159-166 (1970).
3. Kierulf, P., Godal, H.C.: Acta Med. Scand. 190, 185-190 (1971).
4. Bang, N.U., Chang, M.L.: Sem. Thrombos. Hemostas. 1, 91-128 (1974).
5. Hafter, R., Graeff, H.: Progr. Chem. Fibrinolys. Thrombos. 2, 137-149 (1976).
6. Edgar, W., McKillop, C., Howie, P.W., Prentice, C.R.M.: Thrombos. Res. 10, 567-574 (1977).
7. Matthias, F.R., Reinicke, R., Heene, D.L.: Thrombos. Res. 10, 365-384 (1977).
8. Shainoff, J.R., Page, I.H.: J. Exp. Med. 116, 687-707 (1962).
9. Lipinski, B., Wegrzynowicz, Z., Budzynski, A.Z., Kopeć, M., Latallo, Z.S., Kowalski, E.: Thrombos. Diathes. Haemorrh. 17, 65-77 (1967).
10. Blombäck, B., Vestermark, A.: Arkiv Kemi 12, 173-182 (1958).
11. Alkjaersig, N., Fletcher, A., Burstein, R.: Am. J. Obstet. Gynecol. 122, 199-209 (1975).
12. Graeff, H., von Hugo, R.: Thrombos. Diathes. Haemorrh. 27, 610-618 (1972).
13. Krell, W., Mahn, I., Müller-Berghaus, G.: Thrombos. Res. 14, 299-310 (1979).
14. Müller-Berghaus, G., Mahn, I., Krell, W.: Thrombos. Res. 14, 561-572 (1979).
15. Mahn, I., Krell, W., Müller-Berghaus, G.: Thrombos. Res. 14, 651-663 (1979).
16. Mahn, I., Krell, W., Müller-Berghaus, G.: Thrombos. Res. 17, 623-627 (1980).
17. Krell, W., Mahn, I., Heimburger, N., Müller-Berghaus, G.: Thrombos. Res. in press.
18. Mahn, I., Schönbach, F., Müller-Berghaus, G.: Thrombos. Res. 11, 67-80 (1977).
19. Müller-Berghaus, G., Mahn, I.: Thrombos. Haemostas. 38, 220 (1977).

20. Hogg, D.H., Blombäck, B.: Thrombos. Diathes. Haemorrh. 34, 324 (1975).

21. Blombäck, B., Hessel, B., Hogg, D., Therkildsen, L.: Nature 275, 501-505 (1978).

NO COMPLEX FORMATION OF DES-AB FIBRIN WITH FIBRINOGEN INVESTIGATED BY THE HUMMEL AND DREYER PROCEDURE +)

Ingo Mahn, Wolfgang Krell and Gert Müller-Berghaus
Zentrum für Innere Medizin, Justus-Liebig-Universität Giessen

Des-AB Fibrin generated from fibrinogen by the action of thrombin and solubilized in buffered urea can be kept in solution by using plasma as a solvent. Fibrinogen is thought to form complexes with des-AB fibrin, thus preventing fibrin from aggregating with each other and getting insoluble (1, 2, 3, 4).

The Hummel and Dreyer procedure using the gel filtration technique allows to quantify association rates of reversible complex formation between macromolecules. The prerequisite for performing the Hummel and Dreyer procedure is that both molecular components can be separated from each other by gel chromatography (5).

Previous studies have shown that des-AB fibrin can be separated from fibrinogen during gel filtration. Thus, the prerequisite for performing the Hummel and Dreyer method is fulfilled (6).

Assuming that des-AB fibrin forms complexes with fibrinogen, fibrinogen should derivate from the constant elution profile increasing to a peak together with fibrin and subsequently decreasing to a trough in the region where fibrinogen is normally eluted.

Samples of plasma containing I-125 fibrinogen and I-131 des-AB fibrin were prepared and run through columns packed with sepharose CL-6B equilibrated and eluted with plasma contai-

+) Supported by the Deutsche Forschungsgemeinschaft

ning I-125 fibrinogen in the same amount as above. The experiments were performed simultaneously at 20°C and at 37°C using three quantities of I-131 des-AB fibrin related to fibrinogen (8, 12, 16 % fibrin).

I-131 des-AB fibrin was eluted at 20°C and 37°C in a single peak. The elution profile of I-125 fibrinogen indicated neither a peak nor a trough. The elution profile of fibrinogen did not differ between 20°C and 37°C and were uniform at all three concentrations of fibrins (Fig. 1).

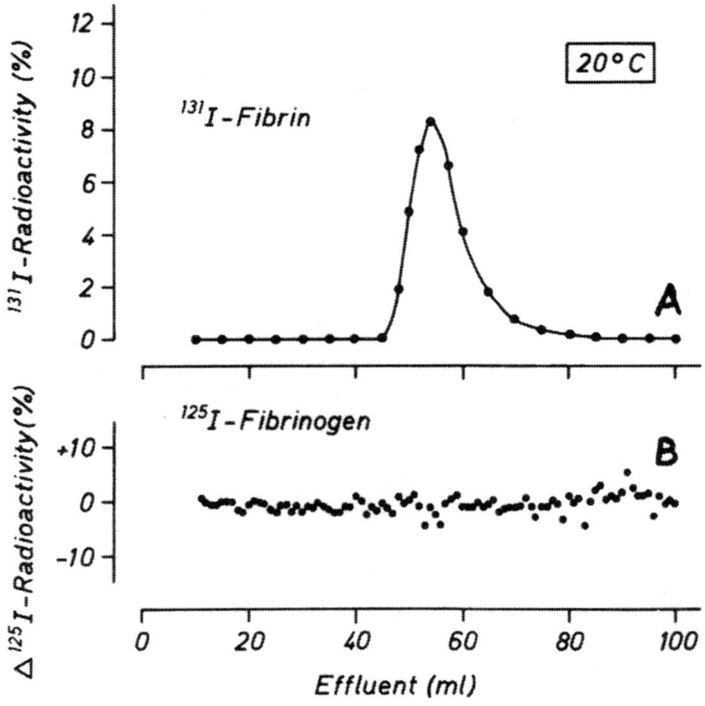

Figure 1: The elution of I-131 des-AB fibrin (A) and I-125 fibrinogen from a sepharose-CL 6B column at 20°C. The ratio of fibrin to fibrinogen was 1:12.5. The I-131 radioactivity is given in % of the total radioactivity applied to the column. The I-125 radioactivity is given in % of deviation of the mean radioactivity of all elution samples.

All the I-131 des-AB fibrin applied to the columns could be recovered in the effluent indicating that precipitation of fibrin on the gel matrix did not occur (Tab. 1).

Table 1.: Recovery rates of I-131-des-AB-fibrin on Sepharose-CL 6B equilibrated and eluted with plasma containing I-125-fibrinogen.

Temperature	Fibrin/Fibrinogen ratio	Recovery of I-131-fibrin* (%)
20 °C	1 : 12.5	111
	1 : 8.3	98
	1 : 6.3	96
37 °C	1 : 12.5	113
	1 : 8.3	93
	1 : 6.3	105

+) % of the radioactivities applied to the columns.

From the findings we conclude that des-AB fibrin does not form complexes with fibrinogen for a longer period of time, most of the fibrin must be present in the plasma in the dissociated form (17). Assuming that fibrin is kept in solution by reversible complex-fibrinogen $\rightleftharpoons$ fibrin + fibrinogen, only small amounts of fibrin are complexed with fibrinogen which can not be detected with this technique.

References

1. Shainoff, J.R. and Page, H.H.: Cofibrins and fibrin-intermediates as indicators of thrombin activity in vivo. Circul. Res. 8, 1013 (1960).

2. Fletcher, A.R., Alkjaersig, N., O'Brien, J. and Tulevski, V.: Blood hypercoagulability and thrombosis. Transact. Assoc. Am Physicians 83, 159 (1970).

3. Graeff, H. and von Hugo, R.: Identification of fibrinogen dirivatives in plasma samples. Thrombos. Diath. Haemorrh. 27, 610 (1972).

4. Matthias, F.R., Reinicke, R. and Heene, D.L.: Affinity chromatography and quantitation of soluble fibrin from plasma. Thromb. Res. 10, 365 (1977).

5. Hummel, J.P., Dreyer, W.J.: Measurement of protein-binding phenomena by gel filtration. Biochim. Biophys. Acta, 63, 530 (1962).

6. Mahn, I., Krell, W. and Müller-Berghaus, G.: Separation of Human des-AB-Fibrin and Fibrinogen by Sepharose-Plasma Chromatography at 20°C and 37°C. Thromb. Res. 14, 651-663 (1979).

7. Mahn, I., Krell, W. and Müller-Berghaus, G.: Equilibrium binding studies of des-AB fibrin and fibrinogen by gel filtration at 20°C and 37°C. Thromb. Res. 17, 623-627 (1980).

VI. FIBRIN(OGEN) DEGRADATION AND CROSSLINKED FIBRIN DERIVATIVES

ON THE STRUCTURE OF D-FRAGMENTS OF HUMAN FIBRINOGEN AND FIBRIN

Willem Nieuwenhuizen, Anton Vermond, Frits Haverkate
Gaubius Institute TNO, Leiden, The Netherlands

Introduction

When fibrinogen is digested exhaustively by plasmin two high molecular weight protein families are formed. These are designated as D and E fragments (1,2). Fully cross-linked fibrin gives rise to the formation of D-dimer and E fragments (3). D fragments are derived from the mid-chain and C-terminal portions of fibrin(ogen) (4,5). E fragments comprise the inner central disulphide knot (6,7) in which the N-terminal regions of the six polypeptide chains of fibrinogen are assembled via disulphide bonds (8,9,10). Under denaturing conditions smaller D fragments (fragments "d") can be formed (11,12). D and E fragments occur in whole digests of fibrinogen as non-covalent complexes called D-E complexes (13-16). In digests of fully cross-linked fibrin the D-dimer-E complex is formed (17,18). In previous publications (19,20) we have shown that after prolonged incubation the structure of the D fragments is greatly influenced by the presence or the absence of Ca in the digestion medium. In the presence of Ca one type of D fragment, Dcate (19) is formed, molecular weight 93,000, which consists of remnants of the $A\alpha$, $B\beta$ and γ-chains with molecular weights of 12,000, 43,000 and 38,000, respectively. D-dimer can be described as two Dcate molecules cross-linked via isopeptide bonds on the C-terminal ends of their γ-chain remnants. In the absence of Ca or in the presence of EGTA and after prolonged incubation a smaller type of D fragment, D EGTA (19), is formed. The molecular weight of D EGTA is 80,000, and it consists of remnants of the $A\alpha$, $B\beta$ and γ-chains with molecular weights of 12,000, 43,000 and 25,000, respectively. Since Dcate and D EGTA have

Fibrinogen - Recent Biochemical and Medical Aspects

identical N-terminal amino acids, we concluded that D EGTA when compared with Dcate lacks a 13,000 molecular weight piece at the C-terminal end of its γ-chain remnant. We demonstrated that the stabilizing effect of Ca on this piece could be explained by the binding of Ca for which this area appeared to be essential (21,22).

As will be described in this paper, we found that the C-terminal end, 13,000 molecular weight, in Dcate and D-dimer becomes extra vulnerable to plasmin digestion in the presence of EGTA. This may be caused indirectly by chelation of Ca (23) or by a direct interaction with fibrinogen (24,25,26). Several lines of evidence suggest that addition of EDTA (an analogon of EGTA) affects the structure of fibrinogen i.e., the prolonged thrombin times (23,24,25), the decreased heat stability (23) and the formation of abnormal clots (23). Experiments will be presented demonstrating that EDTA binds to fibrinogen. This favours the idea of a direct interaction between EDTA and fibrinogen distinct from chelation of Ca by EDTA.

Finally we will show that E fragments have a stabilizing effect on the D fragments in whole digests of fibrinogen.

Materials and Methods

Human fibrinogen was purified as described for rat fibrinogen (27). D-E complexes, D-dimer-E complexes, Dcate, D EGTA and D-dimer were purified as described before (20). Commercial human fibrinogen was bought from KABI (Stockholm, Sweden).

Human plasmin was prepared by urokinase activation of plasminogen, prepared in this laboratory and was 60% active as determined by active-site titration according to Chase and Shaw (28) with p-nitrophenyl-p'-guanidinobenzoate.

Ethylenediaminetetra [2-^{14}C] acetic acid, sodium salt (91 mCi/mmol) batch 19, [1,5-^{14}C] citric acid monohydrate (98 mCi/mmol) batch 22 and [2-^{14}C] acetic acid, sodium salt (59 mCi/mmol) batch 112, were bought from the Radiochemical Centre, Amersham.

Digestion of fibrinogen and D fragments

KABI-fibrinogen was dissolved in distilled water to give a final concentration of 2 mg/ml. Fibrinogen prepared in this laboratory was a stock solution of 2 mg/ml in 0.15 M NaCl. Degradation products were dissolved in distilled water to give a concentration of 1 mg/ml. The solutions were each divided into four portions (a-d).

Portion a, was dialyzed for 24 h against Tris-buffer (0.05 M Tris, 0.15 M NaCl, pH 7.4, "Tris alone") made essentially Ca--free according to refs 29 and 30. Portion b, was first dialyzed overnight against the same Tris-buffer containing 10 mM EGTA and then for 24 h against Tris-buffer without EGTA. Portion c, was dialyzed against the Tris-buffer with 20 mM EGTA. Portion d, was dialyzed against the Tris-buffer with 10 mM $CaCl_2$. After dialysis 4 µg of active plasmin (1 µg of plasmin equals 0.027 CTA units) were added per ml to each of the four portions. Incubation was carried out for 24 h at 37°C and stopped by dilution with 200 µl SDS-buffer (6.7 M urea/5 mM NaH_2PO_4/1.3 mM Na_2HPO_4/0.67% (w/v) SDS) per 25 µl incubation mixture. SDS-polyacylamide gel electrophoresis was carried out according to Weber and Osborn (31) using 100 µl of the diluted mixtures with and without prior reduction. Products were characterized by comparison with purified Dcate, D-dimer and D EGTA samples (20) defined as in the introduction.

Since portions a and b gave identical results, only those for portions a will be given under "results".

Equilibrium dialysis experiments with fibrinogen

Human fibrinogen was dialyzed first against 0.05 M Tris-buffer, pH 7.5, 0.15 M NaCl, freed from Ca as described (29,30). Portions of 250 µl (7.5 mg/ml) of the dialyzed fibrinogen were dialyzed for 2 days at room temperature in pretreated dialysis bags (30) against 10 ml portions of non-radioactive EDTA, citrate or acetate solutions ranging in concentration from 10^{-5} to 10^{-3} M. To each vessel was added a constant tracer amount of ^{14}C-labelled EDTA, citrate or acetate. After dialysis 50 µl samples of inner and outer solutions were taken in triplicate, each mixed with 10 ml scintillation liquid (Packard Dioxan Scintillation Special MI-94) and counted in a Tri-Carb liquid scintillation spectrometer. Averaged counts were taken for calculations. After dialysis protein concentrations were determined by the Lowry (32) method using a calibration curve for human fibrinogen. Dialyzed fibrinogen samples were checked for possible proteolytic damage during the experiment. This was done by SDS-polyacrylamide gel electrophoresis (31). No indication for damage was observed. The dissociation constants and the number of binding sites were calculated by curve fitting (33).

Results and Discussion

Table I clearly demonstrates the protective effect of Ca on the γ-chain C-terminal ends of purified Dcate and D-dimer. These D fragments are plasmin resistant (both against 120 µg of plasmin/ml and against plasmin in 2 M urea) in the presence of 10 mM Ca. In the absence of Ca or in the presence of EGTA they are rapidly degraded to a product resembling "d" (11,12). In the absence of Ca^{2+} D-dimer is relatively stable (as compared with Dcate) against plasmin attack at very low plasmin concentrations i.e., 0.3 µg plasmin/ml. This difference in stability

Table I. Stability of purified D fragments against plasmin digestion under different conditions

Medium	D fragment found after 24 h, starting from:		
	Dcate	D-dimer	D EGTA
Tris + 10 mM Ca	Dcate	D-dimer	"d"
Tris + 20 mM EGTA	"d"	"d"	"d"
Tris alone	"d"	D EGTA + "d"	"d"
Tris + 10 mM Ca + 2 M urea	Dcate	D-dimer	"d"

between D-dimer and Dcate seems to disappear in the presence of EGTA.
On the basis of the results in table I it seems surprising that D EGTA could be found at all in a whole plasmin digest of fibrinogen since the pure D EGTA is so easily attacked by plasmin under all conditions tested. However, in whole digests of fibrinogen D fragments occur in complex with E fragments. As shown in table II none of the D fragments is digested further than D EGTA when they are incubated in complex form with E fragments, even in the absence of Ca.

Table II. Stability against plasmin of D fragments complexed with E fragments

Medium	D fragment found after 24 h, starting from:		
	Dcate-E complex	D-dimer-E complex	D EGTA-E complex
Tris + 10 mM Ca	Dcate	D-dimer	D EGTA
Tris + 20 mM EGTA	D EGTA	D EGTA	D EGTA
Tris alone	Dcate	D-dimer	D EGTA

These findings allow the conclusion that not only Ca but also E fragments stabilize the D fragments.

In agreement with our earlier observations KABI-fibrinogen and fibrinogen prepared in this laboratory yield Dcate and D EGTA in the presence of Ca and EGTA, respectively. However, quite unexpectedly "own" fibrinogen yielded Dcate in "Tris alone", whereas KABI-fibrinogen yielded D EGTA in the same medium.
The KABI-fibrinogen preparations contain, according to the manufacturers' information, citrate ions (concentration about 10 mM when a solution of 2 mg/ml is made) and citrate ions might have an effect comparable to that of EGTA in the digestion of fibrinogen. To check this possibility we removed the ions by dialyzing KABI-fibrinogen against Tris-buffer before

incubation with plasmin; another sample was incubated without prior dialysis. We also added citrate (10 mM) to "own" fibrinogen to mimic the situation in a KABI-fibrinogen solution without dialysis. The results are summarized in table III.

Table III. D fragments in whole digests of KABI and "own" fibrinogen under different conditions

	D fragment found after 24 h, starting from:			
	KABI-fibrinogen without dialysis	KABI-fibrinogen after dialysis	"own" fibrinogen	"own" fibrinogen + 10 mM citrate
Tris + 10 mM Ca	Dcate	Dcate	Dcate	Dcate
Tris + 20 mM EGTA	D EGTA	D EGTA	D EGTA	D EGTA
Tris alone	D EGTA	Dcate	Dcate	D EGTA

As can be seen, KABI-fibrinogen yields Dcate, in "Tris alone", after removal of citrate, whereas "own" fibrinogen yields D EGTA upon addition of citrate. This strongly suggests that EGTA and citrate have a similar effect on the fibrinogen structure. Therefore we decided to test other polycarboxylic compounds for this possible effect on the plasmin susceptibility of the γ-chain C-terminal parts of fibrinogen. The results are shown in table IV.

Table IV. Effect of different compounds on the plasmin susceptibility of γ-chain C-terminal parts of fibrinogen

Effect	No effect
EDTA, EGTA, citrate iminodiacetic acid	oxalic acid, malic acid, succinic acid, adipic acid, glutaric acid, tartaric acid, propane-1,2,3-tricarboxylic acid, acetic acid, murexide

None of the dicarboxylic acids tested, with the exception of iminodiacetic acid, had an affect. As shown EDTA, EGTA and citrate have an effect. However, propane-1,2,3-tricarboxylic acid has no effect although the difference between this compound and citrate is only one hydroxyl group at position 2. The fact that also murexide, a Ca chelator without carboxyl groups has no effect suggests that the minimum structure for an effect is:

$$^{-}OOC-(CH_2)_p-A-(CH_2)_n-COO^{-}$$, where A is CHOH or N.

Although the evidence presented here strongly suggests that EDTA, EGTA, citrate and iminodiacetic acid have a direct effect on the fibrinogen structure, independent of Ca, the possibility still remains that those compounds have this effect by virtue of their Ca-chelating properties. If the compounds have a direct effect on the fibrinogen structure it could be expected that they would bind to fibrinogen. This possibility was investigated by equilibrium dialysis. Analyses of the data show that human fibrinogen binds 3.4 citrate ions with $K_d = 1.4 \times 10^{-4}$ M or 0.4 EDTA ions with $K_d = 2.2 \times 10^{-5}$ M. EDTA and citrate appear to have the same binding sites since an excess of EDTA can remove citrate from fibrinogen and vice versa. Acetate is not bound. Soria et al. (34) have observed a pronounced effect of citrate on the thrombin time of fibrinogen Metz and suggested this to be due to the binding of citrate to this abnormal fibrinogen (35). Fibrinogen is not the only clotting factor capable of binding both Ca and EDTA. Bovine factor V (a Ca binding protein) is capable of binding EDTA (0.5 mole i.e., 1 mole of "functional group" per mole of factor V) (36). The binding of citrate and EDTA is pH-dependent (see figure) and is greatly diminished at high ionic strengths. This indicates that the binding is based mainly on electrostatic interactions.

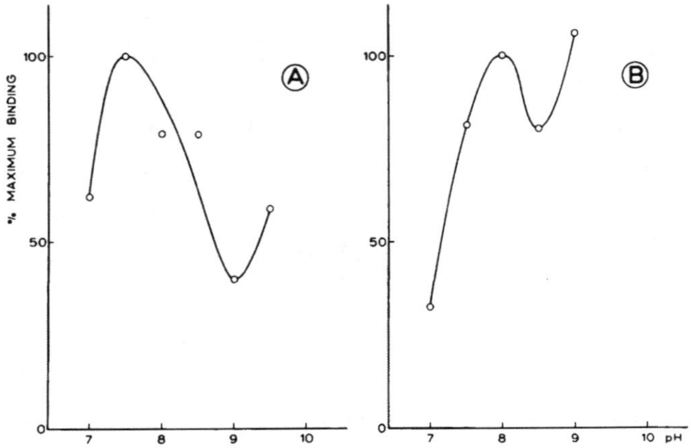

pH-dependence of EDTA (A) and citrate (B) binding to human fibrinogen (details in text)

Conclusions

Based on our observations we conclude that the structure of D fragments is determined by three main factors:

1. Ca protects the γ-chain C-terminal parts of fibrinogen, Dcate and D-dimer, even against high plasmin activities by direct binding to these proteins.
2. Fragments E, complexed with D fragments, also have a protective effect on the D fragments.
3. EGTA, EDTA, iminodiacetic acid and citrate, make the γ-chain C-terminal ends of fibrinogen, Dcate and D-dimer more susceptible to plasmin attack. They do so by direct interaction with the proteins probably by affecting their structure.

These observations should be kept in mind when studying the structure of fibrinogen and degradation products.

References

1. Marder, V.J., Budzynski, A.Z. (1974) in Progress in Hemostasis and Thrombosis (Spaet, T.H., ed.) Vol. 2, pp 141-174, Grune and Stratton, New York

2. Nussenzweig, V., Seligman, M., Pelmont, J., Grabar, P.: Ann. Inst. Pasteur 100, 377-389 (1961).

3. Pizzo, S.V., Taylor, L.M., Schwartz, M.L., Hill, R.L., McKee, P.A.: J. Biol. Chem. 248, 4584-4590 (1973).

4. Marder, V.J., Budzynski, A.Z., James, H.L.: J. Biol. Chem. 247, 4775-4781 (1972).

5. Pizzo, S.V., Schwartz, M.L., Hill, R.L., McKee, P.A.: J. Biol. Chem. 247, 636-645 (1972).

6. Gårdlund, B., Kowalska-Loth, B., Gröndahl, N.J., Blombäck, B.: Thromb. Res. 1, 371-388 (1972).

7. Kowalska-Loth, B., Gårdlund, B., Egberg, N., Blombäck, B.: Thromb. Res. 2, 423-450 (1973).

8. Blombäck, B., Blombäck, M., Henschen, A., Hessel, B., Iwanaga, S., Woods, K.R.: Nature (London) 218, 130-134 (1968).

9. Furlan, M., Beck, E.A.: Biochim. Biophys. Acta 263, 631-644 (1972).

10. Mills, D.A.: Biochim. Biophys. Acta 263, 619-630 (1972).

11. Furlan, M., Beck, E.A.: Biochim. Biophys. Acta 310, 205-216 (1973).

12. Kemp, G., Furlan, M., Beck, E.A.: Thromb. Res. 3, 553-564 (1973).

13. Budzynski, A.Z., Stahl, M., Kopec, M., Latallo, Z.S., Wegrzynowicz, Z., Kowalski, E.: Biochim. Biophys. Acta 427, 313-323 (1967).

14. Matthias, F.R., Hocke, G.: Biochim. Biophys. Acta 427, 569-574 (1976).

15. Plow, E.F., Edgington, T.S.: Thromb. Res. 1, 605-618 (1972).

16. Plow, E.F., Edgington, T.S.: Proc. Natl. Acad. Sci. USA 71, 158-162 (1974).

17. Gaffney, P.J., Lane, D.A., Kakkar, V.V., Basher, M.: Thromb. Res. 7, 89-99 (1975).

18. Hudry-Clergeon, G., Paturel, L., Suscillon, M.: Pathol. Biol. Suppl. 22, 47-52 (1974).

19. Haverkate, F., Timan, G.: Thromb. Res. 10, 803-812 (1977).

20. Van Ruijven-Vermeer, I.A.M., Nieuwenhuizen, W., Haverkate, F., Timan, G.: Hoppe-Seyler's Z. physiol. Chem. 360, 633-637 (1979).

21. Haverkate, F., Nieuwenhuizen, W.: accompanying paper in these proceedings (1981).

22. Nieuwenhuizen, W., Vermond, A., Nooijen, W.J., Haverkate, F.: FEBS Letters 98, 257-259 (1979).

23. Godal, H.C.: Scand. J. Clin. Lab. Invest. Suppl. 53, 12, 3-20 (1960).

24. Bithell, T.C.: Biochem. J. 93, 431-439 (1964).

25. Blombäck, B., Blombäck, M., Laurent, T.C., Pertoft, M.: Biochim. Biophys. Acta 127, 560-562 (1966).

26. Capet-Antonini, F.C.: Biochim. Biophys. Acta 200, 497-507 (1970).

27. Van Ruijven-Vermeer, I.A.M., Nieuwenhuizen, W.: Biochem. J. 169, 653-658 (1978).

28. Chase Jr., T., Shaw, E.: Methods Enzymol. 19, 20-27 (1970).

29. Van Ruijven-Vermeer, I.A.M., Nieuwenhuizen, W., Nooijen, W.J.: FEBS Letters 93, 177-180 (1978).

30. Marguerie, G., Chagniel, G., Suscillon, M.: Biochim. Biophys. Acta 490, 94-103 (1977).

31. Weber, K., Osborn, M.: J. Biol. Chem. 244, 4406-4412 (1969).

32. Lowry, O.H., Rosebrough, N.J., Farr, A.L., Randall, R.J.: J. Biol. Chem. 193, 265-275 (1951).

33. Nieuwenhuizen, W., Vermond, A., Hermans, J.: manuscript submitted (1981).

34. Soria, C., Soria, J., Samama, M., Poirot, E., Kling, C.: Vth Congress of the International Society on Thrombosis and Haemostasis, 225 (abstract) (1975).

35. Samama, M., Soria, J., Soria, C.: in Recent Advances in Blood Coagulation (Pollier, L. ed.) Churchill Livingstone, London, 313-335 (1977).

36. Hibbard, L.S., Mann, K.G.: J. Biol. Chem. 255, 638-645 (1980).

ROLE OF THE C-TERMINUS OF THE FIBRINOGEN γ-CHAIN IN FIBRINOGEN Ca^{2+} BINDING AND IN FIBRIN POLYMERIZATION

Frits Haverkate and Willem Nieuwenhuizen
Gaubius Institute TNO, Leiden, The Netherlands
A.C.M. van der Drift
Medical Biological Laboratory TNO, Rijswijk, The Netherlands

Introduction

Our previous studies (1) and those of others (2,3) showed a protective effect of calcium ions on fibrin(ogen) in the digestion by plasmin (EC 3.4.21.7). Calcium ions appeared to protect the part of the γ-chain located in the D moiety. The present study establishes the essential role of this C-terminal end of the human fibrinogen γ-chain in both the Ca^{2+} binding and the fibrin polymerization.

Materials and Methods

Plasmin digestion of fibrinogen. Human fibrinogen from Kabi (Stockholm, Sweden; 1.5 mg/ml) was digested in 0.04 M veronal buffer, pH 7.4, containing 0.10 M NaCl, 0.25 CU human plasmin/ml and either 2 mM $CaCl_2$ or 10 mM EGTA. Incubations were carried out at 37°C.
SDS polyacrylamide gel electrophoresis (SDS-PAGE). Electrophoresis of 25 µl samples of non-reduced proteins was performed in 5% polyacrylamide gels containing 0.1% (w/v) sodium dodecyl sulphate (SDS) (4). Reduction of proteins was carried out in a medium containing 1 mg/ml protein, 6.7 M urea, 0.67% (w/v) SDS and 0.67% (v/v) 2-mercaptoethanol for 3 h at 37°C. Electrophoresis of reduced samples was performed in 9% gels. The molecular weight of the protein chain, after reduction, was determined using MW standards for gel chromatography (Serva, Heidelberg,

Fibrinogen - Recent Biochemical and Medical Aspects

W. Germany) as marker proteins. Gels were stained in 0.25% (w/v) Coomassie Brilliant Blue R 250 as described earlier (1). The D-β-chain was characterized by a positive carbohydrate staining of gels performed with periodic acid Schiff reagent (5). The γ-chains of the higher MW fragments D were characterized by the presence of crosslink sites. To that purpose the fibrinogen (starting material) was labelled with the fluorescent monodansylcadaverine (MDC) in a medium containing 20 mM $CaCl_2$, 5.0 mM MDC (dissolved in DMSO and diluted with water), 0.9 NIH U thrombin/ml and 1.5 mg fibrinogen (Kabi)/ml containing Factor XIII. After 3 h of incubation at 37°C the formed clots were digested with plasmin, reduced with 2-mercaptoethanol and subjected to SDS-PAGE. Fluorescence of bands in the gels under UV light showed the presence of crosslink sites.

Purification of D fragments. D fragments were isolated from whole fibrinogen digests by anion-exchange chromatography (6) using a column of DEAE cellulose (DE 52 from Whatman), 1.5 x 50 cm. The column was equilibrated with 0.01 M $NaHCO_3$, pH 8.9, containing Trasylol (10 KIU/ml) and eluted in the same solution, followed by an 800 ml linear NaCl gradient (from 0 to 0.2 M in 0.01 M $NaHCO_3$, pH 8.9). Fractions containing mainly D fragments, as revealed by SDS-PAGE, were pooled, dialyzed against distilled water and freeze dried. The freeze-dried material was subjected to gel filtration on a Sephadex G100 column, eluted with 0.025 M Tris/HCl buffer, pH 7.4 containing 0.15 M NaCl. Fractions containing pure fragment D as revealed by SDS-PAGE were pooled. At a later stage in the studies we used the purification procedure described elsewhere (7).

N-terminal amino acids were qualitatively assessed according to Gray (8).

Calcium binding. The binding of calcium ions by protein was determined by equilibrium dialysis using ^{45}Ca as described before (9). The amount of calcium ions bound to protein and the dissociation constants were calculated with a computer by curve fitting (10).

Anticlotting activities of fragments D. Two test systems were used.

(a) An adapted thrombin time test. Normal pooled human plasma (0.1 ml) collected in sodium citrate (final concentration 0.011 mol/l) was mixed with 0.1 ml of saline containing varying amounts of purified fragment D. After 3 min at 37°C, 0.2 ml bovine thrombin (3 NIH/ml) was added and the clotting time assessed using a steel hook which was moved up and down.

(b) A fibrin polymerization test. 0.35 ml of veronal buffer (0.05 mol/l), pH 7.9, containing NaCl (0.15 mol/l) and $CaCl_2$ (0.2 mmol/l) was mixed with 0.10 ml of saline containing varying amounts of purified fragment D. The polymerization time was assessed after addition of 0.15 ml of a solution containing 0.6% (w/v) purified human fibrin monomer (11) in acetic acid (0.02 mol/l). The polymerization time without the addition of fragments D was 26 s.

Results and Discussion

Breakdown of the D-γ-chain from both the N-terminal and C--terminal side has been described (12). We investigated the gradual breakdown of the D-γ-chain as influenced by the presence of calcium ions. To that end we isolated the fibrinogen fragment D with the highest MW observed in the digests. This was obtained from a 45 min digest in the presence of calcium ions. It appeared to have a MW of 96,000 and was designated as D_{Ca}. As revealed by reduced SDS-PAGE, this fragment D_{Ca} consists of a β-chain (MW 43,000), a γ-chain with slightly higher mobility (MW 41,000) and an α-chain of MW 12,000. The β-chain was characterized by a positive carbohydrate staining and the γ-chain by the presence of crosslink sites. Further incubation of fragment D_{Ca} with plasmin in the presence of Ca^{2+} ions resulted in a slight reduction in MW of the D-γ--chain from 41,000 to 38,000, with maintenance of the crosslink sites, which are known to be located near its C-terminus (13).

This indicates a loss of a piece of 3,000 MW cleaved from the N-terminal side of the D-γ-chain. In agreement with this the N-terminal amino acids changed during the digestion from D_{Ca} to D_{cate}, from asp, thr, gly, ser and ala to asp, val and ser, respectively. This cleavage also occurs in the absence of calcium ions, and is similar to the cleavage of a 3,000 MW piece from the N-terminal side of the D-γ-chain described by Collen et al. (14). They found a concomitant change in N--terminal amino acid the γ-chain remnant from ala to ser. Further digestion in the presence of Ca^{2+} was not observed even in 2 M urea which is known to make the substrate more susceptible to plasmin digestion (15). Moreover, it is probable that the N-terminus of the D-α-chain changes from asp to val (15), in agreement with our N-terminal amino acid analysis, but it would appear that the loss of the α-peptide is too small to be detected by SDS-PAGE. We isolated this terminal D fragment (Dcate) formed after prolonged plasmin digestion of fibrinogen in 2 mM $CaCl_2$. This fragment has a MW of 93,000 and a D-γ--chain of MW 38,000. Dcate is very rapidly broken down by plasmin in the presence of EGTA added to sequester Ca^{2+} (16). SDS-PAGE revealed a decrease in the MW of the D-γ-chain from 38,000 to 25,000, through an intermediate of about 28,000. The D fragment formed had a MW of 80,000 and was designated as D EGTA. The crosslink sites disappear during this digestion and the N-terminal amino acids of the D fragments, i.e. asp, val, and ser do not change. Results indicate that the breakdown of the D-γ-chain in the absence of Ca^{2+} ions proceeds from the C-terminus.

These findings enabled us to isolate well-defined fibrinogen fragments D. We purified Dca, Dcate and D EGTA. As the C-terminal parts of their γ-chains are the same as in fibrinogen and in fibrin, but is missing in D EGTA, we were able to study the function of this part of the molecule in fibrinogen with respect to Ca^{2+} binding and fibrin polymerization.

Fibrinogen appeared to have three high-affinity calcium binding sites with a K_d of 19 x 10^{-6}M. Calcium binding studies also

showed that Dcate binds Ca^{2+} in a 1:1 molar ratio with a dissociation constant of 13.10^{-6} M, whereas D EGTA does not bind Ca^{2+}. Consequently, the C-terminal 13,000 piece of the D-γ-chain is essential in Ca^{2+} binding. The results indicate that the two D moieties in the intact fibrinogen have similar Ca^{2+} binding properties for which the C-terminal parts of the fibrinogen γ-chains are essential.
The role of the C-terminal end of the γ-chain in the fibrin--fibrin polymerization was studied by determining the anticlotting activity of the purified fragments D. This approach is based on a competitive inhibition of the polymerization by the D fragments. We assessed the anticlotting activity of purified D fragments in two test systems. It appeared that fragments Dca and Dcate had a strong anticlotting activity. In the presence of fragments D_{ca} and Dcate at a concentration of 5 μM and higher, no clotting was observed within 1 hour, whereas the clotting times without addition of fragments D were 20-25 sec. Addition of D EGTA in concentrations up to 20 μM did not influence the clotting time at all. Results clearly show that only D fragments with an intact C-terminal part of the γ-chain have anticlotting activities. Consequently, the C-terminal end of the fibrin γ-chain is essential in the fibrin-fibrin polymerization.

References

1. Haverkate, F., Timan, G.: Thromb. Res. 10, 803 (1977).
2. Purves, L.R., Lindsey, G.G.: South Afr. J. Sci. 74, 202 (1978).
3. Lawrie, J.S., Kemp, G.: Biochem. Soc. Trans. 6, 769 (1978).
4. Weber, K., Osborn, M.: J. Biol. Chem. 244, 4406 (1969).
5. Fairbanks, G., Steck, T.L., Wallach, D.F.H.: Biochemistry 10, 2606 (1971).
6. Nussenzweig, V., Seligmann, M., Pelmont, J., Grabar, P.: Ann. Inst. Pasteur 100, 377 (1961).

7. Ruijven-Vermeer, I.A.M. van, Nieuwenhuizen, W., Haverkate, F., Timan, G.: Hoppe-Seyler's Z. physiol. Chem. 360, 633 (1979).

8. Gray, W.R.: Methods Enzymol. 25B, 121 (1972).

9. Ruijven-Vermeer, I.A.M. van, Nieuwenhuizen, W., Nooijen, W.J.: FEBS Letters 93, 177 (1978).

10. Nieuwenhuizen, W., Ruijven-Vermeer, I.A.M. van, Nooijen, W.J., Vermond, A., Haverkate, F., Hermans, J.: Thromb. Res. In press (1981).

11. Haverkate, F., Timan, G. (1976) in Progress in Chemical Fibrinolysis and Thrombolysis, Vol. 2, p 73. Raven Press, New York.

12. Hörmann, H., Henschen, A.: Thrombos. Haemostas. (Stuttgart) 41, 691 (1979).

13. Doolittle, R.F.: Adv. Protein Chem. 27, 1 (1973).

14. Collen, D., Kudryk, B., Hessel, B., Blombäck, B.: J. Biol. Chem. 250, 5808 (1975).

15. Furlan, M., Kemp, G., Beck, E.A.: Biochim. Biophys. Acta 400, 95 (1975).

16. Nieuwenhuizen, W., Vermond, A., Haverkate, F.: This volume.

FIBRINOGEN DEGRADATION BY GRANULOCYTE ELASTASE AND CATHEPSIN G - GENERATION OF FIBRINOGEN DEGRADATION PRODUCTS WITH ANTI-COAGULANT PROPERTIES.

Margarethe Gramse and Christine Bingenheimer
Abt. Hämatologie/Onkologie, Zentrum für Innere Medizin, Marburg/Lahn, W. Germany

Introduction

Fibrinogen fragments generated during degradation by plasmin are described to interfere with thrombin mediated fibrinogen coagulation (1,2). The interfering plasmin fibrinogen degradation products (FDP) are X, Y and D; Y was found to be most active (2). Haverkate et al. (3) showed that fragment D generated in a calcium containing buffer was stronger interfering with fibrinogen coagulation than D generated in a calcium-free incubation milieu. Studies of the last few years deal with fibrinogen degradation by granulocyte neutral serine proteases (4,5,6,7) called "alternative pathway of fibrinogen degradation" by Plow & Edgington (8). The elastase-like protease (ELP) was first described by Janoff & Scherer (9), the chymotrypsin-like protease (CLP) or cathepsin G by Schmidt & Havemann (10). Both enzymes are described to be released during frustrated phagocytosis (11) following contact with antigen-antibody complexes (12), C5a (13), and bacterial endotoxin (14). Indications for their in vivo action are given by demonstrations of high leukocyte protease concentration at inflammatory sites (15,16).

This paper describes the in vitro ability of granulocyte protease generated FDP to inhibit fibrinogen coagulation. Comparison is given to the known anticlotting activity of plasmin generated FDP.

The anticoagulant properties are shown to be influenced by calcium ions present during fibrinogen degradation.

Results

Fibrinogen, free of factor XIII and plasminogen, was degraded by granulocyte proteases and plasmin in the presence of 5mM $CaCl_2$ and 5mM ethylene glycolbis (ß-aminoethylether)-N,N'-tetra acetic acid (EGTA), respectively. Granulocyte protease action was stopped by soybean trypsin inhibitor, plasmin by kallikrein inhibitor. Thrombin was added to give clotting times of 17-24 s with undegraded fibrinogen. Addition of the inhibited enzymes did not alter this control value. Fibrinogen was designated as uncoagulable at clotting times more than 180 s.
Two tests were used to show the influence of the granulocyte proteinases on fibrinogen coagulation.
Technique I: Fibrinogen at different degradation stages was precipitated by thrombin to find out the enzyme concentration which makes fibrinogen uncoagulable.
Technique II: Fibrinogen at different degradation stages was given to undegraded fibrinogen. Thrombin was added to estimate the anticoagulant activity of the FDP.
At any time thrombin was allowed to act in a $CaCl_2$-containing buffer by adding an appropriate amount of $CaCl_2$ to the EGTA-degradation mixtures.

Technique I:

Fibrinogen loses its ability to coagulate after degradation with increasing concentrations of ELP, CLP, and plasmin.
Figure 1 shows this process to be dependent on the presence of calcium ions. Fibrinogen was incubated for 5 min with increasing concentrations of ELP and CLP, and of plasmin for comparison. Degradation in $CaCl_2$-buffer requires higher enzyme concentrations to make fibrinogen uncoagulable than degradation in EGTA-buffer. It is remarkable that the CLP action on fibrinogen is

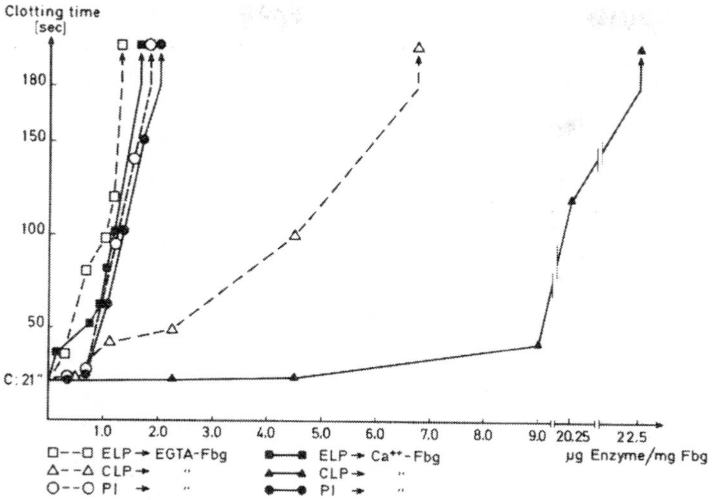

Fig.1: Thrombin clotting time of fibrinogen (Fbg) incubated with different concentrations of ELP, CLP, and plasmin in the presence of calcium ions (——) or EGTA (---).

stronger influenced by calcium ions than ELP or plasmin action.

Technique II:

In a second step the interference of ELP-, CLP-, and plasmin-FDP with fibrinogen coagulation was examined. Starting from those enzyme concentrations which after 5 min makes fibrinogen uncoagulable the enzyme concentrations were further enlarged to get FDP with anticlotting properties. Following enzyme concentrations effective in generation of anticlotting FDP after 5 min incubation were found:

Interfering plasmin-FDP in $CaCl_2$- as well as in EGTA-buffer are produced with 2.0-3.5 µg plasmin/mg fibrinogen.

With granulocyte elastase interfering Ca-FDP arise at higher enzyme concentrations (2.9-3.3 µg ELP/mg fibrinogen) than interfering EGTA-FDP (1.9-2.3 µg ELP/mg fibrinogen).

As expected from experiments in technique I the difference in the concentrations necessary for getting anticlotting Ca- or EGTA-FDP is the highest with CLP as fibrinolytic enzyme. In addition CLP only in EGTA-buffer generates strongly interfering

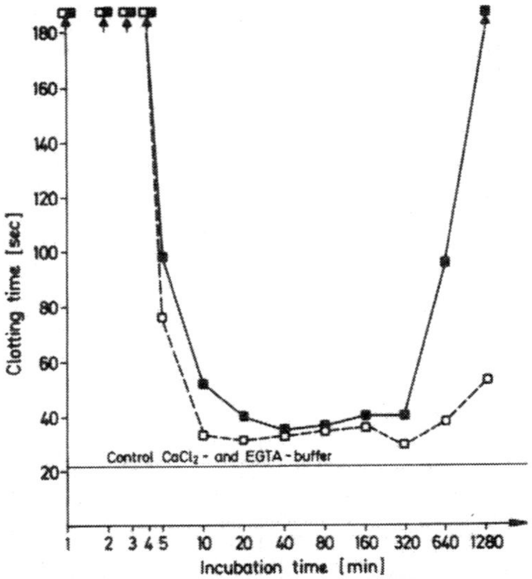

Fig.2: Thrombin clotting times of fibrinogen in the presence of ELP-FDP obtained after 1 to 1280 min degradation in $CaCl_2$- (■—■) or in EGTA-buffer (□--□).

FDP (9.0-13.5 µg CLP/mg fibrinogen). In $CaCl_2$-buffer merely weakly interfering split products are generated (40-60 µg CLP/ mg fibrinogen).

As described by Haverkate (3) fragment D generated during plasmin fibrinogenolysis in $CaCl_2$-buffer stronger interferes with fibrinogen coagulation than fragment D generated in EGTA-buffer. We wanted to examine the same phenomenon for the late ELP- and CLP-FDP. Starting from the enzyme concentrations which after 5 min incubation lead to early strongly interfering FDP incubation times were prolonged up to 1280 min.
Figure 2 shows the time depending generation of interfering ELP-FDP. The incubations in $CaCl_2$-buffer were performed with 3 µg ELP/mg fibrinogen, the incubations in EGTA-buffer were performed with 2 µg ELP/mg fibrinogen. As to be seen strongly interfering early FDP appear after 1 to 5 min degradation in $CaCl_2$- as well as in EGTA-buffer followed by a series of digests with weak anticlotting activity. FDP arising after very long incubation times again strongly interfere with fibrinogen coagulation. All

FDP of later degradation stages in $CaCl_2$-buffer in general more prolong fibrinogen coagulation than the corresponding digests in EGTA-buffer. This difference is distinctly marked after 1280 min degradation.
The ELP-fibrinogen-incubation mixtures described in figure 2 were applied to polyacrylamide gels as shown in the figures 3a&b.

Fig. 3a

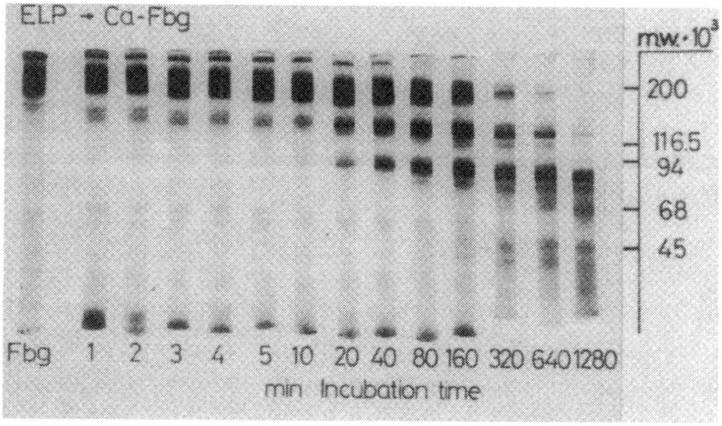

Fig. 3b

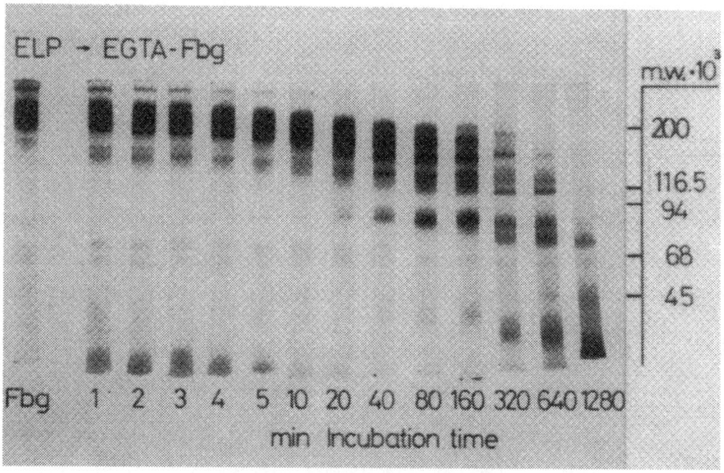

Fig. 3a&b: Sodium dodecylsulfate-polyacrylamide gel electrophoresis (SDS-PAGE) of fibrinogen cleaved by ELP a) in the presence of $CaCl_2$, and b) of EGTA. The distances moved by the marker proteins are indicated. The digests applied to the gels are identical to those examined for anticoagulant properties(see Fig.2).

The short time digests (up to 5 min) strongly interfering with fibrinogen coagulation contain degradation products of high molecular weight. These large FDP obtained in $CaCl_2$- as well as in EGTA-buffer have molecular weights of around 200,000 and 150,000 - 160,000.

The degradation mixtures in $CaCl_2$-buffer with weakly interfering capacity (plateau of clotting times from 20 to 320 min incubation time, see figure 2) contain fibrinogen fragments in 3 main regions of molecular weight:

the first region around 200,000(comparable to X),

the second from 155,000-130,000 (comparable to Y), and

the third from 115,000-70,000.

The appropriate incubation mixtures in EGTA-buffer also contain a variety of fibrinogen fragments with a molecular weight up to 75,000.They cannot be subdivided, however, into molecular weight regions like the calcium dependent ELP-FDP. After 1280 min incubation in $CaCl_2$-buffer the FDP anew strongly interfering (Fig.2) show molecular weights smaller than 90,000 except one fragment around 140,000. The appropriate degradation mixture in EGTA-buffer in a higher concentration only contains a split product

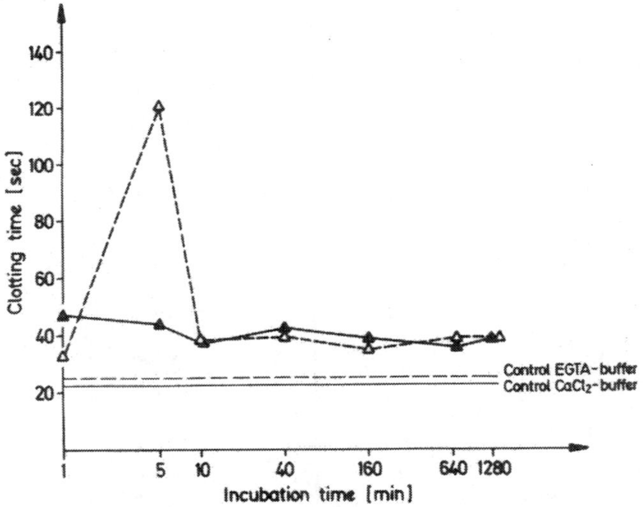

Fig.4: Thrombin clotting times of fibrinogen in the presence of CLP-FDP obtained after 1 to 1280 min degradation in $CaCl_2$- (▲—▲) or in EGTA-buffer (Δ--Δ).

with the molecular weight around 72,000.
To test the interfering capacity of the late FDP generated by CLP time dependent incubations were performed with 50 /ugCLP/mg fibrinogen in $CaCl_2$-buffer, and with 10 /ug CLP/mg fibrinogen in EGTA-buffer. These enzyme concentrations after 5 min yielded early FDP strongly interfering when obtained in EGTA-buffer and weakly interfering with fibrinogen coagulation when obtained in $CaCl_2$-buffer. FDP generated after 10 to 1280 min only slightly prolong the thrombin clotting time of fibrinogen. There is no significant difference in the anticoagulant activity between Ca- or EGTA-FDP (Fig.4).
The described digests were separated by SDS-PAGE shown in the figures 5a&b. The 5 min-EGTA-incubation mixture with strong interfering capacity contains FDP of high molecular weight: around 200,000 and 155,000-140,000. The 1- and the 10-min incubations show the same pattern in polyacrylamide gels. Further degradation of fibrinogen in EGTA-buffer up to fragments lower than 70,000 of molecular weight (1280 min incubation) form the plateau of slightly interfering FDP in figure 4. The 1- to 10-min-incubations in $CaCl_2$-buffer show an additional fragment with a molecular weight around 90,000 compared to the corresponding incubation mixtures in EGTA-buffer. 40- to 1280-min incubations in $CaCl_2$-buffer result in FDP of a larger size than FDP of compara-
Fig. 5a

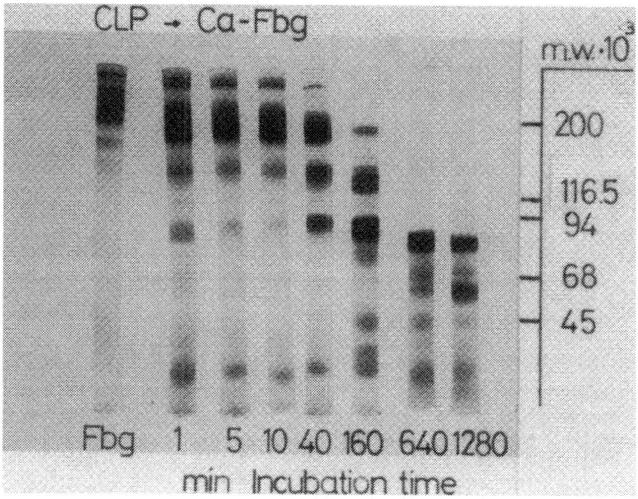

Fig.5b

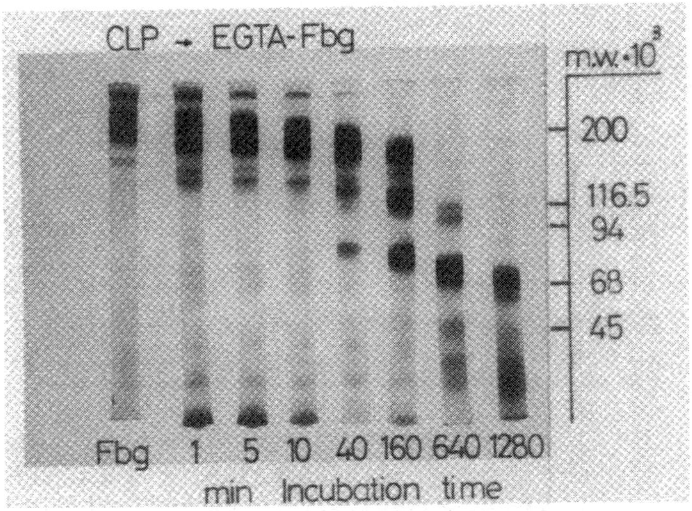

Fig. 5a&b: SDS-PAGE of fibrinogen cleaved by CLP a) in the presence of $CaCl_2$, and b) of EGTA. The distances moved by the marker proteins are indicated.

ble degradation stages.

According to the described tests with ELP and CLP as proteolytic enzymes the interfering capacity of late plasmin derived

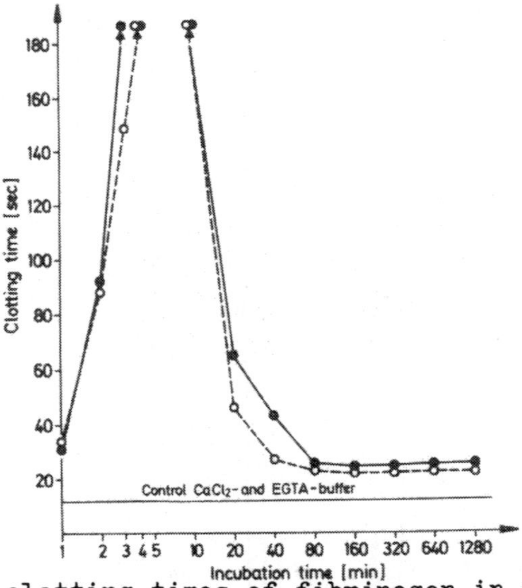

Fig.6: Thrombin clotting times of fibrinogen in the presence of plasmin-FDP obtained after 1 to 1280 min degradation in $CaCl_2$- (●—●) or in EGTA-buffer (○--○).

FDP was shown(Fig.6). The strong coagulation interference of early plasmin-FDP is generated both in $CaCl_2$- and in EGTA-buffer with 2 µg plasmin/mg fibrinogen. This effect is ascribed to the fragments X and Y which are visible in the 2- to 10-min-incubation mixtures (Fig.7a&b). Fragment D that more weakly interferes with fibrinogen coagulation reaches its highest concentration after 20 min incubation. In $CaCl_2$-buffer this split product remains constant up to 1280 min incubation whereas in EGTA-buffer it is degraded to smaller fragments. The $CaCl_2$-stabilized fragment D shows an interference somewhat stronger than the EGTA-dependent D (Fig.6).

Fig.7a

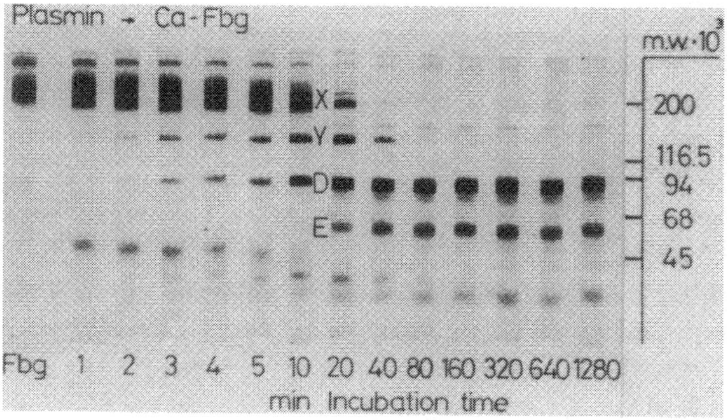

Fig.7b

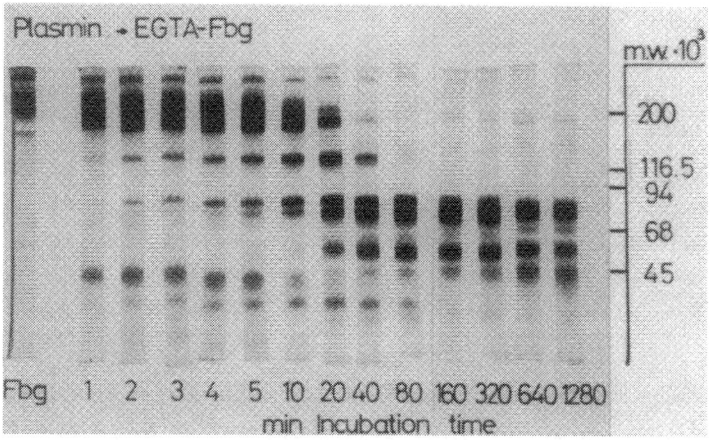

Fig.7a&b: SDS-PAGE of fibrinogen cleaved by plasmin a) in the presence of $CaCl_2$, and b) of EGTA. The distances moved by the marker proteins are indicated.

Discussion

The protective effect of calcium ions on the degradation of fibrinogen and fibrin has been described by Haverkate and Timan (17), Marguerie (18), and Purves et.al.(19). Marguerie localized 3 calcium binding sites of high affinity on the fibrinogen molecule. Haverkate and Timan (17) also studied the influence of calcium ions on the anticlotting properties of fragment D. They found a higher interfering activity of the calcium terminated D. In the presence of EGTA fragment D was further split off into peptides which more weakly interfered with fibrinogen coagulation. Our tests confirm his results.

The early plasmin-FDP X and Y caused both calcium- and EGTA-dependent a strong coagulation interference. A similar result was obtained with ELP as a proteolytic enzyme. Late ELP-FDP, however, showed marked differences in their anticlotting properties depending on the presence of calcium ions during their generation: All $CaCl_2$-FDP stronger inhibited the fibrinogen coagulation than the EGTA-FDP.

Only ELP-digests reached a second maximum of interference at a late degradation stage. Neither plasmin nor CLP degradation of fibrinogen resulted in such strongly interfering fragments of low molecular weight.

Fibrinogen cleavage by CLP yielded fragments which showed a poor anticoagulant activity at nearly all degradation stages. Only when calcium ions were missing strongly interfering early CLP-FDP arose.

The measurement of thrombin activity with the chromogenic substrate S-2238 (H-D-Phenylalanyl-L-pipecolyl-L-arginin-p-nitroanilide dihydrochloride) in the presence of all described anticlotting FDP (not demonstrated in the paper) showed that the phenomenon of coagulation interference by high and low molecular weight FDP is not due to competitive inhibition of thrombin by these FDP. Therefore it can be assumed that these FDP cause a polymerization inhibition as described by Plow and Edgington (8) for leukocyte protease-FDP and by Alkjaersig et.al.(1), Marder

and Shulman (2) for plasmin-FDP.
Both the coagulation experiments and the PAGE's show that fibrinogen degradation in the presence of calcium ions leads to FDP different from those obtained in the presence of EGTA.
As described earlier (20) the CLP only in a calcium free buffer is able to attack the γ-chain of fibrinogen at an early degradation stage. In contrast, calcium ions do not protect the γ-chain against the early cleavage by ELP. The influence of calcium on plasmin action becomes evident later when fragment D is formed (17). This may be an explanation for the strong differences in anticoagulant properties of the early CLP derived Ca- and EGTA-FDP. This difference is much smaller for ELP-FDP and negligible for plasmin-FDP.
In addition, calcium ions delay the proteolysis of the fibrinogen molecule. That means, higher enzyme concentrations are required in the presence than in the absence of calcium ions to receive a definite degradation stage, e.g. to make fibrinogen uncoagulable or to get strongly interfering split products of a comparable size. This fact is especially true for fibrinogen degradation by CLP, to a less extent for cleavage by ELP while plasmin fibrinogenolysis hardly is retarded by calcium ions.
The presented results make clear that the ELP has much stronger anticoagulant properties compared to the CLP. Taking in account that ELP concentration is about twice that of CLP within the granulocytes and that the stimuli for an enzyme release during inflammatory processes also cause an increase of the CLP inhibitor α_1-antichymotrypsin ELP in vivo will play the more important role in coagulation pathology.
Evidences for ELP derived FDP are given at patients with bronchial carcinomas which show a parallel increase in the plasma levels of ELP and FDP (21). Van Mourik (22) found high FPA values in patients with various types of malignancy which were not influenced by heparin suggesting a thrombin independent protease action on fibrinogen. Finally, there are also data on the occurence of FDP in bacterial infections not associated with the parameters of a consumption coagulopathy (23, 24).

References

1. Alkjaersig,N., Fletcher,A.P., Sherry,S.: J.Clin.Invest.42, 917-934 (1962).
2. Marder,V.J., Shulman,N.R.: J.Biol.Chem. 244, 2120-2124 (1969)
3. Haverkate,F., Nieuwenhuizen,W., Brakman,P.: Abstr.51st Sci. Session of Am. Heart Ass., Dallas, Texas (1978).
4. Ohlsson,K., Olsson,I.: Eur. J. Biochem.42, 519-527 (1974).
5. Schmidt,W., Egbring,R., Havemann,K.: Thromb.Res.6, 315-326 (1974).
6. Gramse,M., Bingenheimer,C., Schmidt,W., Egbring,R., Havemann,K.: J. Clin. Invest.21, 1027-1033 (1977).
7. Bilezikian,S.B., Nossel,H.L.: Blood 50, 21-28 (1977).
8. Plow,E.F., Edgington,T.S.: J. Clin. Invest.56, 30-38 (1975).
9. Janoff,A., Scherer,J.: J. Exp. Med.128, 1137-1151 (1968).
10. Schmidt,W., Havemann,K.: Hoppe Seyler's Z. Physiol. Chem. 358, 555-564 (1977).
11. Henson,P.M.: Progression Immunology, ed,by Amos,B., Academic Press, New York, 155-171 (1971)
12. Leffell,M.S., Spitznagel,J.K.: Infect. Immun.12, 813-820 (1974).
13. Goldstein,M, Weissmann,G.: J.Immunol.113, 1583-1588 (1974).
14. Goldstein,M., Wünschmann,B., Astrup,C., Henderson,S.: Blood 33, 447-453 (1971).
15. Ohlsson,K.: FEBS Letters 47, Symp.A6, 11th Meeting, Copenhagen, Regulatory proteolytic enzymes and their inhibitors, ed. by Magnusson,S. et.al., Pergamon Press, Oxford • New York (1977).
16. Saklatvala,J., Barrett,A.J.: BBA 615, 167-177 (1980).
17. Haverkate,F., Timan,G.: Thromb. Res.10, 803-812 (1977).
18. Marguerie,G.: BBA 494, 172-181 (1977).
19. Purves, L.R., Lindsay, G.G., Brown,G., Franks,K.J.: Thromb.Res.12, 473-494 (1978).
20. Gramse,M, Bingenheimer,C., Havemann,K.: Thromb.Res. 19, 201-209 (1980).
21. Gropp,C., Egbring,R.,Havemann,K.: Europ. J.Cancer 16, 79-85 (1980).
22. van Mourik,J.A.: Vth Meeting Europ. and Afric. Div.,Int. Soc.Haematol.,Abstr., Hamburg (1979).
23. Skansberg,P.,Cronberg,S., Nilsson,I.M.: Scand. J. Infect. Dis.6, 197-203 (1974)

24. Egbring,R.,Schmidt,W., Fuchs,G., Havemann,K.: Blood 49, 219-231 (1977).

A SIMPLE METHOD FOR QUANTIFICATION OF FIBRINOGEN-CHAINS AND EARLY SPLIT-PRODUCTS IN SDS-DISC-ELECTROPHORESIS

Thomas Wuppermann, Norbert Folke Rath, Hans Jürgen Zielonka, Matti Seppo Kestilä

Abteilung Angiologie im Zentrum Innere Medizin der Medizinischen Hochschule, Karl-Wiechert-Allee, D-3000 Hannover

In early diagnosis of deep venous thrombosis it is a problem to test the biologic quality of labelled fibrinogen by half-lifetime. In vitro it is possible to separate fibrinogen-chains and their break-down-products by molecular weight by SDS-disc-electrophoresis (1).
By additional labelling of fibrinogen with iodine isotope before plasminolysis and electrophoresis, slicing the whole disc-gel afterwards and analysis of the radioactivity of the disc-gel continuously slice by slice a quantitative analysis of the split products in terms of their molecular weight and additionally by their quantitative distribution is possible.
By comparison of four steps of plasminolysis the method can be further differentiated.

Method

Fibrinogen Kabi is isolated by affinity chromatography (3) and dissolved in 0,05 M phosphate buffer pH 5,3. After precipitation with saturated ammonium-sulfate-solution fibrinogen is redissolved in the same buffer and brought to a concentration of 3 mg fibrinogen/ml. 2 ml of this protein solution are labelled by Chloramin T technique (2) with 300 µCi carrier free 131 iodine isotope.
On a Sephadex G-25 column the salts and the iodine labelled fibrinogen are separated.
Standard-plasminolysis of 400 µg fibrinogen is achieved with

0.5 CTA-units of streptokinase activated plasminogen at 37°C. To obtain six steps of fibrinolysis aliquots of 40 µg of fibrinogen are separated at 10 sec, 50 sec, 2 min, 7 min, 15 min and 30 min after beginning of the fibrinolysis and further plasminolysis is stopped in each probe by 100 KIE Trasylol.
The probes are then boiled with denaturation buffer at the same time as calibration substances (cytochrome C, chymotrypsin A, ovalbumine, bovinealbumine and phosphorylase).
For SDS-disc-electrophoresis (1) the start-gel is cast with 4% polyacrylamid pH 8.3 and the separation-gel with 9% pH 8.8. The gels are charged with 20 µg of protein and every lysis step was separated electrophoretically twice.
SDS-disc-electrophoresis is done with an electrode-buffer pH 8.3 and 4 mAmp/gel over 4 1/2 hrs with bromphenol-blue as a marker, stained with coomassie blue and destained over 24 hrs.
Afterwards the first gel is photographed in natural scale and the molecular weight of the proteins is determined by help of the calibration substances.
The frozen second gel is cut by gel-slicer in 130 slices each slice 1 mm thick. The radioactivity of each slice is measured in a gammaspectrometer and the individual counts converted to a total activity of 100 000 counts/gel.
The entire process is repeated three times so that six gels are obtained for each lysis step.

Results

131 iodine labelled reference fibrinogen Kabi is demonstrated in 4 of the 6 lysis steps (Fig. 1). The four steps are: before plasminolysis, 50 sec after, 7 min after and 30 min after beginning of plasminolysis. The stained gels are demonstrated below with the protein fractions characterized by their molecular weight. For each gel slice the corrected activity

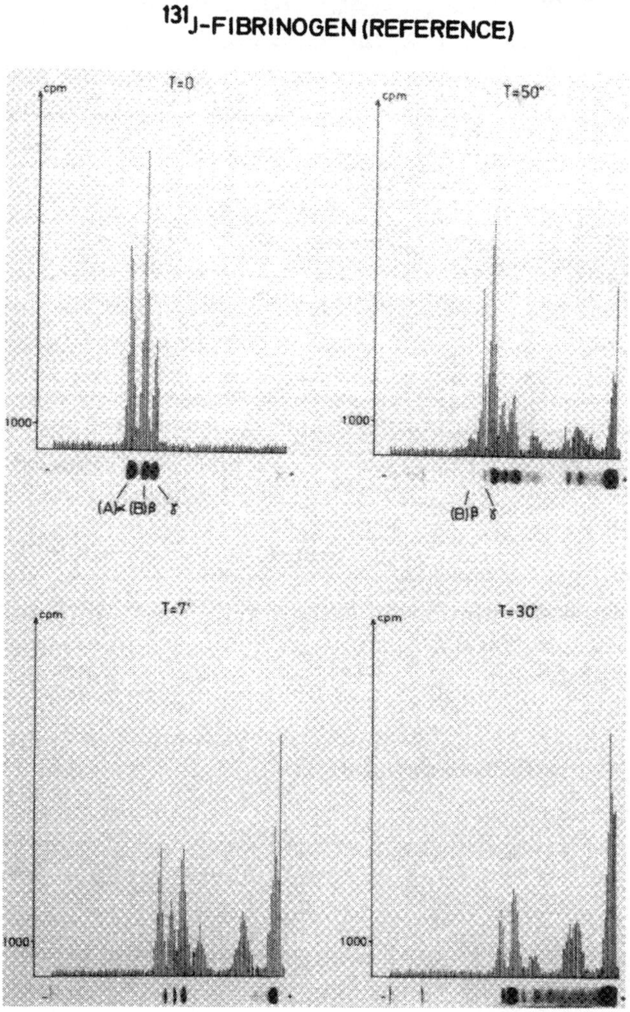

Fig. 1: Lysis behaviour of 131-J-Fibrinogen

^{131}J-FIBRINOGEN (REFERENCE)

<table>
<tr><th rowspan="2">$\overline{MG}$</th><th rowspan="2">S_{MG}</th><th colspan="2">Lysis: T_0</th><th colspan="2">10"</th><th colspan="2">50"</th><th colspan="2">2'</th><th colspan="2">7'</th><th colspan="2">15'</th><th colspan="2">30'</th></tr>
<tr><th>$\bar{x}$ [cpm]</th><th>S_x</th><th>$\bar{x}$</th><th>S_x</th><th>$\bar{x}$</th><th>S_x</th><th>$\bar{x}$</th><th>S_x</th><th>$\bar{x}$</th><th>S_x</th><th>$\bar{x}$</th><th>S_x</th><th>$\bar{x}$</th><th>S_x</th></tr>
<tr><td>67000</td><td>1400</td><td>29070</td><td>2180</td><td>6470</td><td>3520</td><td></td><td></td><td></td><td></td><td></td><td></td><td></td><td></td><td></td><td></td></tr>
<tr><td>57000</td><td>2200</td><td>33030</td><td>1570</td><td>23030</td><td>9960</td><td>2870</td><td>1180</td><td></td><td></td><td></td><td></td><td></td><td></td><td></td><td></td></tr>
<tr><td>49000</td><td>1700</td><td>13000</td><td>1540</td><td>24570</td><td>6530</td><td>10070</td><td>7200</td><td>3130</td><td>670</td><td></td><td></td><td></td><td></td><td></td><td></td></tr>
<tr><td>46000</td><td>1900</td><td></td><td></td><td>5670</td><td>1210</td><td>28520</td><td>12500</td><td>23700</td><td>3630</td><td>12500</td><td>3700</td><td></td><td></td><td></td><td></td></tr>
<tr><td>43000</td><td>1600</td><td></td><td></td><td>5700</td><td>3220</td><td>5770</td><td>2450</td><td>4150</td><td>610</td><td>7600</td><td>3080</td><td>7250</td><td>1120</td><td>6780</td><td>1630</td></tr>
<tr><td>40000</td><td>1600</td><td></td><td></td><td>2870</td><td>1170</td><td>8550</td><td>4960</td><td>9450</td><td>1050</td><td>16070</td><td>2970</td><td>11950</td><td>910</td><td>10980</td><td>2070</td></tr>
<tr><td>33000</td><td>1200</td><td></td><td></td><td></td><td></td><td>4130</td><td>1100</td><td>4670</td><td>150</td><td>6970</td><td>1070</td><td>4970</td><td>900</td><td>3270</td><td>550</td></tr>
<tr><td>26000</td><td>1100</td><td></td><td></td><td>2730</td><td>310</td><td></td><td></td><td></td><td></td><td></td><td></td><td></td><td></td><td></td><td></td></tr>
<tr><td>22000</td><td>500</td><td></td><td></td><td></td><td></td><td>2300</td><td>1060</td><td>2830</td><td>210</td><td rowspan="2"></td><td rowspan="2"></td><td rowspan="4">12770</td><td rowspan="4">1750</td><td rowspan="4">15430</td><td rowspan="4">1720</td></tr>
<tr><td>20000</td><td>600</td><td></td><td></td><td>3930</td><td>1960</td><td></td><td></td><td></td><td></td></tr>
<tr><td>18500</td><td>300</td><td></td><td></td><td></td><td></td><td>6170</td><td>2170</td><td>5300</td><td>660</td><td rowspan="2">11730</td><td rowspan="2">1540</td></tr>
<tr><td>17000</td><td>600</td><td></td><td></td><td></td><td></td><td>2670</td><td>810</td><td>3630</td><td>250</td></tr>
<tr><td><14000</td><td></td><td></td><td></td><td>7430</td><td>4610</td><td>12900</td><td>6750</td><td>23130</td><td>11600</td><td>24770</td><td>3170</td><td>35070</td><td>4900</td><td>28000</td><td>12350</td></tr>
</table>

MG = molecular weight of the sulfitolysized fibrinogen split products

cpm = counts per minute

T = length of lysis

S_{MG} = S_x of molecular weight

Table 1: Correlation of molecular weight and activity in different lysis steps of fibrinogen by SDS-disc-electrophoresis in counts/min is demonstrated correspondingly the molecular weight. The three chains of fibrinogen molecule are recorded in different quantities before the beginning of lysis. This probably depends on a unequal labelling by chloramin-T-technique.

After 50 sec of plasminolysis the Aα-chain disappears completely and only traces of Bβ-chain are found, while the γ-chain is nearly intact.

At the same time, several lysis products appear of which the greatest quantity is found at a molecular weight of 46 000. It is lysed only 7 min after beginning of plasminolysis together with the γ-chain. It may be identical with a known, highly lysis resistant part of Bβ-chain in fragment D (4).

The two last lysis steps (15 min, 30 min) seem to be of lower interest as it can not be decided which chain the fragments belong to.

The molecular weights and the corresponding mean counts with standard deviation (Tab. 1) in the first three lysis steps demonstrate for example the γ-chain with a molecular weight of 49 000 and 13 000 ± 1500 cpm.

After 10 sec of plasminolysis the counts corresponding to the molecular weight 49 000 increase to 24 570 ± 6530 cpm. This demonstrates, that the increase of counts at a certain molecular weight cannot be related to a certain product i. e. not the γ-chain has really increased, but break down products of Aα- or Bβ-chain have to be the cause.

Summary

By a simple variation of SDS-disc-electrophoresis and radio-iodinated fibrinogen it is possible to get some information on the quantitative distribution of the chains and early lysis products of fibrinogen.

The main limitations of this method are:

1. in progress of lysis and appearance of low molecular weight

products the origin of these individual products cannot be determined.

2. unequal iodination of the three chains falsifies the real quantitative proportion of the split products to a certain degree.

Furthermore the comparison of fibrinogen labelled by different methods may be difficult.

However we believe the method to be more exakt in control of the quality of labelled fibrinogen used for early diagnosis of deep vein thrombosis than biologic half-life-time.

References

1. Laemmli, U. K.: Nature 227, 680-685 (1970).
2. McConahey, P. J., Dixon, J. F.: Int. Arch. Allergy 29 185-189 (1966).
3. Stemberger, A., Hörmann, H.: Hoppe-Seyler's Zschr. Physiol. Chem. 357, 1003-1005 (1976).
4. Tagaki, K., Kawei, T.: Thrombos. Haemostas. 37, 464-470 (1977).

HOMOLOGUES OF FIBRIN X OLIGOMERS

Reimar Hafter, Henner Graeff

I. Frauenklinik der Universität München, Maistrasse 11
8000 München 2, Germany

Introduction

Crosslinked fibrin derivatives are assumed to signalize intravascular coagulation (1,2). The D dimer (3,4), the Y-D fragment (5) and fibrin X oligomers (6) were observed in plasma samples from mainly obstetric patients with severe coagulation disorders (consumption coagulopathy). Subunit structure analysis revealed that X oligomers with a mean molecular weight up to 2 million daltons are composed of: γ-γ, γ-γ^1, ß, $ß^2$, α^1 and α^2 chains (2).

X oligomers, Y-D and D dimer are related to each other in their subunit structure in a way which reflects the progress in degradation of crosslinked fibrin. Therefore, if X oligomers are present in plasma or body fluids, Y-D and D dimer should also be present. This concept could be substantiated in this study for ascitic fluid obtained from patients with ovarian cancer (7,8).

The same derivatives are formed *in vitro*, if thrombin, F XIII and plasmin react simultaneously on fibrinogen, which indicates that this reaction scheme might also be responsible for the formation of these derivatives *in vivo* (9,10).

It was the aim of this study to evaluate the subunit structure of single molecular entities of crosslinked fibrin derivatives. The *in vitro* formed derivatives are ideally suited as models with which the *in vivo* formed derivatives can be compared.

Fibrinogen - Recent Biochemical and Medical Aspects

Materials and Methods

Plasma samples from patients with intravascular coagulation were prepared as described earlier (6). Ascitic fluid from patients with advanced ovarian cancer was anticoagulated immediately after collection (1:10) (anticoagulant ref. in 6) and centrifuged for 30 min at 3,000 x g. The sediment containing cellular material was discarded.

Crosslinked soluble fibrin derivatives were prepared by simultaneous action of thrombin, factor XIII, and plasmin on human fibrinogen (Deutsche Kabi GmbH, Munich) in presence of calcium as described elsewhere (6).

Immunoprecipitation of fibrin derivatives (11), gel filtration chromatography of ß-alanine precipitated protein, enrichment of protein in fractions of gel filtration by acid precipitation (12), and sodium dodecylsulfate polyacrylamide gel electrophoresis (SDS-PAGE) of unreduced and reduced samples were performed as outlined in (6). To evaluate the subunit structure of electrophoretically separated fragments, single bands of unreduced samples in 4 or 5 % PAA-gels were cut out at the respective positions. The gels had been stained beforehand with Coomassie Brillant Blue G-250 in methanol - 7 % acetic acid (1:1, v/v), and destained in the same solvent. The sliced gel cylinders were treated for about 30 min with 0.05 N Tris-HCl buffer, pH 7.4, containing 2 % mercaptoethanol. The gel slices were then placed on top of 7.5 % SDS-PAA gels and reelectrophoresed for 12 hours (4 mA/ gel). The protein staining with this particular dye is completely reversible on electrophoresis, and protein migration is not altered.

Analogously treated gel slices containing fibrinogen and reduced crosslinked fibrin were used as references.

Subunit polypeptide chains and chain remnants were identified according to their migration position on SDS-PAGE (6,13).

Molecular weights of subunits and fragments were calculated from the amino acid compositions of the respective subunit polypeptide chains according to their amino acid sequences, as published by Henschen et al (14),under consideration of the most likely plasmic cleavage sites, proposed and reviewed by Furlan (15), and Lottspeich and Henschen (16). An additional molecular weight of 2,200 was taken into account for each single carbohydrate side chain in the Bß-(position 364) and γ-chain (position 52) (14). The chain remnants were designated following their decreasing molecular size with elevated arabic numbers. We abandoned the nomenclature of Pizzo et al (13) with dash symbols which we had used before (6) in order to avoid confusion with dash symbols used by Mosesson et al (17) and Martinez (18) for γ chain variants.

Results

Simultaneous action of thrombin, factor XIII, and plasmin on fibrinogen in the presence of calcium gives rise to a whole spectrum of soluble low and high molecular weight derivatives as shown in figure 1, gel c. The band pattern in the top region of the gel refers to the presence of crosslinked X oligomers, the strongly expressed band in position of the X fragment (see gel d) to the Y-D fragment. The following bands represent D dimer, Y, D, E, and smaller fragments, most likely α chain remnants.

The pattern of gel a (Fig. 1) represents the urea soluble portion of a crosslinked fibrin clot, partially lysed by plasmin. It shows a similar pattern as gel c, but with less bands in the top region (X oligomers), and a smaller ratio of D-Y to D-D. There is also relatively less Y and D compared to gel c , which indicates that crosslinking had taken part to a higher degree than in the sample shown in gel c.

Gel b represents the terminal lysis product of the cross-

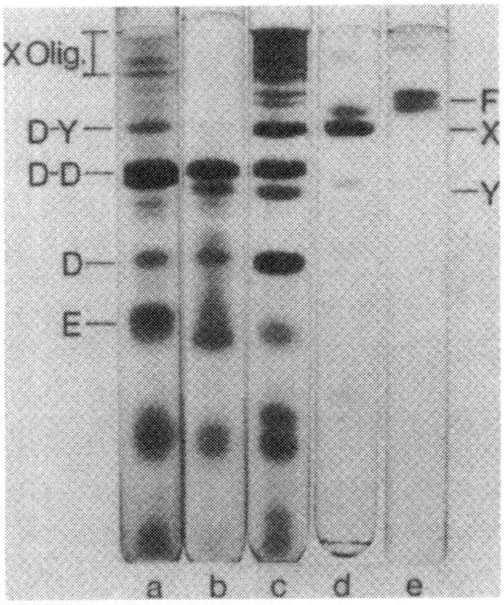

Fig. 1

SDS-PAGE of fibrin derivatives (5 % gel). a: Crosslinked fibrin clot which was partially lysed with plasmin; the residual clot was dissolved in 6 M urea. b: Crosslinked fibrin clot which was completely lysed with plasmin. c: Crosslinked fibrin derivatives obtained by simultaneous action of thrombin, F XIII, and plasmin on fibrinogen. d: Early FDP, mainly X fragment. e: Fibrinogen

linked fibrin clot. The absence of X oligomers and D-Y indicates that these derivatives are further degraded to D-dimer.

Figure 2 shows the patterns of subunit chains and chain remnants of the samples c and a from figure 1 after reduction with mercaptoethanol. Three dimeric γ chains are visible, designated γ-γ, γ-γ^1 and γ^1-γ^1 respectively. The additional bands were identified as ß chain and $ß^2$, α^1 and α^2 chain remnants. The nature of the band in position between $ß^2$ and α^1 is uncertain.

Figure 3 shows an experiment in which the bands of D-D and

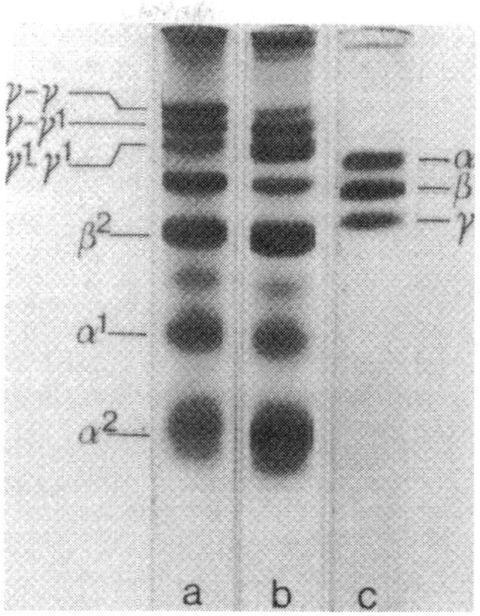

Fig. 2

SDS-PAGE patterns of reduced samples (7,5 % gels). a: Fibrin lysis product as shown in fig. 1c. b: Fibrin lysis product as shown in fig. 1 a. c: Fibrinogen as reference

Y-D were cut out from a 4 % gel (shown in the center) and reelectrophoresed after mercaptolysis. D-D gave rise to the chain remnants $\gamma^1-\gamma^1$, β^2 and α^2, and Y-D to the chain remnants $\gamma-\gamma^1$, β^2, α^1 and α^2.

The Y-D band further gave rise to small but equal amounts of β, β^1 and γ subunits. This may be caused by a contaminating X fragment, which has the same migration behaviour as Y-D (see fig. 1).

In another experiment the top part of gel c (shown in fig. 1 and 3), i.e. the migration zone of the X oligomers was sliced into three parts of equal length and reelectrophoresed after reduction (fig. 4). The three fractions gave similar patterns when compared with each other. One máin difference to Y-D is

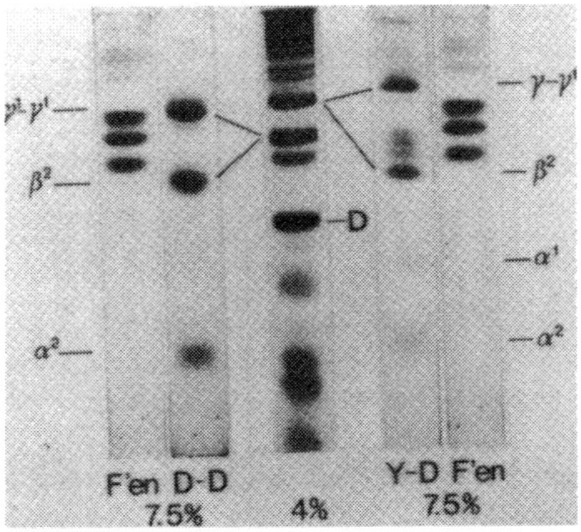

Fig. 3

SDS-PAGE patterns (7,5 % gels) of subunit chain remnants of the D-D and the Y-D fragments cut out from the gel in the center (4 % gel). Aα, Bß and γ (reduced fibrinogen) as reference

the presence of two dimeric γ chains (γ-γ and γ-γ^1) instead of only one (γ-γ^1). Another difference, which is best expressed in the sample from the very top (first gel from the right), is the presence of an undegraded ß chain. In addition there are differences between the three samples in respect to different molar ratios of γ-γ to γ-γ^1.

Figure 5 shows an analogous experiment. A fibrin fraction obtained by acid precipitation from ascitic fluid (second gel from the right) was sliced in correspondence to D-D, Y-D and X oligomers (two slices) respectively. After reduction the slices were subjected to reelectrophoresis (left side panel). The subunit patterns obtained from fragments D-D and Y-D are in agreement with the ones shown before from the <u>in vitro</u> sample,and confirm the presence of D-D and Y-D in ascitic fluid. The two fractions corresponding to X oligomers reveal

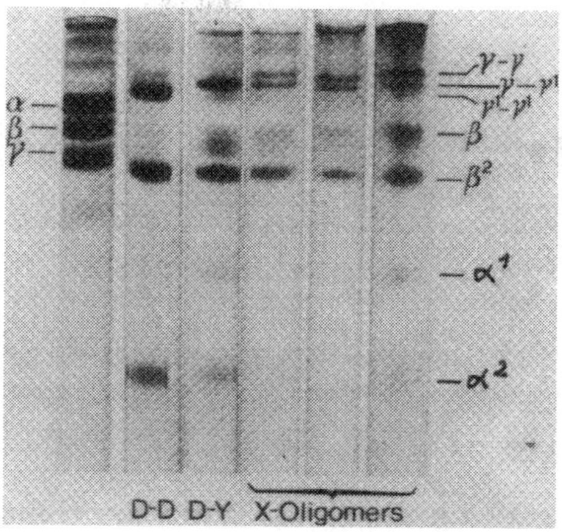

Fig. 4

SDS-PAGE patterns (7,5 % gels) of subunit chains and remnants of D-D, Y-D and X oligomers cut out from gel c (shown in fig. 1). Reduced fibrinogen is given as reference.

the typical subunit patterns of $\gamma-\gamma$, $\gamma-\gamma^1$, ß, $ß^2$, α^1, and α^2. In addition different molar ratios of the γ dimers of about 2:1 and 1:1 are observed.

In figure 6 an experiment is shown in which a high molecular weight fibrin fraction, obtained by gel filtration chromatography after ß-alanine precipitation of ascitic fluid, was electrophoresed on 4 % PAA gel. The indicated bands were sliced out of the gel and reelectrophoresed after reduction. The subunit patterns reveal the presence of $\gamma-\gamma_1$, $\gamma-\gamma^1$, ß, $ß^1$, $ß^2$, and α^1. The patterns of $\gamma-\gamma$ and $\gamma-\gamma^1$ again show different molar ratios between each other. In one case (fig. 6a) the ratio is about 1:1 and in the other 2:1 (fig. 6b) which is consistent with the fragments D-X-Y and D-X-X-Y respectively.

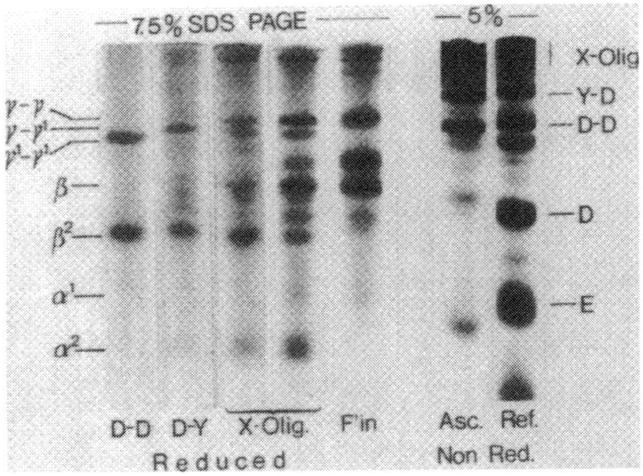

Fig. 5

Reduced samples (SDS-PAGE, 7,5 %) of fragments D-D, D-Y and of X oligomers which were cut out from the gel Asc. (5 % SDS-PAGE pattern of non reduced fibrin derivatives from ascitic fluid). Reduced fibrin (F'in) and non reduced fibrin derivatives from in vitro material (Ref.)(also shown in figures 1 and 3) are given as references.

According to the subunit chain compositions as described above (see figs. 3-5) structural models of D-D, D-Y, and X oligomers are proposed as given in figure 7. D-D and D-Y can be described by the formulas $\gamma^1 - \gamma^1$, $(\alpha^2, ß^2)2$ and $\gamma - \gamma^1$, α^1, α^2, $2ß^2$, α^E, $ß^E$, γ^E, respectively.

The next two structural models being proposed are X oligomers which can be described as D-X-Y and D-X-X-Y respectively. Structures with D and Y at the ends are in favour of the asymmetric cleavage of fibrinogen by plasmin and are therefore more likely to occur than symmetric structures with either Y or D moieties on both ends. According to the observations made to now, X oligomers always contain two dimeric γ chains, i.e the undegraded $\gamma - \gamma$ and the partially degraded

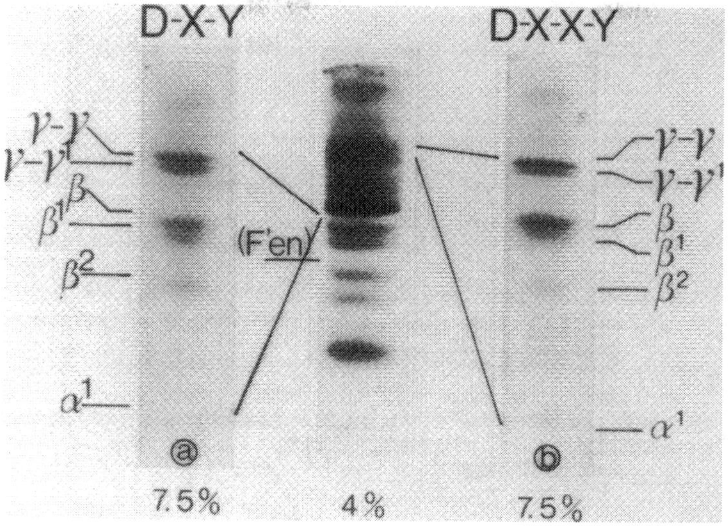

Fig. 6

SDS-PAGE patterns of high molecular weight fibrin fraction of non reduced sample (4 % gel) from ascitic fluid, and reduced bands (7,5 % gels) after being cut out at the indicated positions. As reference the migration position of fibrinogen is indicated

$\gamma - \gamma^1$. The molecular weights given in figure 7 are calculated values. They were calculated from the molecular weights of the respective subunits (Table 1) involved in the proposed structures.

Discussion

The *in vitro* experiments presented show that crosslinked fibrin derivatives exist as a series of homologous structures which can be described as D-D, D-Y, D-X-Y, and D-X-X-Y and so forth. Following this concept, a molecular weight difference of approximately 240,000 daltons (equal to one X moiety)

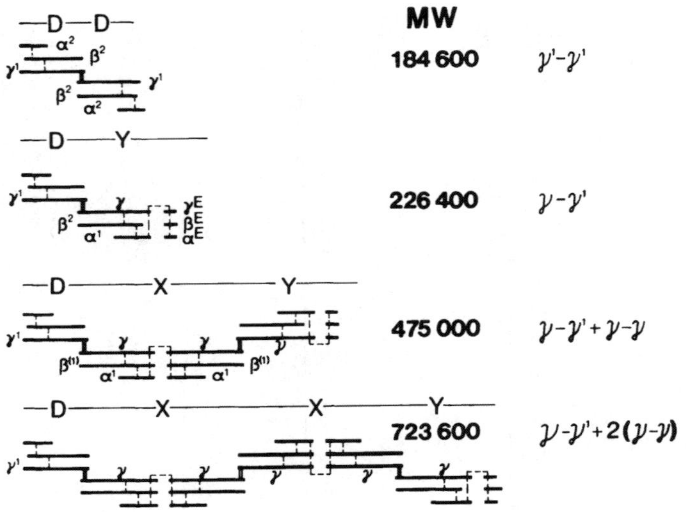

Fig. 7

Structural models of γ chain crosslinked D-D, D-Y, and X oligomers (D-X-Y, D-X-X-Y). The calculated molecular weights and the respective chain compositions are demonstrated on the right.

should exist between following X oligomer homologues. However, there is a rather dense pattern of X oligomers in the molecular range of about one million daltons and more as shown in figure 1 (gel c). It must therefore be assumed that additional structures exist. At least some of them might be rather stable complexes with X fragments or others. One possible explanation could also be the occurrence of larger γ chain variants as recently described by Wolfenstein-Todel and Mosesson (19) and Francis et al (20). However, the rather small differences in the molecular weight between normal and variant γ chains (~ 2000 daltons) do rather unlikely add up to the observed variety in the million dalton range.

Symmetrical structures in which the ends bear equal moieties, like D-X-D or Y-X-Y, might also be considered as a cause of

Table 1

Calculated Molecular Weights of Subunit Chains and Chain Remnants Involved in Crosslinked Fibrin Derivatives

DESIGNATION	POLYPEPTIDE	APPROXIMATE MOL. WEIGHT	ORIGIN
α^1	Aα17-208	22,600	X, Y
α^2	Aα79-208 OR104-208 OR111-208	15,200 12,300 11,500 MEAN: 13,000	D
α^E	Aα20-78	7,000	E
β	Bβ15-461	53,000*	
β^1	Bβ54-461	49,000*	X, Y
β^2	Bβ134-461	40,000*	D
β^E	Bβ54-122	7,500	E
γ	γ 1-411	48,700*	
γ^1	γ 63-411	39,300	D
γ^E	γ 1- 53	8,300	E
$\gamma-\gamma$	$(\gamma\ 1\text{-}411)_2$	97,400**	
$\gamma-\gamma^1$	(γ1-411)(γ63-411)	88,000*	D-Y
$\gamma^1-\gamma^1$	$(\gamma 63\text{-}411)_2$	78,600	D-D

* INCLUSIVE 1 OR 2(**) CARBOHYDRATE SIDE CHAINS (MW~2,200 EACH)

the observed variety. This would however imply a symmetrical cleavage which is less likely. Nevertheless fragments like D-X-D and Y-Y have recently been reported to occur in lysis products of crosslinked fibrin (22). The existence of the symmetrical D dimer fragment can easily be explained by the fact that it is a terminal lysis product. A further plausible explanation for the variety of high molecular weight derivatives could well be the different degree of degradation of certain molecular entities (comparable to the D- and X- families).

Our suggestions that X oligomers exist as homologues to the structure D-X-Y and not of a symmetric structure like D-X-D or Y-X-Y is confirmed by our findings that X oligomers revealed always two dimeric γ chains. This is incompatible with the latter structures. They would contain only one type of dimeric chain of either $\gamma - \gamma^1$ or $\gamma - \gamma$ respectively.

Disulfide bond splitting of the <u>in vitro</u> sample revealed that three dimeric γ chains exist in crosslinked derivatives (see fig. 2a) and that these three dimeric γ chains are equally distributed. At this stage of clot lysis X oligomers have reached a size in which the largest still soluble entity might have a molecular weight of about 2 million daltons, while the smallest crosslinked derivative has reached the terminal lysis stage, i.e. the D dimer with a molecular weight of approximately 185,000 daltons. A distribution of molecular entities in the range between 185,000 and 2 million daltons can therefore be expected.

The three dimeric γ chains obtained were designated $\gamma - \gamma$, $\gamma - \gamma^1$, and $\gamma^1 - \gamma^1$. Strikingly similar observations concerning the three dimeric γ chains were made by P.J. Gaffney (see same volume).The first two are constituents of X oligomers, the last one of D dimer. The $\gamma - \gamma^1$ is also a constituent of the D-Y and thus indicates the intermediary state of the D-Y fragment between X oligomers and D dimer. A shift in the distribution of γ dimers towards the terminal dimer $\gamma^1 - \gamma^1$ is observed when lysis has proceeded, or when an already crosslinked fibrin clot was subjected to lysis and the urea soluble part is investigated (fig. 2, gel b).

Additonal dimeric γ chains which are related to the presence of γ chain variants (19-21) could not be demonstrated by our technique.

The evaluation of single fragments out of a mixture became possible by gel slicing technique. By this technique the nature of the D-Y (or Y-D) fragment could be elucidated. It could be shown (fig. 3) that D-Y contains $\gamma-\gamma^1$, $ß^2$, α^1, and α^2. From this evaluation the structural model depicted in figure 6 is proposed. The molecular weight was calculated to be 226,400 which is roughly the same as for the X fragment with which it shares a corresponding electrophoretic behavior. An interesting feature is the finding that the ß chain is degraded to the $ß^2$ remnant, the terminal ß chain plasmin lysis product. In contrast X oligomers still contain the undegraded ß chain. When the degradation of D-X-Y proceeds, the molecule is split into two D-Y fragments (see fig. 7). At this stage the ß chain becomes exposed and hence susceptable to ß chain degradation.

It was also possible to demonstrate that X oligomers, D-Y, and D dimer are formed in vivo. In an earlier paper (6) we had already shown typical patterns of X oligomers from patients with intravascular coagulation. The findings were obtained from ß-alanine precipitated plasma samples after gel filtration and acid precipitation (12) of fractions from the void volume and the shoulder volume prior to the fibrinogen peak. SDS-PAGE of the non reduced fractions revealed bands which were identified as X oligomers. Less pronounced bands compatible with D-Y and D-D (these derivatives respond poorly to ß-alanine precipitation and remain therefore mainly in the supernatant) were also present. The reduced fractions revealed the typical subunit composition of X oligomers such as $\gamma-\gamma$, $\gamma-\gamma^1$, ß and $ß^2$. Furthermore, different ratios of $\gamma-\gamma$ to $\gamma-\gamma^1$ favouring the proposed structural models were observed. The ratios were approximately 2:1 (or even 3:1) in the sample from the void volume (Vo) and 1:1 in the sample from the shoulder volume (Vs) in agreement to the structural models D-X-X-Y (Vo) and D-X-Y (Vs) respectively.

These earlier findings correspond very well to the observations concerning the occurrence of X oligomers, D-Y and D-D in ascitic fluid, as presented in this paper. It can therefore be assumed that the combined action of thrombin, factor XIIIa and plasmin on the fibrinogen molecule may occur as well intravascularly as extravascularly under certain conditions. This pathobiochemical process leads to a rather consistent pattern of crosslinked fibrin derivatives which may have a marker function for certain disease states.

Summary

D dimer, Y-D, and X oligomers are observed in plasma from obstetric patients with severe coagulation disorder. They are also found in ascitic fluid from patients with advanced ovarian cancer and can be produced in vitro by simultaneous action of thrombin, plasmin, and factor XIII with fibrinogen. The study was aimed to evaluate the subunit structure of separated molecular entities. The derivatives were separated by 4 % SDS-PAGE preceded in case of the in vivo products by gel filtration and/or by immunoabsorption technique. The gels were sliced at the respective migration positions, and derivatives therein reelectrophoresed on 7,5 % gels after reduction. Subunit characterisation revealed that D dimer is composed of the chain remnants $\gamma^1-\gamma^1$, β^2, α^2, and Y-D of $\gamma-\gamma^1$, β^2, α^1, α^2, besides α^E, β^E and γ^E. Crosslinked X oligomers are composed of $\gamma-\gamma$, $\gamma-\gamma^1$, β, (β^1), β^2, α^1, and α^2 besides α^E, β^E, and γ^E. Three possible combinations of plasmin degraded and undegraded dimeric γ-chains were observed in vivo and in vitro: $\gamma-\gamma$, $\gamma-\gamma^1$ and $\gamma^1-\gamma^1$. The ratio of degraded (γ^1) to undegraded γ-chains in dimeric γ-chain patterns reflects the molecular structure of the respective derivative. Two X oligomers could be demonstrated in which the ratio of $\gamma-\gamma$ to $\gamma-\gamma^1$ in terms of stain intensity was either 1:1 or 2:1. Their subunit compositions are in accordance with structures describable as D-X-Y and D-X-X-Y. Their molecular weights calculated from the subunit compositions are 475,000 and 723,600 respectively. It is proposed that crosslinked X oligomers exist as a homologous family with increasing X fragment content.

Acknowledgements

This study was supported by the Deutsche Forschungsgemeinschaft, Bonn-Bad Godesberg, FRG (Sonderforschungsbereich 51, München, grant no. B/6). The technical assistance of Miss B. von Brunn is gratefully acknowledged.

References

1. Hafter, R., Graeff, H.: Thrombosis Research 7, 391-399 (1975).

2. Graeff, H., von Hugo, R., Hafter, R.: in: Progress in Chemical Fibrinolysis and Thrombolysis, vol. 3, J.F. Davidson, R.M. Rowan, M.M. Samama, P.C. Desnoyers (eds.) Raven Press, New York 1978, 435-447.

3. Gaffney, P.J.: Clin. Chim. Acta 65, 109-115 (1975).

4. Marder, V.J., Budzynski, A.Z.: Thrombos. Diathes. haemorrh. 32, 49-56 (1974).

5. Whitaker, A.N., Rowe, E.A., Masci, P.P., Gaffney, P.J.: Thrombosis Research 18: 453-459 (1980).

6. Graeff, H., Hafter, R., Bachmann, L.: Thrombosis Research 16, 313-328 (1979).

7. Svanberg, L., Astedt, B.: Cancer 35, 1382-1387 (1975).

8. Murr, H., Grünst, J., Eisenburg, J., Hiller, E., Zumtobel, V.: Klin. Wochenschr. 58, 85-90 (1980).

9. Hafter, R., von Hugo, R., Graeff, H.: Thrombos. Haemostas. 42, 275 (1979).

10. Graeff, H., Hafter, R.: in: Progress in Fibrinolysis and Thrombolysis, vol. 5, J.F. Davidson (ed.), Churchill Livingstone, London, in press.

11. Lane, D.A., Scully, M.F., Kakkar, V.V.: Thrombosis Research 9, 191-200 (1976).

12. Hafter, R., von Hugo, R., Graeff, H.: Hoppe Seyler's Z. Physiol. Chem. 359, 759-763 (1978).

13. Pizzo, S.V., Schwartz, M.L., Hill, R.L., McKee, P.A.: J. Biol. Chem. 247, 638-645 (1972).

14. Henschen, A., Lottspeich, F., Southan, C., Töpfer-Petersen, E.: in: Protides of Biological Fluids, vol. 28, H. Peeters (ed.), Pergamon Press, Oxford-New York 1980, 51-60.

15. Furlan, M.: in: Fibrinolysis, P.J. Gaffney, S. Balkuv-Ulutin (eds.), Academic Press, London 1978, 97-108.

16. Lottspeich, F., Henschen, A.: in: Fibrinogen, Fibrin und Fibrinkleber, K. Schimpf (ed.), F.K. Schattauer-Verlag, Stuttgart-New York 1980, 61-63.

17. Mosesson, M.W., Finlayson, J.S., Umfleet, R.A.: J. Biol. Chem. 247, 5223-5227 (1972).

18. Martinez, J.: Blood 56, 417-420 (1980).

19. Wolfenstein-Todel, C., Mosesson, M.W.: Proc. Natl. Acad. Sci. 77, 5069-5073 (1980).

20. Francis, C.W., Marder, V.J., Martin, S.E.: Blood 56, 456-464 (1980).

21. Francis, C.W., Marder, V.J., Martin, S.E.: J. Biol. Chem. 255, 5599-5604 (1980).

22. Francis, C.W., Marder, V.J., Barlow, G.H.: J. Clin. Invest. 66, 1033-1043 (1980).

FIBRIN FRAGMENTATION BY PLASMIN: MOLECULAR CONSIDERATIONS AND THEIR RELEVANCE TO DIC AND THROMBOSIS

Patrick J. Gaffney

National Institute for Biological Standards and Control, Holly Hill, Hampstead, London NW3 6RB, England

Introduction

The occurrence of disseminated intravascular coagulation (DIC) is associated with a wide variety of clinical conditions. An understanding of the mechanisms involved during DIC may allow new concepts in treatment to emerge. That thrombin is the trigger for the initiation of DIC is widely accepted, while the involvement of the fibrinolytic system in generating fibrin degradation products (FDP) has been supported by both the consumption of plasminogen and fibrinogen. During this brief review we shall endeavour to highlight the fact that forming fibrin rather than fibrin deposits generates most of the FDP fraction. Details of the fragmentation of various types of fibrin will be discussed and the relevance of various structural entities to the clinical assessment of DIC and thrombosis will be outlined.

Background Information

Considerable data have been accumulated concerning the interaction of plasmin with fibrinogen (for review, see 1) and these data have acted as a guide to experimental procedures and nomenclature used to examine the nature of fibrin-plasmin interactions. The initial detailed examinations of fibrinogen-plasmin interactions were performed by the group of Nussenzweig and colleagues (2). They concluded that two large plasmin-resistant

Fibrinogen - Recent Biochemical and Medical Aspects

fragments, D and E make up 70% of exhaustively digested fibrinogen, while a variety of peptide material (A, B and C in these authors' chromatographically-based nomenclature) comprised the other 30%. They also observed degradation products of fibrinogen early in the digestion process, which have since been noted by others and named as fragments X and Y (3). Much of the work since then has been concentrated on interrelating the structural details of fragments X, Y, D and E and the other peptides generated during plasmin-mediated digestion with the structure of the originating fibrinogen.

The scheme in Figure 1 shows that one mole of fibrinogen generates two moles of fragment D and one mole of the dimeric E, while one mole each of fragments X and Y are found in intermediate plasmin-mediated fibrinogen digests. A variety of peptides is also released from fibrinogen by plasmin (A, B, C in the terminology of Nussenzweig et al., 2). These may have significant biological activities in vivo, and work on their structures and activities is developing. Most of these peptides derive from the carboxy-terminal ends of the Aα chains, and a radioimmunoassay has been developed in plasma for certain of these fragments (20,000-30,000 mol. wt.). These types of assay have been suggested as sensitive means of detecting elevated fibrinolytic activity in vivo (10-12).

In order to appreciate the interactions of plasmin with fibrin it is necessary to describe the various types of fibrin which can be formed by the interactions of fibrinogen with thrombin and plasma transglutaminase (Factor XIII). Fibrin formation is initiated by the thrombin-mediated cleavage of fibrinopeptides A and B from fibrinogen (13). Only the removal of fibrinopeptide A is necessary for the polymerisation process to begin. The des-AA fibrin monomer $(\alpha\ B\beta\ \gamma)_2$ polymerises spontaneously in an end-to-end manner while the cleavage of polypeptide B seems to signal an additional lateral polymerisation process either prior to or on visualisation of fibrin strands (14). These fibrin strands become stabilised or crosslinked in the presence of calcium and Factor XIII (activated by

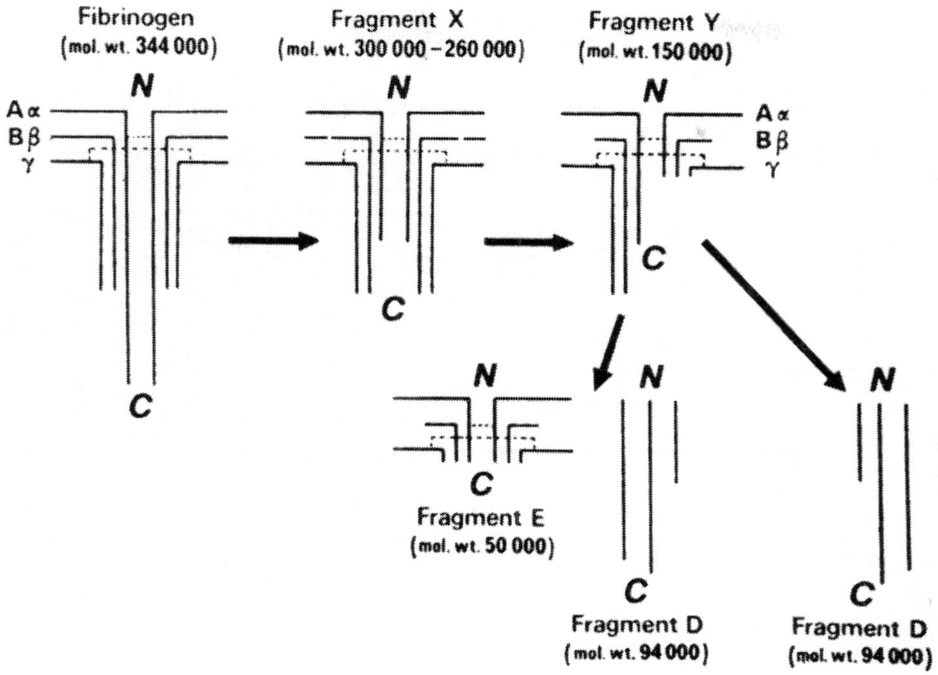

Figure 1

Schematic diagram of the plasmin-mediated conversion of fibrinogen to its core fragments D and E, showing the intermediate fragments X and Y. The polypeptide chain origin of the constituent subunits of each fragment can be related to those of the preceding fragment and in turn to chains of the originating fibrinogen molecule - Aα, Bβ and γ. The amino (N)- and carboxy (C)-terminal ends of the fragments are shown, to aid in visualizing more easily the site of structural breakdown. The range of molecular weights of fragment X depends on the amount of peptide material removed from the carboxyl ends of the Aα chains and whether or not the peptides have been hydrolysed from the N-terminal ends of the Bβ chains. Taken from Gaffney (66).

thrombin from an inactivated precursor (15). These reactions are summarised in Figure 2. The crosslinking process seems to begin immediately prior to the visualisation of stranded fibrin polymers and continues after deposition to enhance the stabilisation of the deposited fibrin (16). The γ, glutamyl-ε,

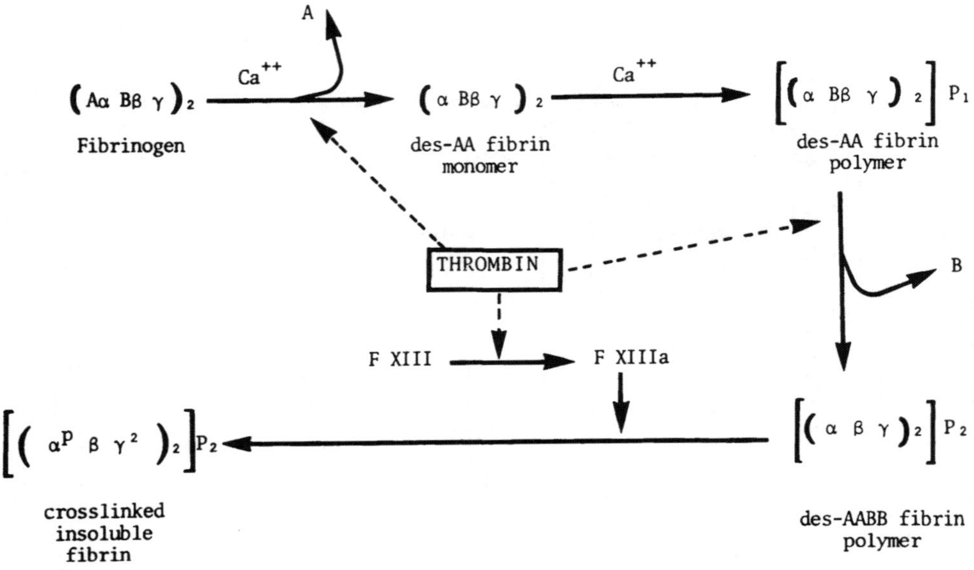

Figure 2

A proposed sequence of events outlining the effects of thrombin on fibrinogen and Factor XIII. The release of fibrinopeptides A and B takes place in two distinct steps, the former stimulating the initial polymerisation process. The Factor XIII-induced formation of γ-γ cross-links occurs first, followed by the formation of α-α crosslinks. Soluble γ-γ crosslinked fibrin (not shown in this diagram) can be an intermediate between soluble des-AABB fibrin and insoluble crosslinked fibrin. Taken from Gaffney (67).

lysine (17,18) crosslinks form rapidly between the γ chains to form γ chain dimers (γ-γ) and more slowly between the α chains to form α chain polymers (α^P) of molecular weights 400,000 and above (19). In all, six moles of crosslinks can exist per mole of fibrin subunit in totally crosslinked (TXL) fibrin, two moles of which are involved in γ chain crosslinkage and four moles in α chain interconnections (20). The upper part of Figure 3 shows a) that the γ chain crosslinking amino acids

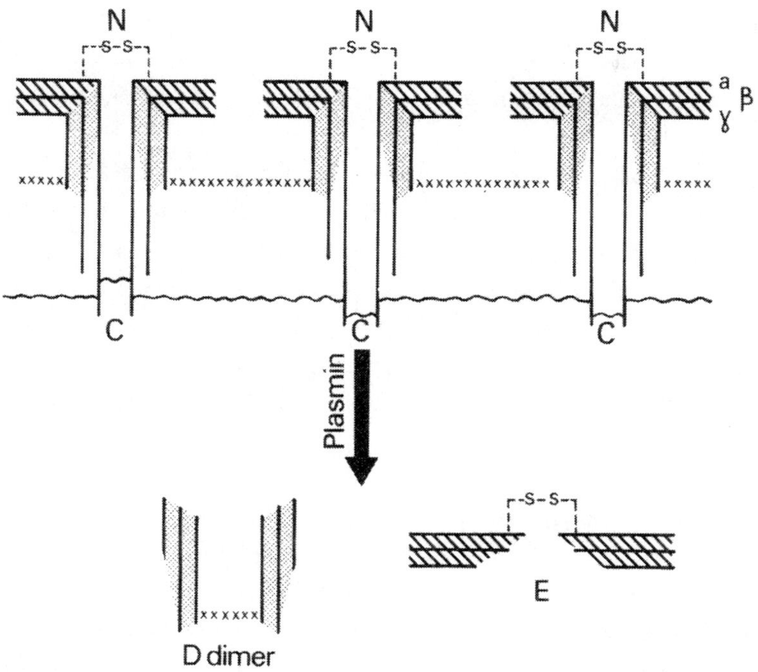

Figure 3
Schematic representation of the polypeptide chains of crosslinked human fibrin and its plasmin-mediated degradation products D dimer and fragment E. The shaded area marks the D domain while striations show the E domain. The crosslinked γ-chain dimers (γ-γ crosslinks shown by xxxx), the crosslinked α-chain polymers (α_p crosslinks shown by ∿∿), and the non-crosslinked β chains, are here related to each other schematically. Taken from Gaffney (61).

(glutamine and lysine) are located near the carboxy-terminal ends of the γ chains (21) and b) that the γ chain crosslinks are intermolecular (22). The antiparallel arrangement of the crosslinked γ chains (23) is not shown for diagrammatic convenience. The α chain crosslinks are shown near the carboxy terminal ends of the α chains also for convenience, though the precise locations are now known and are associated with the mid section of the α chain (24,25). Whether some or all the crosslinked α chains are organised in an antiparallel manner is as yet unknown. The clinical significance of fibrinopeptide A release and the formation and lysis of crosslinked fibrin will be discussed later in this presentation.

From Figures 2 and 3 and the above description it is evident that a number of fibrins can be formed from the interaction of thrombin and fibrinogen. Soluble fibrin oligomers are mostly non-crosslinked (NXL) though some forms may be γ-γ crosslinked. Insoluble fibrin can be classified into partly crosslinked (PXL) having only γ-γ crosslinks in the fibrin, and totally crosslinked (TXL) containing both γ-γ and α-α crosslinks.

Fibrin-Plasmin Interactions

Figure 2 shows that thrombin can mediate a variety of forms of fibrin, differing both in their degree of polymerisation and of Factor XIII-mediated crosslinking. While considerable investigation has been devoted to fibrinogen-plasmin interactions (for review, see 1), it seems that plasmin-fibrin interactions may be more relevant clinically. Only on rare occasions is fibrinogen directly attacked by plasmin; the commonest occurrence is during streptokinase therapy, when the fast-acting α_2-antiplasmin level is exceeded by the production of plasmin. The initial formation of fibrin and its subsequent digestion by plasmin is the commonest expression of systemic fibrinolysis and is usually expressed during disseminated intravascular coagulation (DIC). Thus an understanding of the molecular mechanisms involved during the plasmic digestion of the various forms of fibrin is paramount to any interpretation of the clinical relevance of the different molecular forms of the fibrin degradation products (FDP) found in plasma.

Chemically non-crosslinked (NXL) fibrin differs from fibrinogen only in that the latter contains the fibrinopeptides A and B. Thus it is not surprising that, like fibrinogen (4), the α chain is first digested by plasmin to an amino-terminal fragment of about 25,000 molecular weight, with the β and γ chains being digested in that order (26). Likewise, fragments similar in size to fibrinogen fragments X, Y, D and E (Fig. 1) are released from NXL-fibrin (27,28). While the chemical composition of NXL-fibrin and fibrinogen is similar, they are known to differ quite dramatically in their conformational states. The overall primary, secondary and tertiary structures of fibrinogen have been argued to dictate its degradation by a variety of different proteases, rather than being affected by the various specificities of these enzymes (29). Since the 'core' degradation fragments D and E do not differ significantly in molecular size between fibrinogen and its conformationally-different analogue, fibrin, it seems that these stable disulphide bonded regions do not take part in the fibrinogen → fibrin conformational change and primary sequences in the Aα and Bβ chains are the deciding influences affecting the specificity of the various proteases which degrade fibrinogen and fibrin. It has been argued that there is a need to distinguish the fragments derived from fibrinogen and from NXL fibrin in clinical material. This has been claimed to have been achieved using their respective binding specificities to various antisera in a radioimmunoassay (30). However, since fibrinogenolysis is arguably a very rare occurrence in vivo and since nearly all episodes of fibrinolysis involve the digestion of fibrin, the more relevant distinctions which we now seek are those involving the degradation of various forms of fibrin, notably those from crosslinked (XL) fibrin and NXL-fibrin. It is this theme which will be discussed for the remainder of this paper, examining the molecular interactions involved and introducing a hypothesis to cover their clinical relevance. From Figure 2 it is evident that XL fibrin can contain either partly crosslinked (PXL), e.g. crosslinked γ chains (γ-γ) only, or totally crosslinked (TXL) containing both crosslinked γ chains and crosslinked α chains (α polymers, α^{p}). While this discussion is mainly concerned with the reaction of plasmin with these various fibrins, it is appropriate to outline briefly the probable role which crosslinking plays in the behaviour of fibrin. Gormsen and colleagues (16) first pointed out that XL fibrin was less susceptible to plasmin digestion than NXL-fibrin and this resistance to digestion has been shown to be associated not only with α chain crosslinkage (31) but also with the complexity and completeness of this type of crosslinking (32). While γ chain crosslinking has been shown not to affect digestion rate by plasmin (32), it however endows fibrin with its insolubility in detergents (31). The latter authors examined aqueously insoluble fibrin containing varying amounts of γ chain crosslinks and found that the ratio of crosslinked to non-crosslinked γ chains was directly related to the amount of fibrin which did not dissolve in 1% monochloroacetic acid. While these data are acceptable and experimentally valid in vitro, they may not reflect the dynamic interrelationship existing between fibrin polymerisation and γ chain crosslinking in vivo. It has been shown (33) that γ chain crosslinked fibrin can exist in plasma in an aqueously soluble state while low molecular weight ($< 10^6$) fibrin aggregates (probably dimers and trimers), containing γ chain crosslinks, have been solubilised by detergent from aqueously insoluble fibrin clots (34). Thus it seems that our understanding of fibrin polymerisation and its relationship to γ chain crosslinking is quite incomplete. Certainly soluble γ chain crosslinked fibrin can exist in vivo and the insolubility of γ chain crosslinked fibrin in detergents must depend not only on crosslinking but also on both the length and complexity of the various associations of this polymer.

While the necessity of introducing the above ambiguity may become apparent during the subsequent discussion of fibrinolysis in clinical material, it is now appropriate to outline our understanding of the interactions of plasmin with XL-fibrin. The spectrum of fragments obtained from PXL- and TXL-fibrins is quite similar (35). It seems that α chain crosslinks affect fibrin lysis rate rather than having any influence on specificity. All XL-FN digests contained a major distinct fragment not found in NXL-fibrin digests (36,37,38). This was called D dimer (27) since it was composed of two D domains from adjacent fibrin molecules, joined by their crosslinked γ chain remnants (36). A scheme showing the generation of the D dimer and E fragments from XL-FN is shown in Figure 3. Later it was found that some D dimer and E molecules could be present in a digestion mixture as an electrophoretically stable but non covalent complex, known as the D dimer-E complex (39,40). Subsequent experiments in which the lysis of XL-FN was performed in serum of plasma, indicated that all the D dimer and E molecules were present as a stoichometrically-satisfied complex, having the general formula (D-D)E (41,42). The schematic representation of the D dimer-E complex, shown in Figure 4, indicates that each domain originates in a distinct fibrin molecule. More important is that the D dimer-E complex is held together by non-covalent associations present in the originating fibrin; while these associations may be synonymous with the end-to-end polymerisation sites (43) located in the D and E domains, recent work (44) has shown that they are more probably related to other polymerisation sites exposed by fibrinopeptide B release (45) and implicated in lateral fibrin aggregation. That one mole of D dimer and one mole of E form one mole of the complex has been shown (46) while the same workers have shown that conversion of the complex to free D dimer and E fragments is mediated by digestion of the E domain (47). The D dimer-E complex can dissociate in 8 M urea and reforms following removal of the detergent (41). Specific antigenic markers have been recognised in D dimer and the D dimer-E complex (48), but since these crosslink related markers represent only a small proportion of the total antigenicity of the D dimer fragment, it was difficult to devise a useful test for these fragments in vivo.

Being satisfied that the D dimer-E complex is a major product of crosslinked fibrin lysis in vivo (49,50), a variety of lytic regimens was investigated to determine if this complex was the earliest soluble expression of XL-fibrin lysis. Reganon and

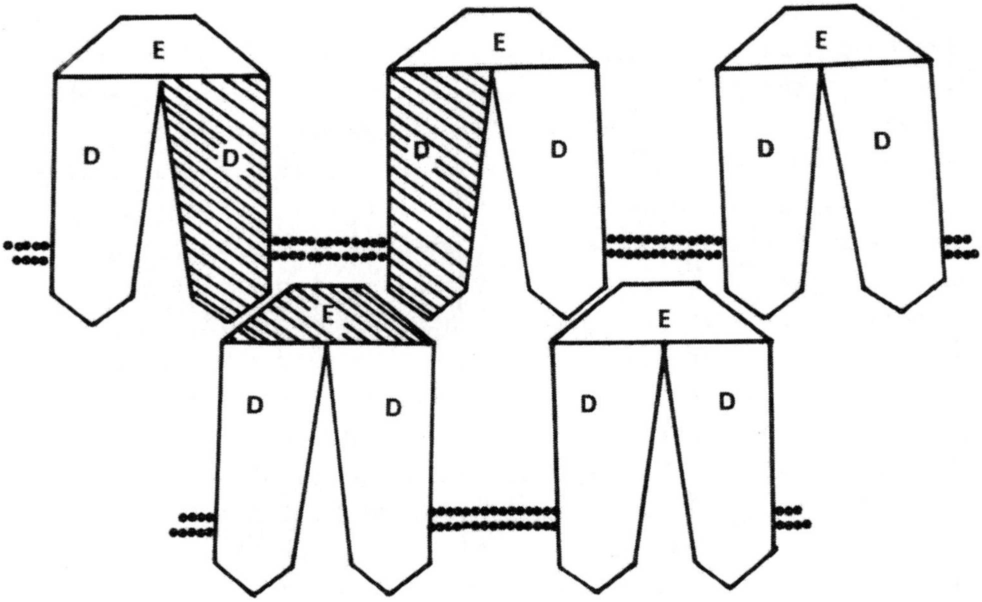

Figure 4

Schematic view of probable fibrin subunit origins of the three component parts of the D dimer-E complex. The intermolecular γ chain crosslinks are shown as dotted lines. The suggestion is inherent that the COOH-terminal ends of the γ chains of fibrin which contain the crosslinking residues are also involved in the binding of one E fragment between the two D fragments.

coworkers (51) had already shown that crosslinked Y-Y and Y-D complexes are present in crosslinked fibrin digests, besides the D-D fragment. It transpired that, besides the D dimer-E complex, higher molecular weight complexes are released from XL-FN which have been classified as crosslinked fragment X oligomers, Y-D and D dimer complexes (52). Figure 5 shows schematic structures of these fibrin degradation products, elucidated from sodium dodecyl sulphate polyacrylamide gel electrophoresis, carbohydrate staining and ^{125}I-label distribution (52). Thus it would seem that crosslinked X, Y and D

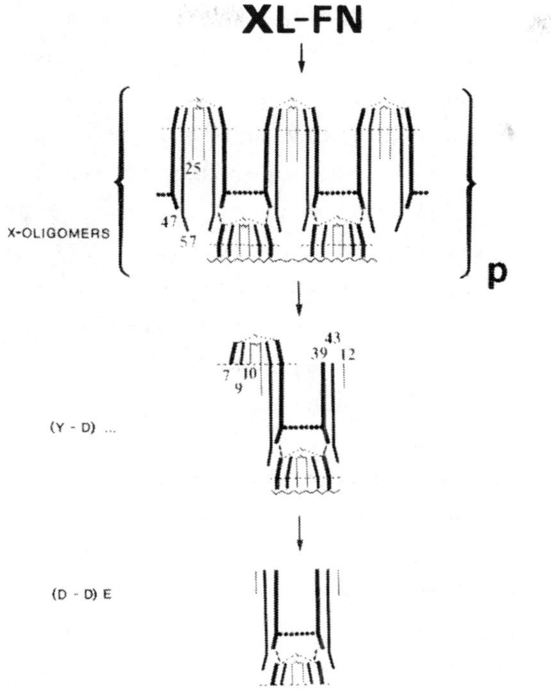

Figure 5

Proposed schematic structures of the crosslinked fragments found in XL-FN digests following a lytic regimen similar to that expected under physiological conditions. The molecular weights (x 10^{-3}) are shown only once to identify each chain and the key which is included should adequately explain the covalent (Factor XIII-mediated) and non-covalent (polymerisation) bonds holding these crosslinked aggregates together. Taken from Gaffney (52).

fragments can be released from XL-FN as soluble fragments, as had been suggested earlier (51) and in contradiction to our earlier opinions (27). This author's data (52) indicate that the crosslinked X-oligomers are the first fragments derived from XL-fibrin by plasmin, are the precursors of the D dimer-E complex, and are present during lysis of fibrin performed under near-physiological conditions in vitro. These crosslinked X oligomers have also been isolated from patient plasma during episodes of fulminant DIC (50) and other forms of intravascular coagulation (53).

The high molecular weight XL-fibrin derivatives have also been examined in detail by Francis and colleagues (54). Their digests were prepared by direct plasmin interaction, whereas the digests described above and in Figure 5 were prepared from XL-fibrin enriched with plasminogen and immersed in streptokinase solutions. This may account for the reported differences in the compositions of the digests. Francis et al. (54) conclude that, besides the (D-D)E complex, there exist the following associations in a XL-fibrin digest: (DY) (YD); (DXD) (YY) and, possibly, (YXD) (DXY). In each case the domains within parentheses are covalently linked while the association between, for instance (DXD) and (YY), is maintained by non-covalent polymerisation sites originating in the crosslinked fibrin polymer. As mentioned earlier, the end-to-end D:E associated polymerisation sites (43) may be involved but the arguments concerning the relevance of fibrinopeptide B release and secondary set of polymerisation sites may provide a more satisfactory explanation of these large non-covalent associations (44,45). Whether the high molecular weight X oligomers described in Figure 5 are compatible with some of the structures described by Francis et al. (54) remains to be seen. However, there is agreement that high molecular weight crosslinked and polymerised structures represent the earliest form of crosslinked fibrin degradation in vivo (50,53,54) and their detection in blood may further our understanding of the hypercoagulable state, whether it is a florid systemic condition as in DIC or a low grade, local episode predisposing to thrombosis.

Clinical Relevance of FDP Structures

It is agreed that disseminated intravascular coagulation (DIC) involves the generation of thrombin which converts circulating fibrinogen to fibrin. Other consequent features of thrombin generation are the adsorption of plasminogen activator and

plasminogen to the surface of forming fibrin where plasmin converts the fibrin to FDP (for review, see 55). During DIC it seems that as fibrin is formed it orchestrates its own destruction as FDP's (56). The haemorrhagic consequence of DIC may thus be associated with a) elevated FDP's and their inhibitory effects on thrombin action and platelet aggregation, b) the consumption of coagulation factors during the generation of thrombin, and c) the lack of intact circulating fibrinogen.
A number of workers have examined the subunit composition of the FDP fraction of patient plasma during DIC. During a case of amniotic fluid embolism, D dimer and the crosslinked Y-D complex were observed (49). At the time the Y-D complex was described as an X-like component. Most patients with florid DIC evidence both D dimer-E, Y-D and some crosslinked high molecular weight X-oligomers (50), while Graeff et al. (53) have described X-oligomers in the plasma of patients with various forms of obstetric disorders, clinically expressed as DIC. High molecular weight crosslinked fragments have also been described by Francis et al. (57) during both DIC and thrombosis and these fragments were suggested to be elevated during thrombolytic therapy. This latter observation is difficult to harmonise with the insensitive methodology used and our understanding of the amount of fibrin in various forms of occlusive thrombosis. It has been established that no crosslinked FDP's are formed during the DIC-like condition induced during defibrination by ancrod (coagulant from the venom of Agkistrodon rhodostoma), while patients bitten by the snake, Echis carinatus, evidence significant amounts of D dimer in their plasma (60). The direct conversion of prothrombin to thrombin by Echis venom allows the formation of Factor XIII mediated crosslinked fibrin and its subsequent digestion to crosslinked D fragments. The latter clinical condition bears some of the hallmarks of advanced DIC in contradistinction to the benign effect of ancrod as a defibrinating agent.
Whereas the crosslinked high molecular weight fibrin fragments, be they crosslinked as X-oligomers (52), or as (YXD) (DXY), etc.

complexes (54), have been elaborated mainly during the in vitro digestion of XL-fibrin, their in vivo expression seems to have been mostly described during DIC. Figure 6 suggests pathways by which crosslinked X-oligomers may be formed during the competitive interactions of thrombin and plasmin for fibrinogen and fibrin. It is possible that these fragments have been only observed during DIC because the latter is a systemic effect and the amount of FDP in the circulation is great. In Figure 6 is also shown the production of crosslinked X-oligomers from the XL-fibrin, which is the known (61) condition of fibrin in an occlusive thrombus. It is expected that very sensitive assays are needed to detect the small amounts of X-oligomers present in the circulation following disintegration of a local thrombus.

Possibly a more important clinical issue is the detection of the hypercoagulable or prethrombotic state and to query whether the high molecular weight crosslinked fibrin fragments described above can play any diagnostic role. It seems reasonable to suggest that the thrombin-initiated effects described above during systemic DIC must also apply at a local level in the initiation of thrombosis, especially at the venous side of the circulation. Thus one of the more popular diagnostic tools in use at the moment is the measurement of fibrinopeptide A (FPA) by radioimmunoassay (62). As in DIC, the localised formation of fibrin stimulates the fibrinolytic system and it is probable that low levels of crosslinked X-oligomers are formed. As Figure 7 suggests, the more thrombin is produced locally the greater is the possibility of thrombus formation. It is here suggested that the detection of crosslinked X oligomers may auger the onset of thrombosis at some point in the vascular tree. It can also be argued that this form of diagnosis may be far more relevant to the prethrombotic state than fibrinopeptide A assay. Indeed, it has been suggested (53) that crosslinked X-oligomers may act as a focus for thrombus formation but this would pre-suppose that certain forms of DIC predispose necessarily to thrombosis.

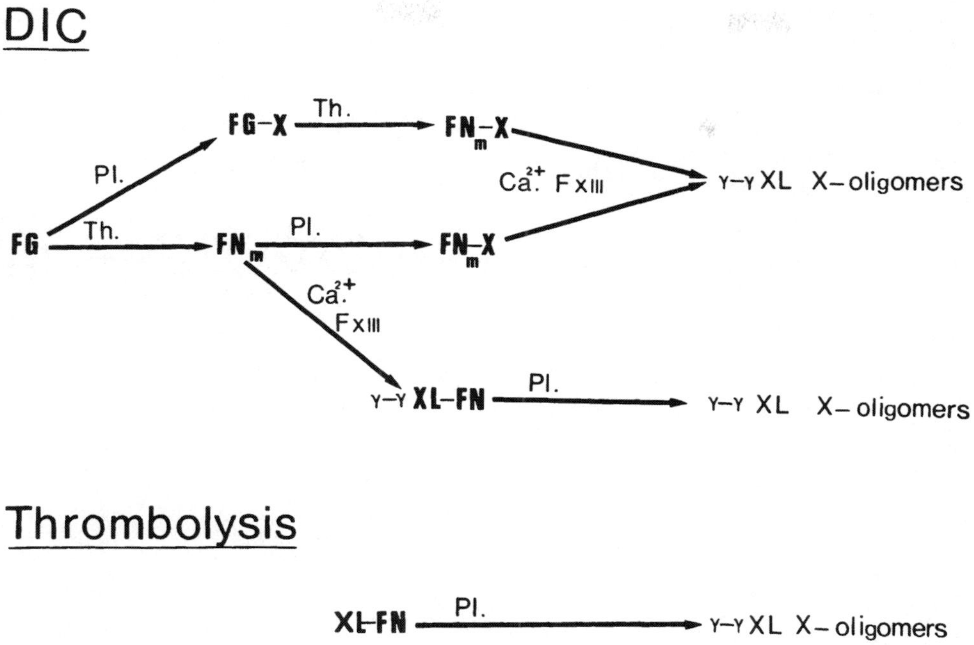

Figure 6

A scheme suggesting a variety of pathways by which crosslinked X-oligomers might be found in blood during disseminated intravascular coagulation (DIC). Similar crosslinked fragments would be expected during thrombolysis of preformed fibrin-containing thrombi, as has been demonstrated in vitro and referred to in the accompanying text. Taken from Gaffney (52).

Since it has been observed that X-oligomers (or indeed any high molecular weight crosslinked complexes of FDP) seem to accompany the onset of severe DIC and since similar fragments have been proposed as diagnostic tools for the prethrombotic state, assays are being developed for their specific and sensitive detection in plasma. Antisera raised in rabbits to human X oligomers crossreacted poorly with human fibrinogen in a novel form of radioimmunometric assay (63) while non-crosslinked FDP's (X, Y, D and E) interfered with the competitive assay of

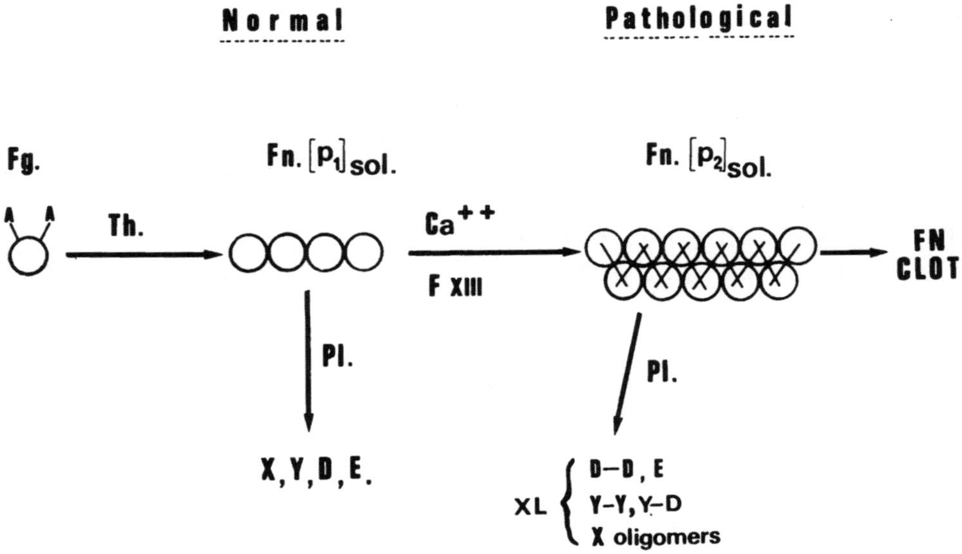

Figure 7

Diagram outlining possible sequences of events in vivo during the hypercoagulable or prethrombotic state. After the removal of fibrinopeptide A from fibrinogen (Fg) by thrombin (Th) soluble fibrin polymers (Fn_{p1}) form which can be converted by plasmin (P1) to fibrin degradation products (X, Y, D, E). Such a balanced series of reactions is thought to occur at a local level regularly in healthy subjects. Should either the available Th production be excessive or the level of plasmin depressed, large fibrin polymers (Fn_{p2}) may form and crosslink under the influence of Ca^{++} and Factor XIII. These crosslinked fibrin polymers can act as a focus for clot formation or can be digested by plasmin to generate crosslinked fragments - for example, (D-D)E, Y-Y, Y-D, and X-oligomers. The diagram proposes that the detection in plasma of XL fibrin polymers or their plasmin-mediated fragments defines a 'pathological' prethrombotic state.

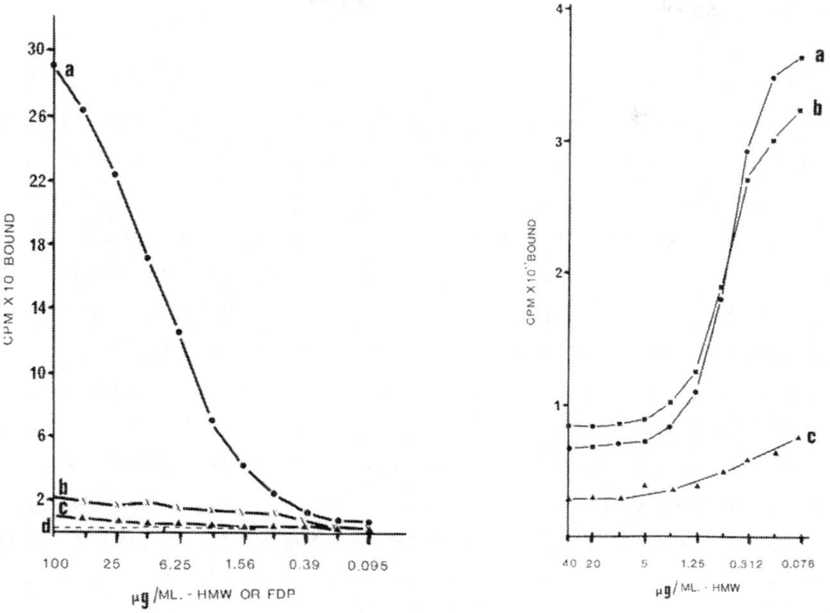

Figure 8

Left: binding curves of an antiserum to HMW fibrin fragments to various concentrations (100-0.95 μg/ml) of immobilised a) HMW fraction, b) fibrinogen FDP, c) fibrin FDP and d) fibrinogen. Right: competitive radioimmunometric assay curves between HMW fraction and its antiserum in a) plasma, b) serum and c) in the presence of 2 mg/ml fibrinogen-FDP in plasma.

X-oligomers in serum and plasma (64) (Fig. 8). Attempts to raise specific antisera to (D-D)E which do not react with fibrinogen have not been successful in this author's laboratory while specific antigenic markers have been recognised by others (48) in D dimer and the D dimer-E complex; however, these have not been found appropriate to establish a worthwhile radioimmunoassay system.

Concluding Remarks

While there is some dispute (compare 52 with 54) about the precise structure of the earliest fragments derived from crosslinked fibrin, there is a general consensus concerning their importance in the diagnosis of impaired haemostasis. Specific and sensitive assays are yet to be developed. It seems that DIC is nearly always evidenced by measurable levels of D dimer in plasma (65) by quite insensitive methods. It is possible that the severity of DIC may be measured by the level and size of the high molecular complexes (e.g. X-oligomers) in plasma but this has as yet not been shown. Based on the acceptable proposition that the mechanisms associated with the prethrombotic local event are similar to those which take place during systemic DIC, the measurement of nanogram levels of crosslinked X-oligomers may represent a far more reliable diagnostic test for thrombosis than the current FPA assay. Such assays need to be developed, introduced into a simple form of radioimmunoassay, and tested on a variety of clinical material.

References

1. Gaffney, P.J.: In: The Biochemistry, Physiology and Pathology of Haemostasis, Ed. D. Ogston and B. Bennett, pp 105-108, John Wiley, New York (1977).
2. Nussenzweig, V., Seligman, M., Pelmont, J., Grabar, P.: Ann.Inst.Pasteur 100, 377-387 (1961).
3. Marder, V.J., Shulman, N.R., Carroll, W.R.: J.Biol.Chem. 244, 2111-2119 (1969).
4. Gaffney, P.J., Dobos, P.: FEBS Lett. 15, 13-16 (1971).
5. Shainoff, J.R., Lahiri, B., Bumpus, F.M.: Thromb.Diath. Haemorrh. Suppl. 39, 203-217 (1970).
6. Mosesson, M.W., Finlayson, J.S., Umfleet, R.A., Galanakis, D.: J.Biol.Chem. 247, 5210-5219 (1972).
7. Mosesson, M.W., Finlayson, J.S., Galanakis, D.K.: J.Biol. Chem. 248, 7913-7929 (1973).

8. Budzynski, A.Z., Marder, V.J., Shainoff, J.R.: J.Biol.Chem. 249, 2294-2302 (1974).

9. Gaffney, P.J.: Biochim.Biophys.Acta 263, 543-458 (1972).

10. Gollwitzer, R., Timpl.: R. FEBS Lett. 58, 269-272 (1975).

11. Harfenist, E.J., Canfield, R.E.: Biochemistry 14, 4110-4117 (1975).

12. Blombäck, M., Blombäck, B., Holmqvist, H.: Thrombos.Res. 8, 567-571 (1976).

13. Bettelheim, F.R., Bailey, K.: Biochim.Biophys.Acta. 9, 578-579 (1952).

14. Laurent, T.C., Blombäck, B. Acta.Chem.Scand. 12, 1875-1877 (1958).

15. Lorand, L., Konishi, K.: Arch.Biochem.Biophys. 105, 58-67 (1964).

16. Gormsen, J., Fletcher, A.P., Alkjaersig, N., Sherry, S.: Arch.Biochem.Biophys. 120, 654-665 (1967).

17. Matacic, S., Loewy, A.G.: Biochem.Biophys.Res.Commun. 30, 356-362 (1968).

18. Pisano, J.J., Finlayson, J.S., Peyton, M.P.: Science 160, 892-893 (1968).

19. McKee, P.A., Mattock, P., Hill, R.L.: Proc.Natl.Acad.Sci. USA 66, 738-744 (1970).

20. Pisano, J.J., Finlayson, J.S., Peyton, M.P., Nagai, Y.: Proc.Natl.Acad.Sci.USA 68, 770-772 (1971).

21. Chen, R., Doolittle, R.F.: Proc.Natl.Acad.Sci.USA 66, 472-479 (1970).

22. Doolittle, R.F., Chen, R., Lau, F.: Biochem.Biophys.Res. Commun. 44, 94-100 (1971).

23. Chen, R., Doolittle, R.F. Biochemistry 10, 4486-4491 (1971).

24. Finlayson, J.S., Mosesson, M.W.: Thrombos.Res. 2, 467-478 (1973).

25. Takagi, T., Doolittle, R.F.: Biochemistry 14, 940-946 (1975).

26. Gaffney, P.J.: Nature(New Biol.) 234, 281-283 (1971).

27. Gaffney, P.J.: Thrombos.Res. 2, 201-218 (1973).

28. Pizzo, S.V., Schwartz, M.L., Hill, R.L., McKee, P.A.: J.Biol.Chem. 248, 4574-4583 (1973).

29. Nakamura, S., Iwanaga, S., Suzuki, T.: Biochim.Biophys. Acta. 336, 86-99 (1974).

30. Edgington, T.W., Plow, E.: In: Haemostasis and Thrombosis, Eds. G.G. Neri-Serneri, C.R.M. Prentice, pp 27-43, Academic Press, London (1979).

31. Schwartz, M.L., Pizzo, S.V., Hill, S.L., McKee, P.A.: J.Clin.Invest. 50, 1506-1513 (1971).

32. Gaffney, P.J., Whitaker, A.N.: Thrombos.Res. 14, 85-94 (1979).

33. Kierulf, P.: Thrombos.Res. 1, 527-532 (1972).

34. Moroi, M., Inoue, N., Yamasaki, M.: Biochim.Biophys. Acta. 379, 217-226 (1975).

35. Gaffney, P.J., Lane, D.A., Brasher, M.: Clin.Sci.Mol.Med. 49, 149-156 (1975).

36. Gaffney, P.J., Brasher, M.: Biochim.Biophys.Acta. 295, 308-313 (1973).

37. Pizzo, S.V., Taylor, L.M., Schwartz, M.L., Hill, R.L., McKee, P.A.: J.Biol.Chem. 248, 4584-4590 (1973).

38. Kopec, M., Teisseyre, E., Dudek-Wojciechowska, G., Klocewiak, M., Pankiewicz, A., Latallo, Z.S.: Thrombos. Res. 2, 283-291 (1973).

39. Hudry-Clergeon, G., Patural, L., Suscillon, M.: Pathol. Biol. 22 (Suppl.), 47-52 (1974).

40. Gaffney, P.J., Lane, D.A., Kakkar, V.V., Brasher, M.: Thrombos.Res. 7, 89-99 (1975).

41. Gaffney, P.J.: In Progress in Chemical Fibrinolysis and Thrombolysis, Eds. J.F. Davidson, V. Cepelak, M.M. Samama, P.C. Desnoyers, Vol. 4, pp 424-432, Churchill Livingstone, Edinburgh (1979).

42. Gaffney, P.J., Joe, F.: Thrombos.Res. 15, 673-687 (1979).

43. Kudryk, B.J., Collen, D., Woods, K.R., Blombäck, B.: J.Biol.Chem. 249, 3322-3325 (1974).

44. Olexa, S.A., Budzynski, A.Z.: Biochemistry 19, 647-651 (1980).

45. Shainoff, J.R., Dardik, B.N.: Science 204, 200-202 (1979).

46. Olexa, S.A., Budzynski, A.Z.: Biochemistry 18, 991-995 (1979).

47. Olexa, S.A., Budzynski, A.Z., Marder, V.J.: Biochim. Biophys.Acta. 576, 39-50 (1979).

48. Budzynski, A.Z., Marder, V.J., Parker, M.E., Shames, P., Brizuela, B.S., Olexa, S.A.: Blood 54, 794-804 (1979).

49. Gaffney, P.J.: Clin.Chim.Acta. 65, 109-115 (1975).

50. Whitaker, A.N., Rowe, E.A., Masci, P., Gaffney, P.J.: Thrombos.Res. 18, 453-459 (1980).

51. Reganon, E., Vila, V., Aznar, J.: Thrombos.Haemos. 40, 368-376 (1978).

52. Gaffney, P.J., Joe, F., Mahmoud, M.: Thrombos.Res. 20, 647-662 (1980).

53. Graeff, H., Hafter, R., Bachmann, L.: Thrombos.Res. 16, 313-328 (1979).

54. Francis, C.W., Marder, V.J., Barlow, G.H.: J.Clin.Invest. 66, 1033-1043 (1980).

55. Collen, D.: Thrombos.Haemos. 43, 77-89 (1980).

56. Gaffney, P.J.: In: Haemostasis and Thrombosis, Ed. A.L. Bloom and D.P. Thomas, pp 198-224, Churchill Livingstone 1981

57. Francis, C.W., Marder, V.J., Martin, S.E.: Blood 54, 1282-1295 (1979).

58. Prentice, C.R.M., Edgar, W., McNicol, G.P.: Brit.J. Haematol. 27, 77-87 (1974).

59. Gaffney, P.J., Brasher, M.: Nature 251, 53-54 (1974).

60. Edgar, W., Warrell, M.J., Warrell, D.A., Prentice, C.R.M.: Brit.J.Haematol. 44, 471-481 (1980).

61. Gaffney, P.J., Brasher, M., Lord, K., Strachan, C.J.L., Wilkinson, A.R., Kakkar, V.V., Scully, M.F.: Cardiovasc. Res. 10, 421-426 (1976).

62. Nossel, H.L., Ti, M., Kaplan, K.L., Spanondis, K., Soland, T., Butcher, V.P. Jr.: J.Clin.Invest. 58, 1136-1144 (1976).

63. Gaffney, P.J., Joe, F., Mahmoud, M., Fossati, C.A., Spitz, M.: Thrombos.Res. 19, 815-822 (1980).

64. Gaffney, P.J., Joe, F., Mahmoud, M., Spitz, M.: Proc. Vth Int.Conf. on Synthetic Fibrinolytic and Thrombolytic Agents (Malmö, June 17-20, 1980). Abstr. 143 (1980).

65. Lane, D.A., Kakkar, V.V.: In: Progress in Chemical Fibrinolysis and Thrombolysis, Eds. J.F. Davidson, V. Cepelak, M.M. Samama, P.C. Desnoyers, Vol. 4, pp 433-438, Churchill Livingstone, Edinburgh (1979).

66. Gaffney, P.J.: Haemostasis 6, 2-25 (1977).

67. Gaffney, P.J.: Haematologica 65, 612-625 (1980).

FIBRINOGEN IN LIVER CIRRHOSIS. FIBRIN MONOMER COMPLEXES AND DYSFIBRINOGENAEMIA WITH HIGH SIALIC ACID CONTENT

Sergio Coccheri, Gualtiero Palareti and Mario Poggi

Department of Angiology and Blood Coagulation of the University Hospital S. Orsola, Bologna, Italy

The disturbance of blood coagulation and haemostasis in chronic liver disease is based on multifactorial mechanisms. The decreased synthetic ability for proteins involves both Vitamin K and/or liver-dependent clotting factors and the main coagulation inhibitor, antithrombin III. The increase of factor VIII, the activation of fibrinolysis, the decreased production of natural fibrinolysis inhibitors, and frequently thrombocytopenia complicate the coagulation picture.

In recent years, special attention has been focused on the pathophysiology of fibrinogen in cirrhosis of the liver.

Evidence of a shortened survival time of fibrinogen has been supplied by some Authors (1). The reported correction with heparin of this accelerated catabolism (2) suggested the possibility of discrete or low-grade intravascular coagulation phenomena. However, the increased catabolism of fibrinogen in liver cirrhosis may be explained by alternative mechanisms as extravascular loss and/or coagulation (3), presence of an abnormal fibrinogen population with a faster catabolic rate, and even physical degradation of the protein due to altered hemodynamics in the microcirculation (4).

In previous investigations (5) we measured soluble fibrin monomer complexes (SFMC) in moderate and severe liver cirrhosis by

Fibrinogen - Recent Biochemical and Medical Aspects

means of gel filtration of plasma on a sepharose column (5), followed by determination of the fibrinogen related material (FRM) in the eluates by means of the staphylococcal clumping test and rocket immunoelectrophoresis. Our results showed that the relative amount of high m.w. SFMC was significantly increased in liver cirrhosis, especially in severe liver disease. In a further series of experiments (7), fibrinogen of plasma of patients with severe cirrhosis was purified by means of β-alanine and glycine precipitation (8). After filtration on sepharose gel, the eluates were concentrated, dialysed and submitted to polyacrylamide gel electrophoresis with sodium dodecylsulfate (SDS PAGE), with and without reduction with mercaptoethanol.

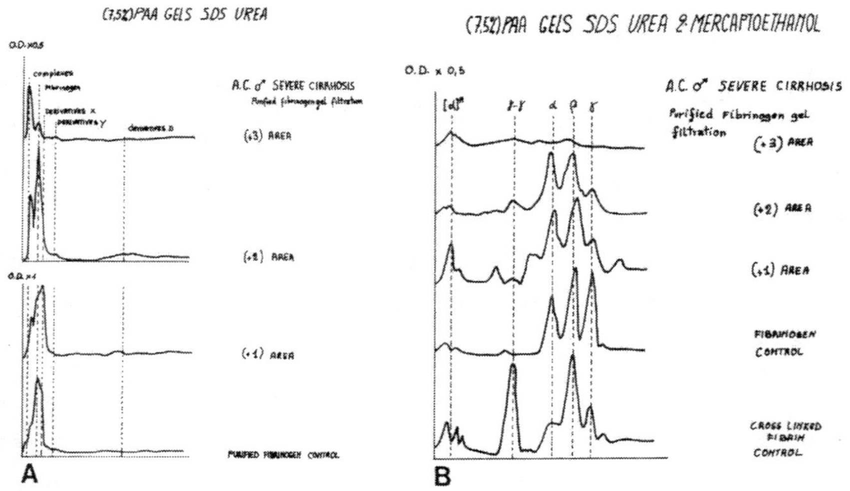

Fig. 1 - SDS PAGE of chromatographic eluates of plasma of a patient with severe liver cirrhosis (see text). 1 A: non-reduced, and 1 B reduced PAGE.

With SDS PAGE we have studied the composition of FRM collected from three equal aliquots of eluates, corresponding to three "areas" comprised between the void volume and the fibrinogen peak in the fibrinogen elution curve. As seen in fig. 1 A, in "area 3" which is adjacent to the void volume, only high molecular weight polymers could be detected, while in areas 2 and 1 SFMC were composed also by fibrinogen and X derivatives. After reduction (fig. 1 B), it was possible to detect in 3 examined cases of severe liver cirrhosis the presence of γ-γ dimers and α oligomers in SFMC, thus showing that cross-linked FRM may be present in this condition, as a marker of F XIII activation.

Recently, we have concluded a collaborative study (9) in cooperation with the IIIrd Medical Clinic of the University of Milan, in which 19 patients with liver cirrhosis classified into moderate and severe cases were simultaneously investigated with sepharose gel chromatography and radioimmunological determination of Fibrinopeptide A (10). We were able to confirm the existence of a significant increase of high m.w. FRM (HMWF) and also of low m.w. FRM (LMWF) (Table 1), indicating that the molecular weight distribution of FRM populations in liver cirrhosis is less concentrated around the median m.w. of the protein in comparison to the controls. FPA was definitely increased ($\bar{m}$ + 2 SD of the control group) in 6 out of 19 cases, with peaks up to 94 ng/ml; the median value found in cirrhotic patients was higher than that of the control group, but the difference between the two groups and in the subgroups was not significant using a non-parametric test (Wilcoxon test) (Table 2). FPA correlated positively ($p < 0.01$) with fibrinogen level, but did not correlate with HMWF of LMWF. In turn, HMWF showed a trend

	HMWF %	LMWF %
A) All Cirrhotics (n = 19)	10.4±2.8	6.8±1.8
M) Moderate (n = 11)	10.8±3.2	6.6±1.6
S) Severe (n = 8)	9.9±2.4	6.7±2.2
C) Controls (n = 16)	6.9±1.4	5.3±1.3
STATISTICAL EVALUATION		
	HMWF	LMWF
A versus C	$p < 0.01$	$p < 0.05$
M versus C	$p < 0.01$	$p < 0.05$
M versus S	n.s.	n.s.

Table 1 - HMWF and LMWF in Liver Cirrhosis (n = 19)

	MEDIAN	RANGE
All cirrhotics (n = 19)	2.8	< 0.6 to 94
Moderate (n = 11)	3	< 0.6 to 94
Severe (n = 8)	2.7	< 0.6 to 62
Controls (n = 16)	2.2	< 0.6 to 4.1

Statistical evaluation: Wilcoxon test
non-significant for all comparisons

Table 2 - FPA (ng/ml) in patients with Liver Cirrhosis

to a negative correlation ($p = 0.05$) with AT III values in the subgroup of moderate cirrhosis (Table 4).
The lack of correlation between HMWF and FPA could be merely a consequence of a different catabolic rate of the two materials especially in liver cirrhosis. However, it may rather suggest that mechanisms other than a direct intravascular thrombin effect could be involved, like intravascular reflections of extravascular coagulation phenomena (3), or alterations in the fibrinogen molecule (11), or facilitation of complex formation due to dysproteinaemia (12).
Another part of our investigations is related to the polymerisation defect of fibrinogen in liver cirrhosis (7). Polymerisation curves as determined photometrically after thrombin or thrombin-like enzymes addition to plasma are grossly altered in liver cirrhosis (7,13,14). This alteration is often , but not always, corrected in purified fibrinogen, showing that antipolymerizing agents or an altered protein environment certainly play a role in the polymerisation defect (7). However,thrombin times are never completely corrected after purification, and there are cases in which the polymerisation curve remains altered even in purified fibrinogen, thus suggesting co-existence of a qualitative defect in the fibrinogen molecule. The presence of "very low" m.w. fibrinogen detected by some Authors (15,16) as indicating a population of fibrinogen more soluble and less prone to polymerisation, derived from partial proteolysis of α chains, was confirmed by us only in 2 out of 10 patients with severe liver cirrhosis.
Starting from reported observations (17,18) of an increase in sialic acid in the carbohydrate moiety of the fibrinogen molecule in a number of patients with liver disease, as being pro-

bably connected with the polymerisation defect, we have measured the sialic acid content of purified fibrinogen (FSA) with the thiobarbiturate method (19) in a group of 22 patients with liver cirrhosis, classified into "moderate" and "severe" according to standardized criteria. FSA has been found increased with high statistical significance in the subgroup of "severe" cirrhosis both in comparison to the controls and the "moderate" patients (Table 3). FSA was inversely correlated with the fibrinogen level (Table 4), was not correlated with FPA, neither with the prolongation of Thrombin time, thus confirming that the polymerisation defect in these conditions is a multifactorial phenomenon.

Increased FSA is therefore a feature of severe liver cirrhosis with hypofibrinogenaemia.

		FSA mg per 100 mg fibrinogen	Thrombin Time Ratio
(A) All cirrhotics	(n = 22)	0.79±0.17	1.40±0.3
(M) Moderate	(n = 11)	0.69±0.10	1.41±0.2
(S) Severe	(n = 11)	0.90±0.16	1.39±0.4
(C) Controls	(n = 10)	0.70±0.13	1.00±0.12

STATISTICAL EVALUATION	FSA	Thrombin Time
A versus C	n.s.	$p < 0.001$
M versus C	n.s.	$p < 0.001$
S versus C	$p < 0.01$	$p < 0.001$
M versus S	$p < 0.005$	n.s.

Correlations between FSA and Thrombin Time: n.s. in any comparison

Table 3 - Fibrinogen Sialic Acid (FSA) and Thrombin Time in Liver Cirrhosis.

(n = 19)	FPA	to	Fibrinogen (FPT)	0.59^^
(n = 19)	FPA	to	HMWF,LMWF, AT III	n.s.
(n = 19)	HMWF	to	Fibrinogen (FPT)	n.s.
(n = 11)	HMWF	to	AT III	-0.42(^) in "moderate" cirrh.
(n = 22)	FSA	to	Thrombin Time	n.s.
(n = 22)	FSA	to	Fibrinogen (Imm)	-0.59^^
(n = 22)	FSA	to	Fibrinogen (Photom.)	-0.55^
(n = 19)	FSA	to	FPA	n.s.
(n = 22)	Thrombin Time	to	Fibrinogen (Imm, Photom.)	n.s.

n.s. = non significant

(^) = $p \approx 0.05$

^ = $p < 0.05$

^^ = $p < 0.01$

Table 4 - Correlations in Liver Cirrhosis

The increase in FSA might be attributed to different mechanisms, on which interesting working hypotheses may be developed. The first possibility to be considered is that the increase of FSA might be relative rather than absolute, due to partial degradation of Fibrinogen mainly involving the non-sialilated α chain. This seems to be unlikely according to our forementioned experience and also that of Soria (17). Possible mechanisms for a true increase of FSA are: a) production of hypersialilated fibrinogen in similarity to production of other hypersialilated proteins in similar conditions (20); b) selection of hypersia lic fibrinogen populations due to faster consumption, catabo -

lism, or complexation of the non-hypersialic fibrinogen (21); c) sialilation of circulating fibrinogen due to increased production of sialil-transferases (22).

We are presently investigating on the clinical significance of the high FSA, and especially on: possible relations to liver cell regeneration (23), and consequences of the association of hypofibrinogenaemia and polymerisation defect for prediction and prognosis of haemorrhage in liver cirrhosis (13).

References

1. Tytgat G.N., Collen D., Verstraete M. et al.: J. Clin. Invest. 50, 1690 (1971).

2. Coleman M., Finlayson N., Bettigole R.E. et al.: Ann. Int. Med. 83, 79 (1975).

3. Rüegg R., Straub P.W.: J. Lab. Clin. Med. 95, 842 (1980).

4. Charm E.S., Wong B.L.: Science, N.Y. 170, 466 (1970).

5. Coccheri S., Palareti G., Dalmonte P.R., et al.: Haemostasis 8, 8 (1979).

6. Fletcher A.P., Alkjaersig N.: Clin. Res. 18, 531 (1970).

7. Palareti G., Poggi M., Coccheri S.: Proceedings V Int. Congress on Thromboembolism Bologna 1978. Published by "Qua — derni della Coagulazione", Pisa, page 186 (1980).

8. Graeff H., von Hugo R.: Thrombos. Diathes. Haemorrh. 27,610 (1972).

9. Coccheri S., Mannucci P.M., Palareti G. et al.:in print (1981).

10. Nossel H.L., Yudelman I., Canfield R.E., et al.: J. Clin. Invest. 54, 43 (1974).

11. Soria J., Soria C., Samama M. et al.: Coagulation 3, 37 (1970).

12. Matsuda M., Saida T., Hasegawo R.: Thrombosis Research 7, 361 (1965).

13. Green G., Thomson J.M., Dymock I.W. et al.: Brit. J.Haematol. 34, 425 (1976).

14. Lane D.A., Scully M.F., Thomas D.P. et al.: Brit. J.Haematol. 35, 301 (1977).

15. Lipinski B., Lipinska I., Nowati A., Gurewich V.: J. Lab. Clin. Med. 90, 187 (1977).

16. Gurewich V., Hyde E., Mobarhan S. et al.: Proceedings V International Congr. on Thromboembolism, Bologna 1978; Quaderni della Coagulazione, Pisa, page 177 (1980).

17. Soria J., Soria C., Samama M. et al.: Proceedings V Int. Congress on Thromboembolism Bologna 1978. Published by"Quaderni della Coagulazione", Pisa, page 191 (1980).

18. Martinez J., Palascak J.E., Kwasniak D.: J. Clin. Invest. 61, 535 (1978).

19. Warren L., J. Biol. Chem. 8, 234 (1959).

20. Burger R.L., Waxman S., Gilbert M.S. et al.: J. Clin. Invest. 56, 1262 (1965).

21. Morell A.G., Gregoriadis G., Scheinberg I.H. et al.: J. Biol. Chem. 246, 1461 (1970).

22. Serafini-Cessi F.: Biochem. J. 166, 381 (1977).

23. Barr R.D., Allardyce M., Brunt P.W. et al.: J. Clin. Path. 31,89 (1978).

INFLUENCE OF VARIOUS THROMBIN FORMATION RATES ON THE FIBRIN STABILIZATION PROCESS

Monika Barthels

Division of Haematology, Medical School Hannover, D-3000 Hannover, Germany

Introduction

Plasma factor XIII is activated by thrombin and consumed during the coagulation process, in normal serum only low concentrations of factor XIII being present. The amount of activated factor XIIIa obtained is directly related to the amount of thrombin used in the assay (7, 9). We asked ourselves the question if the activation of factor XIII or the kinetics of fibrin stabilization representing factor XIII activation were also influenced by the rate of thrombin formation.

Looking for a test system, the biochemical tests with C^{14} incorporation were not suitable, because the necessary heat defibrination destroys several components of thrombin formation in plasma. A factor XIII consumption test parallel to prothrombin consumption also was not useful, as factor XIII is consumed not only by thrombin but also by other factors (9). We therefore modified an assay for quantitative estimation of factor XIII (8) measuring the increasing stability of fibrin in monochloroacetic acid in relation to the thrombin formation rate.

Method

Two-stage assay for the fibrin stabilization rate.
In the first step a constant fibrin clot was formed as a substrate by

Fibrinogen - Recent Biochemical and Medical Aspects

adding ancrod (Arwin® KNOLL AG/Ludwigshafen) or thrombincoagulase® (BOEHRINGER/Mannheim) to citrated plasma or to a purified test system consisting of placenta factor XIII (Fibrogammin® BEHRINGWERKE/Marburg), fibrinogen free of factor XIII (BEHRINGWERKE/Marburg), thrombinum purum (BEHRINGWERKE/Marburg). In the second step $CaCl_2$ was added, thereby starting thrombin formation and factor XIII activation in plasma or factor XIII activation alone in the purified system. At certain times up to 120 min monochloroacetic acid was added and the stability of the clot judged visually after 90 min incubation (fig. 1).

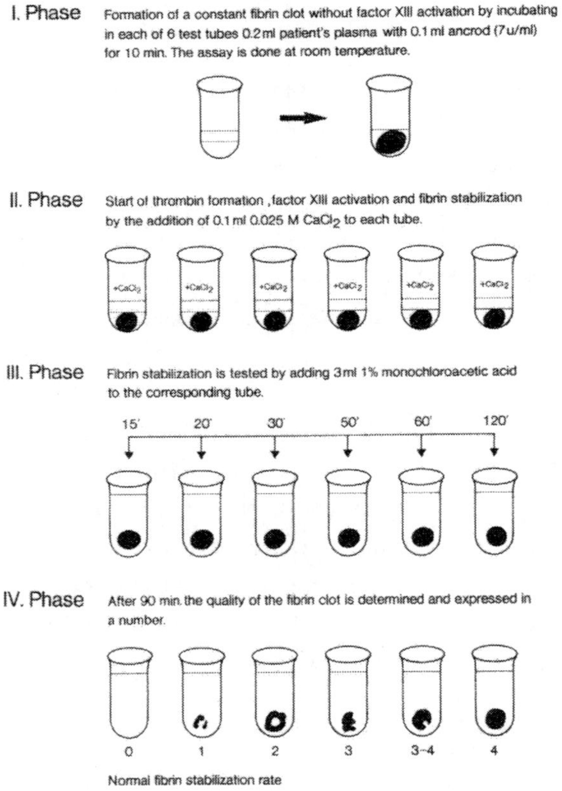

Fig. 1 Fibrin stabilization rate - two-stage assay.

Results

Checking the influence of the various partners on the test system, it could be seen that fibrinogen concentrations between 0.5 and 4.0 g/l did not influence the stabilization rate, neither did platelets at a concentration of more than 30 000/µl. In plasma of patients with severe thrombocytopenia fibrin stabilization was delayed, but could be normalized by adding phospholipids (2). Ancrod at a concentration of more than 20 u/ml improved the fibrin stabilization process. But with the concentration of 7 u/ml, generally used here, the absence of factor XIII in the purified test system led to a complete dissolution of the ancrod-fibrin clot even after 2 h incubation (see fig. 2, left side). Using thrombincoagulase

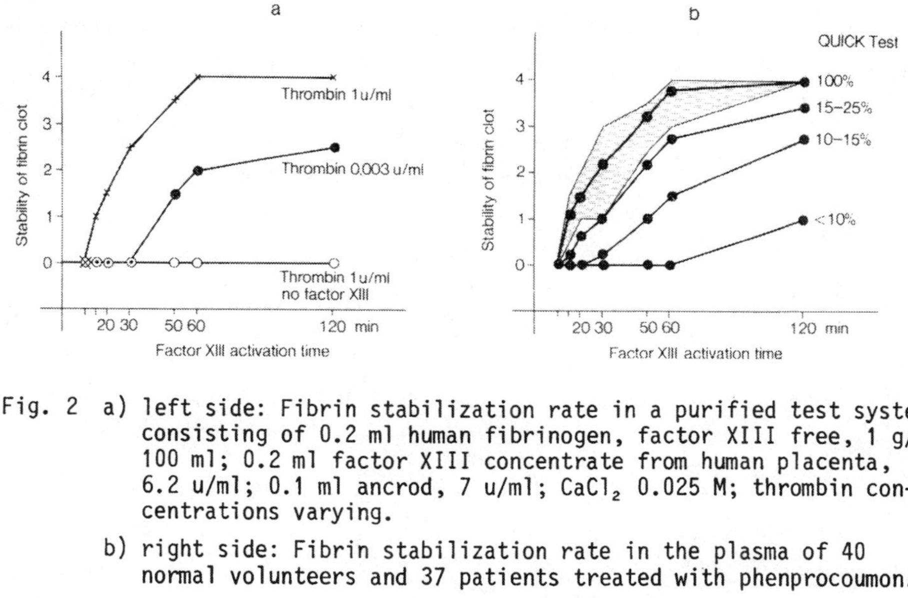

Fig. 2 a) left side: Fibrin stabilization rate in a purified test system consisting of 0.2 ml human fibrinogen, factor XIII free, 1 g/ 100 ml; 0.2 ml factor XIII concentrate from human placenta, 6.2 u/ml; 0.1 ml ancrod, 7 u/ml; $CaCl_2$ 0.025 M; thrombin concentrations varying.

b) right side: Fibrin stabilization rate in the plasma of 40 normal volunteers and 37 patients treated with phenprocoumon.

without factor XIII the clot was equally unstable. On the other hand, with optimal concentrations of factor XIII, thrombin and calcium ions the ancrod-fibrin clot became increasingly stable during the 2 h incubation with $CaCl_2$. A lower thrombin concentration led to a less stable fibrin clot. Between the upper and the lower curve the test system was established.

In citrated plasma the fibrin stabilization rate of normal volunteers reached the optimum in most cases after 60 min incubation with $CaCl_2$, demonstrating slight variations in the normal range when using an ancrod-fibrin clot (see fig. 2, right side).

Investigating the plasma of patients treated with phenprocoumon it could be demonstrated that the fibrin stabilization was related to the concentration of active prothrombincomplex in the patient's plasma, expressed by QUICK's test. In the so called 'therapeutical range' the fibrin stabilization rate was nearly normal, whereas it was severely delayed in plasmas with QUICK's test below the therapeutical range (see fig. 2, right side).

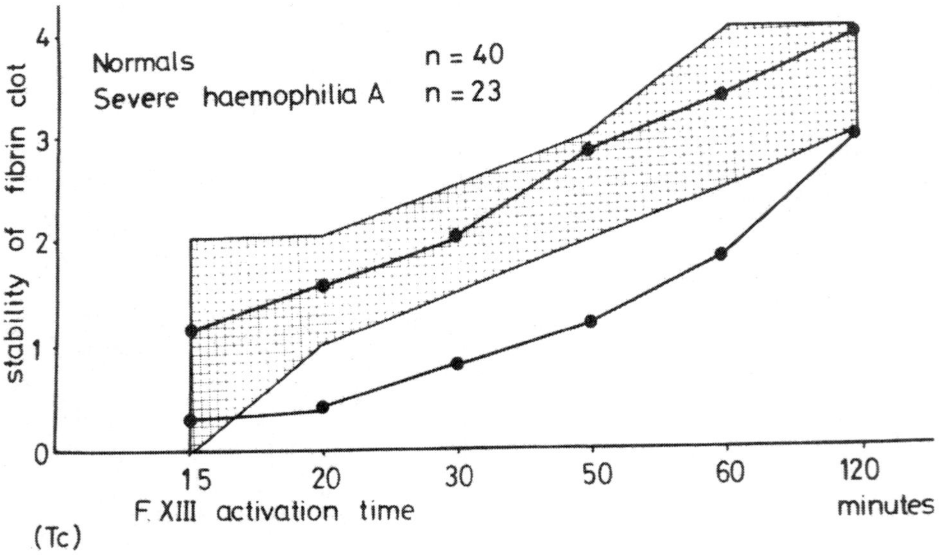

Fig. 3 Fibrin stabilization rate in the plasma of normal volunteers and 23 patients with severe haemophilia A. The patients had received no factor VIII for at least 10 days.

In the plasma of patients with severe haemophilia A the fibrin stabilization was severely delayed compared to the normal range. These observations could be made with clots obtained with thrombincoagulase as well as with ancrod (see fig. 3). But already in moderate haemophilia A and more so in mild haemophilia A the fibrin stabilization rate was normal.

Fig. 4 demonstrates that after substitution of factor VIII the fibrin stabilization rate in the plasma of patients with severe haemophilia A improved or normalized the fibrin stabilization process for 2 to 3 days, depending on the amount of factor VIII given to the patient.

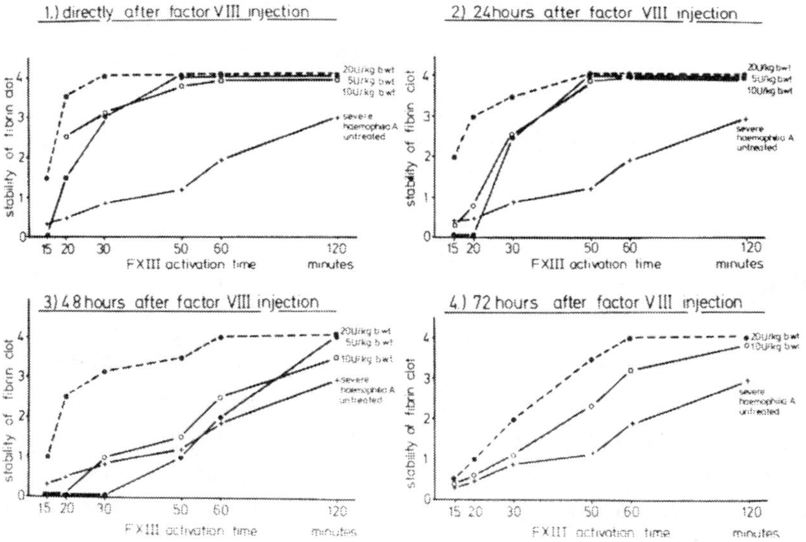

Fig. 4 Fibrin stabilization rate in the plasma of patients with severe haemophilia A after factor VIII substitution therapy.

But in the plasma of 7 patients with severe congenital factor XII deficiency the fibrin stabilization rate was nearly normal although thrombin generation was rather poor, as observed by the abnormal prolonged activated partial thromboplastin time and an abnorme thrombin generation test (see fig. 5).

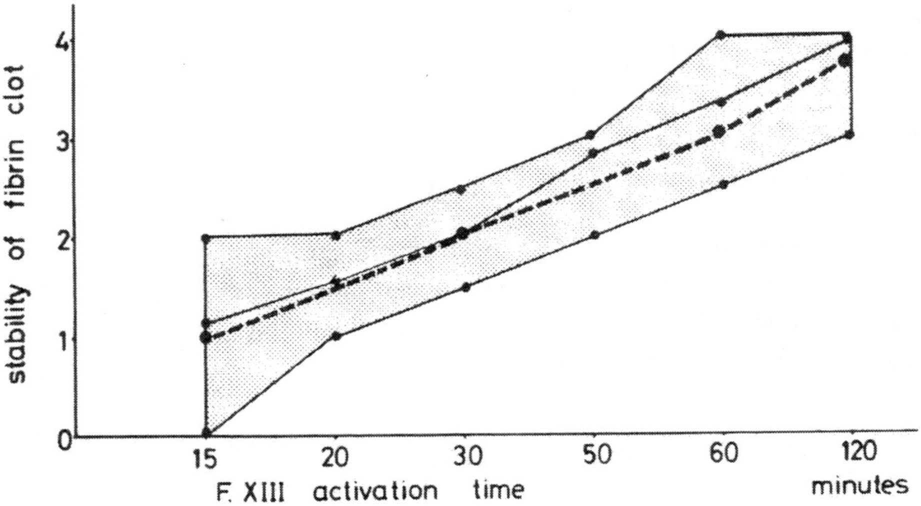

Fig. 5 Fibrin stabilization rate in the plasma of 7 patients with congenital factor XII deficiency using an thrombincoagulase induced clot.

Discussion

Our investigations show, that the kinetics of fibrin stabilization in an in-vitro-system are correlated to the thrombin formation rate, fibrin stabilization being rather poor in severe haemophilia A and in patients with QUICK's test below the therapeutical range. The method used here is a global test, and although our investigations make it probable, that the process observed depends primarily on factor XIII activation, other influences cannot be excluded. This became apparent, when investigating the plasma of patients with congenital factor XII deficiency, who had a nearly normal fibrin stabilization rate in spite of a prolonged activated partial thromboplastin time.

There is evidence that during the 2 h incubation with $CaCl_2$ other proteolytic enzymes, such as e.g. plasmin, start degrading the fibrin clot and therefore influence the visible stability. Investigations published elsewhere (1) could demonstrate, that the fibrin clot in the plasma of patients treated with phenprocoumon was easier lysed by streptokinase than the clot in normal plasmas. Addition of epsilonaminocaproic acid to the test system improved the fibrin stabilization rate. The increased lysability of an incomplete stabilized fibrin clot has already been observed by other authors (3, 5, 6, 10).

To check our results with the global test we investigated the fibrin obtained after adding calcium ions to plasma of patients with severe haemophilia A. Preliminary results show, that the cross-linking of the γ-chains occurs as in normal plasma. This was to be expected, as the normal slope of the fibrin stabilization curve already indicated, that it cannot be connected to the rapid cross-linking of γ-chains. On the other hand the cross-linking of α-chains was less complete in plasmas from patients with severe haemophilia A than in normal plasmas. This corresponds to the investigations of GAFFNEY, who found that the lysability of fibrin clots was more related to the cross-linking of α-chains than to those of the γ-chains (4).

So far our investigations indicate that the bleeding tendency in severe haemophilia A and in patients with a QUICK's test below the therapeutical range may be due not only to delayed fibrin formation but also to an incomplete stabilized fibrin.

References

1. Barthels, M., Borchers, H., Sommer, H.G.: in Progress in Chemical Fibrinolysis and Thrombolysis IV, 229-234. Churchill Livingstone, Edinburgh · London · New York 1979.

2. Barthels, M., Borchers, H., Poliwoda, H., Sommer, H.G.: Fortschr. Med. 95, 1281-1284 (1977).

3. Beck, E., Duckert, F., Ernst, M.: Thrombos. Diathes. Haemorrh. 6, 485-491 (1961).

4. Gaffney, P.J., Whitaker, A.N.: Thrombos. Res. 14, 85-94 (1979).

5. Gormsen, J., Fletcher, A.P., Alkjaersig, N., Sherry, S.: Arch. Biochem. Biophys. 120, 654 (1967).

6. McDonagh, R.P., McDonagh, J., Duckert, F.: Br. J. Haemat. 21, 323-332 (1971).

7. McDonagh, J., McDonagh, R.P.: Brit. J. Haemat. 30, 465-477 (1975).

8. Sigg, P.E., Duckert, F.: Schweiz. Med. Wschr. 93, 1455 (1963).

9. Triantaphyllopoulos, D.C.: Thrombos. Res. 3, 241-250 (1973).

10. Turowski, G., Schaadt, M., Barthels, M., Diehl, V., Poliwoda, H.: in Fibrinogen, Fibrin und Fibrinkleber. Verhandlungsbericht Deutsche Arbeitsgemeinschaft für Blutgerinnungsforschung, 23. Tagung, 227-237. F.K. Schattauer Verlag, Stuttgart · New York 1980.

SOME OBSERVATION ON CONCENTRATION AND COAGULABILITY OF FIBRINOGEN IN PATIENTS WITH CANCER, WITH AND WITHOUT METASTASIS

Chieko Miyashita, Elena Dumitrescu, Ernst Wenzel
Abteilung für Klinische Haemostaseologie und Transfusionsmedizin,
Universitätskliniken, Fb 4/3, D-665o Homburg-Saar, BRD

Richard Johannes Meiser
Medizinische Klinik und Polyklinik, Universität des Saarlandes,
Universitätskliniken, Fb 4/3, D-665o Homburg-Saar, BRD

Introduction

Aside from the typical symptoms of chronic DICFS (disseminated intravascular coagulation and fibrinolysis syndrome) with secondary fibrinolysis, an obviously higher concentration of circulating fibrinogen was proven in patients with carcinoma (1). Because fibrinogen reacts on one side as an acute-phase protein in inflammatory disease and on the other side is subject to more increase in metabolism as a main substrate in DICFS, we controlled coagulation functions in patients suffering from carcinoma, and examined how quality and concentration of clottable fibrinogen changed dependent on the stage of the cancer and whether these various changes in clottable fibrinogen were connected to other changes in the coagulation system in terms of type of carcinoma.

Patient

During about 1 1/2 years we controlled coagulation functions in 41 patients out of 154 with carcinoma of lung, stomach and interstines. At the time of examination the patients had not been treated with anticoagulants, immun-suppressants or with blood transfusions for at least 4 weeks.

Fibrinogen - Recent Biochemical and Medical Aspects

Table 1. Cohorts of patients suffering from various types of cancer with and without metastasis investigated in the studies.

	PATIENTS				
	N	F	M	METASTASES 0	METASTASES +
VARIOUS TYPES OF CANCER	154				
EVALUATED PRIMARY ORGAN INVOLVED	41	14	27	18	23
LUNG	11	1	10	5	6
STOMACH	13	3	10	5	8
COLON/ RECTUM	17	10	7	8	9

* HODGKIN'S DISEASE: 9 PATIENTS

The collective of 41 patients was comprised of 11 patients with lung carcinoma, 13 with stomach carcinoma and 17 with carcinoma of the intestine (Table 1). Out of these 41 patients 23 cases with metastasis were proven and in 18 there was no clinical proof. These patients suffered neither from thrombolic complication nor manifest or latent bleeding tendencies.

Method

The concentration of clottable fibrinogen was measured according to the Clauss-method. The time of fibrin generation was registered by thrombin time (5 NIH U/ml), thrombin-coagulase time and reptilase time (standardized reagents of Boehringer Mannheim). The influence of the amount clottable fibrinogen on the time of reptilase, thrombin-coagulase or thrombin time, is well known from the investigation of Kopeć's and Latallo's group (2, 3), and from findings of Wenzel and coworkers (4). Moreover, specimens of platelet function, partial volume analysis, antithrombin III concentration and activity, coagulation and fibrinolysis parameters were determined;

these methods are described elsewhere (5). In order to relate an increased concentration of fibrinogen to a delayed fibrin polymerization by means of reptilase, thrombin-coagulase or thrombin time indices were calculated for each parameter as quotients.

Results and discussion

As shown in the table 2 and figure 1, among the metastatic and nonmetastatic subgroups significant differences (s = 95%) could be proven between the mean values of clottable fibrinogen concentration and the ratio of the fibrinogen to the reptilase time. In terms of the types of carcinoma, each other tests given, significantly differed between the subgroups with metastasis and without metastasis. The mean value for thrombin time and thrombin-coagulase time significantly differed in patients with bronchial cancer.

Table 2. Observation of significant differences in laboratory parameters in patients suffering from different types of cancers, with and without metastasis.

	CANCER PATIENTS	CANCER OF		
		LUNG	STOMACH	COLON
	0 +	0 +	0 +	0 +
FIBRINOGEN	s!		s!	
THROMBINTIME		s!	s!	
STAPHYL.COAG.		s!		
FBG/REPTILASE	s!		s!	
FBG/REPTILASE + STAPHYL.			s!	
N	18 23	5 6	5 8	8 9

STUDENT TEST, s = 2 p = 0.05

In the groups of patients with stomach cancer, significant differences were found in the mean values for fibrinogen concentration, thrombin time and the quotients for fibrinogen concentration versus reptilase time. Moreover, the thrombin-coagulase time could be differentiated in their mean values in both subgroups. When all patients with and without metastasis were considered alltogether, not taking the localisation of the primary tumor into account, only the clottable fibrinogen concentration differed significantly. In analysing all 41 patients the mean values could not longer be proven to be significantly different. In a small group of patients with Hodgkin's disease increased fibrinogen concentration and prolonged thrombin-coagulase time were observed (table 3). The group of 9 patients examined in the laboratory was too heterogenous to be evaluated statistically on account of varied grades and stages of the disease.

Table 3. Pattern of laboratory parameters observed in patients suffering from Hodgkin's disease.

9 PATIENTS SUFFERING FROM HODGKIN DISEASE	MEAN VALUE	RANGE
FIBRINOGEN-CONC.	368.1 MG%	209 - 665
THROMBINTIME	17.7 SEC	13.8-20.3
REPTILASETIME	24.5 SEC	16.6-43.7
STAPHYLOCOAGULASE-TIME	31.3 SEC	22.2-49.9

According to investigations by other authors (6), in a higher number of cases statistical differences can be proven in the behaviour of fibrinogen and circulating degradating products. Other tested parameters such as the degree of thrombocytic disfunction, behaviour of Antithrombin III, e.g. functionally and immunologically, could not be correlated to those parameters for fibrinogen concentration and its quality. There was also no correlation to the disease (table 4).

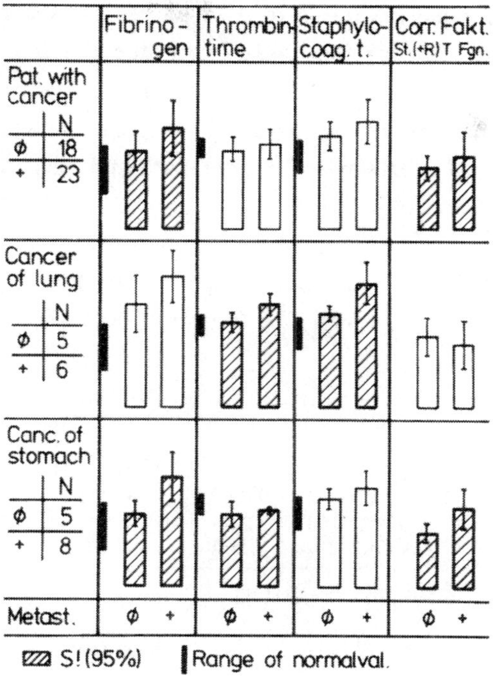

Figure 1. Diagram showing the pattern of laboratory parameters discussed in the table 2.

As mentioned in the introduction and as reflected in the clinical data, it seems that according to the simple physiological parameters for coagulation, the behaviour of fibrinogen (that is its increased synthesis) perhaps due to the deformities or abnormally functioning fibrinogen together with DICFS could not be differenciated. According to our in vitro investigation circulating degradated products with inhibitoring effects on polymerization will mathematically correlate to fibrinogen concentration.

According to our results (4) fibrinogen concentration and the concentration of polymerization inhibiting degradated products are mutually and systematically effecting how distinct thrombin-coagulase, reptilase-time and thrombin time are prolonged in vitro. Evenmore, so a relationship of both parameters independent of the stage of the carcinoma could be proven in randomised groups. On the other hand, there is no relationship with respect to the concentration of circulating degradating products and to the clottable fibrinogen concentration as well as to the stage of the disease. Therefore, we have to discuss that aside from the circulating degradated products inhibiting polymerization, also dysfibrinogenemia resulting perhaps from an excessive fibrin synthesis which could explain the prolonged fibrin generation time in plasma of cancer-patients, should be taken into account, too.

Table 4. Correlationship between antithrombin activity of the plasma and the concentration of antithrombin III (measured by Mancini method).

ANTITHR.ACT.(X)/A-III-CONCENTR.(Y)

N		R
56	HEALTHY BLOOD DONORS	0.94
36	PAT.W.OCCLUS.ART.DIS.	0.73
14	PAT.W.DIC (VARIOUS D.)	0.65
9	PAT.W.CARCIN.(VARIOUS)	0.68

References

1. Hagedorn, A.B., Bowie, E.J.W., Elveback, L.R., Owen, Ch.A., Jr.: Mayo Clinic Proceedings 49, 647 - 653 (1974).

2. Kopeč, M., Latallo, Z.S., Stahl, M., Wegrzynowicz, Z.: Biochim. Biophys. Acta 181, 437 - 445 (1969).

3. Latallo, Z.S., Teisseyre, E.: Scand. J. Haemat. Suppl 13, 261 - 266 (1971).

4. Wenzel, E., Holzhüter, H., Muschietti, F., Angelkort, B., Ochs, H.-G., Pusztai-Markes, S., Nowak, H., Stürner, H.: Dtsch. med. Wschr. 99, 746 - 756 (1974).

5. Wenzel, E., Pfordt, L., Wenig, Ch., Nienhaus, K.H.: in New Methods for the Analysis of Coagulation using Chromogenic Substrates (Ed. Witt, I.), Walter de Gruyter, Berlin 1977, p. 69-89.

6. Sun, N.C.J., Bowie, E.J.W., Kazmier, F.J., Elveback, L.R., Owen, Ch.A., Jr.: Mayo Clinic Proceedings 49, 636 - 641 (1974).

BIOLOGICAL EFFECTS OF LOW MOLECULAR WEIGHT PEPTIDES AND INTERACTIONS WITH OTHER PROTEINS

EFFECTS OF PEPTIDES DERIVED FROM FIBRINOGEN PROTEOLYSIS ON IMMUNEREACTIONS AND TUMOR GROWTH

Maria Kopeć

Department of Radiobiology and Health Protection, Institute of Nuclear Research, Warsaw, Poland

Waldemar Roszkowski

Department of Immunology, National Institute of Tuberculosis, Warsaw, Poland

Bengt Gerdin, Tom Saldeen

Faculty of Forensic Meidcine, University of Uppsala, Sweden

Introduction

Low molecular weight fibrin(ogen) degradation products (LMW-FDP) have been reported to suppress the humoral immune reaction in vivo in mice /1/ and to inhibit the proliferation of lymphocytes in cultures /1, 2/. The evidence for suppression of cell mediated immunoreactivity by LMW-FDP under the conditions in vivo is however lacking.

Results

Delayed cutaneous hypersensitivity (DCH) to oxazolone /3/ in BALB/c mice was applied as a model of cell mediated immune reaction. Concentrates of LMW-FDP (containing peptides of mw $>$ 3,500) separa-

Fibrinogen - Recent Biochemical and Medical Aspects

ted by dialysis from the digest of human fibrinogen by plasmin /4/ were injected in four daily doses of 200 µg (0.1 ml) before sensitization and once again 3 days after. Controls were treated with phosphate buffered saline (PBS) instead of LMW-FDP.

DCH was suppressed moderately by LMW-FDP administered intraperitoneally (i.p) and strongly when peptides had been injected subcutaneously (s.c) in the area drained by the same lymph nodes as the site of sensitization (Table 1).

TABLE 1

Effect of LMW-FDP on DCH to oxazolone, mean increament in ear thickness $\pm$ SD (10^{-3} cm)

Treatment	Intraperitoneal	Subcutaneous
PBS	21.8 ± 2.4	21.4 ± 2.8
LMW-FDP	15.5 ± 2.1^{x}	7.6 ± 1.4^{xx}

means from values obtained in 5 mice; x: $p < 0.05$; xx: $p < 0.001$

Such s.c treatment with LMW-FDP abolished nearly completely an increase of ^{125}I-UdR incorporation and of weight (Fig. 1) provoked in regional lymph nodes by a sensitizing dose of oxazolone. This indicates that LMW-FDP inhibit the primary proliferative phase of response to sensitization.

Further experiments aimed to establish whether LMW-FDP influence the progression of L-1 sarcoma in BALB/c mice. L-1 sarcoma, a tumor that arose spontaneously in the lung of BALB/c mice, can be maintained by s.c. passages in this strain. L-1 sarcoma is immunogenic /6/ and exhibit a potent plasminogen activator activity /7/.

LMW-FDP were injected in 200 ug (0.1 ml) daily doses during 4 con-

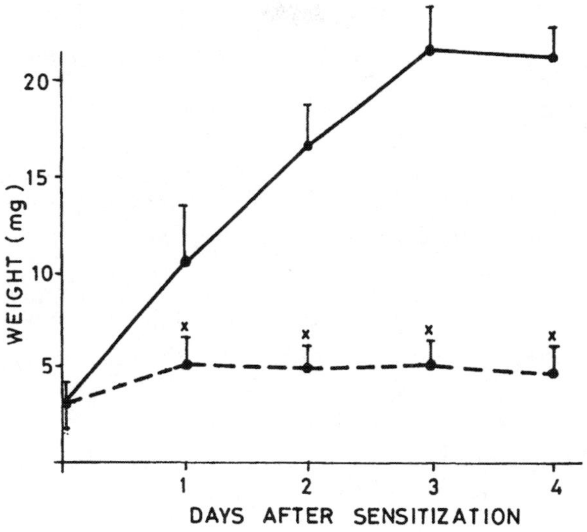

Figure 1

Effect of local treatment with LMW FDP on stimulation of regional lymf node. Results are expressed as mean values ± S.D. —— control; ----- LMW-FDP treated animals
^{x} $p < 0.01$

secutive days before s.c. tumor cell implantation and than at 3 day intervals either locally in the site of tumor implantation (i.t.) or i.p. The data listed in Table 2 indicate that i.t. administration of LMW-FDP provoked a pronounced increase in the rate of the primary tumor growth and in numbers of lung metastases. This was associated with a shortening of survival as compared with PBS treated controls (Table 2). Similar effect were by i.p. administered LMW-FDP.

From the mixture of LMW-FDP three peptides of established primary structure have been isolated namely so called peptide 6A: Ala-Arg--Pro-Ala-Lys, 6D: Thr-Ser-Glu-Val-Lys. /8/.

Peptides 6A and 6D have been already reported to suppress proliferation of stimulated mice spleen cell cultures in concentrations

TABLE 2

Effect of i.t. administered LMW-FDP on progression of L-1 sarcoma in BALB/c mice

Treatment	Weight of primary tumor, mg 16th day after implantation	Median survival days	Number of lung metastases on day ( )
PBS	536 ± 94^{x}	37°	10.9 ± 3.6$^{\bullet}$ /37/
LMW-FDP	1050 ± 230^{x}	28°	21.8 ± 4.1$^{\bullet}$ /28/

x: means ± SD from values obtained in 5 mice
o: 20 mice in each group
•: means ± SD from values obtained in 10 mice

of 100 nmol/ml or higher /9/. Peptide 6E appeared less effective inducing a significant suppression at a close of 400 nmol/ml.

Table 3 presents the results of a study on the influence of peptides

TABLE 3

Effects of peptides on phagocytosis of ^{3}H-thymidine labelled S. aureus by peritoneal macrophages

Peptide nmol/ml	% change related to control Tuftsin	6A	6E
1.0	+ 30.4	+ 28.9	0
2.5	+ 47.0	+ 37.0	0
5.0	+ 56.8	+ 61.4	0
10.0	+ 55.0	+ 86.3	-17
20.0	+ 51.9	+ 50.1	-16
40.0		+ 61.1	-50

Results are expressed as mean values obtained in 5 samples

on phagocytosis of ^{3}H-thymidine labelled S. aureus by peritoneal macrophages of BALB/c mice. Tuftsin a tetrapeptide (Thr-Sys-Pro--Arg) cleaved by trypsin as well as by granulocyte protease(s) from a particular fraction of serum immunoglobulins was included into investigation for comparison. Tuftsin is considered to be a strong natural stimulant of phagocytotic activity of macrophages and blood granulocytes /10, 11/. It can be seen that phagocytosis stimulating activity of peptide 6A is of the same degree as that of tuftsin.

Peptide 6E in concentrations above 10 nmol/ml produced slight and at a dose of 40 nmol/ml pronounced inhibition of phagocytosis. Peptide 6D remained without any effect.

The results presented indicate that peptides cleaved by plasmin from fibrin(ogen) can modulate the immune reactivity and the dynamics of tumor progression in mice. Deposition of fibrin in the cell mediated immune reactions /12/ and at the periphery of various malignant tumors /13/ makes the pathophysiological significance of these LMW--FDP properties not unlikely. Further studies are however required for recognition of mechanism underlying the LMW-FDP action of their components responsible for immune suppression and for enhancement of tumor progression as well as the possible links between these two phenomena.

References

1. Girmann, G., Pees, H., Schwarze, G., Scheurlen, P.G.: Nature 259, 399-401 (1976).

2. Krzystyniak, K., Stachurska, J., Ryżewski, J., Bykowska, K., Kopeć, M.; Thrombos. Res. 12, 523-530 (1978).

3. Asherson, G.L., Ptak, W.: Immunology 15, 405-416 (1968).

4. Stachurska, J., Łopaciuk, S., Gerdin, B., Saldeen, T., Korościk, A., Kopeć, M.: Thrombos. Res. 15, 663-672 (1979).

5. Hughes, W.L., Commerford, S.L., Gitlin, D., Krueger, R.C., Shah, V., Reilly, P.: Fed. Proc. 23, 640-648 (1964).

6. Roszkowski, W., Wrembel, J., Roszkowski, K., Janiak, M., Szmigielski, S.: Int. J. Cancer 25, 289-292 (1980).

7. Chmielewska, J., Rączka, E., Teisseyre, E., Latałło, Z.S.: Thrombos. Haemost. 42, 141 (1979).

8. Belew, M., Gerdin, B., Porath, J., Saldeen, T.; Thrombos. Res. 13, 983-994 (1978).

9. Gerdin, B., Saldeen, T., Roszkowski, W., Szmigielski, S., Stachurska, J., Kopeć, M.: Thrombos. Res. 18, 461-468 (1980).

10. Najjar, V.A., Nishioka, K.: Nature 228, 672-673 (1970).

11. Nishioka, K., Constantoponlos, A., Satok, P.S., Najjar, V.A.: Biochem. Biophys. Res. Commun. 47, 172-179 (1972).

12. Colvin, R.B., Dvorak, H.F.: J. Immunol., 114, 377-387 (1975).

13. Zacharski, L.R., Henderson, W.G., Richles, F.R., Forman, Cornell, C.: J. Jr. et al. Cancer, 44, 732-741 (1979).

PLASMIN AND ELASTASE INDUCED FIBRIN DEGRADATION PRODUCTS AND THEIR ACTION ON VASCULAR PERMEABILITY: AN EXPERIMENTAL STUDY

Marie-Luise Schmeller, Reimar Hafter*, Axel Stemberger, Henner Graeff* and Günther Blümel
Institut für Experimentelle Chirurgie, Technische Universität München, 8000 München 80,
*I.Frauenklinik der Universität München, 8000 München 2, FRG

Introduction

In recent years various groups have shown that plasmin induced fibrin(ogen) degradation products posess numerous biological activities. Effects like inflammatory reactions, chemotactic activity, stimulation of phagocytosis, immunsuppressive action, inhibition of H_3-thymidin incorporation in lymphocytes, inhibition of platelet aggregation and increase in microvascular permeability were described (1).
Histological pictures e.g. in shock lungs or inflammatory processes which combine deposition of fibrin and presence of granulocytes, and the known fact that neutral proteases from leukocytes digest fibrin(ogen), led us to start some research activities in order to elucidate the action of such fibrin(ogen) degradation products on microvascular permeability.
The following contribution is a preliminary report on the comparison of biological activities of fibrinogen degradation products either by plasmin or by elastase harvested from neutrophils.

Materials and Methods

Plasmin degradation products of fibrinogen (p-fdp) were prepared by streptokinase incubation of fibrinogen for 20 minutes

or 3 hours. Early (X and Y) and late (D and E) degradation products were separated and dialysis of the respective p-fdp was subsequently performed.
Elastase degradation products (e-fdp) were prepared by incubation of fibrinogen with leucocyte elastase in a ratio of 500 : 1. Subsequently heat precipitation was performed. The supernatant was further fractionated using ultrafiltration techniques. Fractions with the following molecular weights were prepared: MW $<$ 500, 500 to 1000, 1000 to 5000, 5000 to 10 000, $>$10 000.
In order to characterize biological effects with clinical relevance enhancement of permeability was chosen as parameter. Vascular permeability may be measured by using vital dyes which are bound to plasma proteins particularly to albumin.

According to the modified model of Udaka (2) the v. jugularis of anaesthetized rats was canulated and Evans blue injected: After 5 minutes 0.1 ml p-fdp or e-fdp were given intradermally in the shaved abdominal skin in a concentration range from 1 µg to 1 mg calculated as fibrinogen equivalent. 30 minutes later the injection sites were evaluated macroscopically and afterwards the coloured dermis was punched out and extracted by formamide. The optical density was measured at 620 mµ. In these experiments isotonic saline as blank and bradykinin in a concentration of 1 or 10 µg as positive controls were used.

Results

All peptides tested reacted in a dose-dependent manner. Low molecular weight fibrinogen degradation products induced by plasmin showed a high permeability increasing activity as described in literature (1, 3, 4). In our experiments early (X,Y) and late (D,E) fragments were less effective than the dialysed low molecular weight p-fdp (MW $<$ 20 000) or e-fdp (MW $<$ 40 000) (Fig. 1).

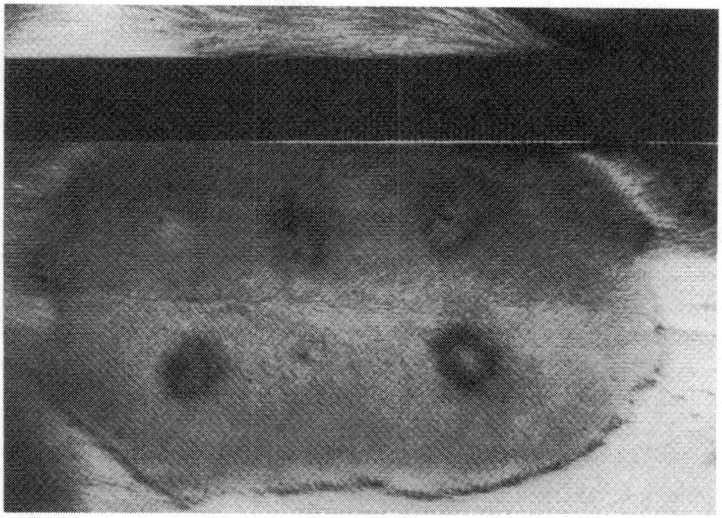

Fig. 1: Enhancement of rat-skin-permeability with different fdp-fractions (0.1 ml i. derm.):

left to right:

top: 1. early p-fdp (X and Y),

2. 20 h dialysate of early p-fdp

3. 40 h dialysate of early p-fdp

bottom: 4. e-fdp

5. NaCl

6. bradykinin

After ultrafiltration the e-fdp fraction below 1000 d was inactive whereas the fractions from 1000 d to 40 000 d showed distinct activity (Fig. 2).

The latter results are of interest because the most effective plasmin induced peptides are a pentapeptide and an undecapeptide (5) which both are split off early.
Our preliminary data may therefore indicate that different fragments are responsible for the biological activities of either p-fdp or e-fdp.

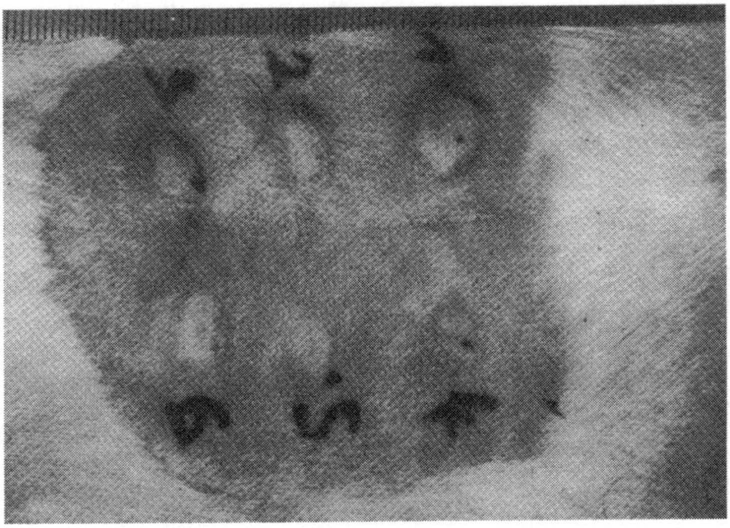

Fig. 2: Enhancement or rat-skin-permeability with e-fdp fractions (0.1 ml i. derm.):

1. MW > 10 000	4. MW 500 - 1 000
2. MW 5 000 - 10 000	5. MW < 500
3. MW 1 000 - 5 000	6. NaCl

References

1 Kopec, M.: The Microembolism Syndrome. (Ed.: Saldeen,T.) Almquist and Wiksell International Stockholm, pp. 193 - 196, 1979.

2 Udaka,K.,Takeuchi,Y.; Movat,H.Z.: Proc.Soc.Exp.Biol.Med. 133,1384-1387 (1970).

3 Gerdin,B.; Saldeen,T.: Thromb.Res.13, 995-1006 (1978).

4 Buczko,W., DeGaetano,G., Franco,R., Donati,M.B.: Thrombos. Haemostas. 35, 651-657 (1976).

5 Belew,M., Gerdin,B., Porath,I., Saldeen,T.: Thrombos. Res. 13,983-994 (1978).

THE EFFECT OF REPEATED INJECTIONS OF FIBRINOPEPTIDES A AND B ON EXPERIMENTAL ALLERGIC ENCEPHALOMYELITIS IN RATS AND GUINEA PIGS.

Gerhard Ruhenstroth-Bauer, Rainer Scherer, M. Abd-El-Fattah
Max-Planck-Institut für Biochemie
D-8033 Martinsried, Germany

Jiři Simon
Max-Planck-Institut für Psychiatrie
D-8000 München 40, Germany

Introduction

In humans with an elevated erythrocyte sedimentation rate there exists a disease specific correlation between this elevation and the changes in the concentration of different plasma proteins (1,2). Analogous changes have been observed in rats (3,4,5). As a result of this and other observations a systemic feed back mechanism has been postulated (6) which operates in different kinds of tissue damage in a very similarly way (Fig.). Individual steps of this feed back mechanism have been confirmed in vitro. The final step of this proposed mechanism is the depressing effect of fibrinopeptides A and/ or B on the edema in a damaged area. This has been shown in the carrageenan edema of the rat (6), in the inflammatory response after X-ray irradiation (7), in a test of delayed immunological reaction, and in experimental allergic encephalomyelitis (EAE) (8).

Fibrinogen - Recent Biochemical and Medical Aspects

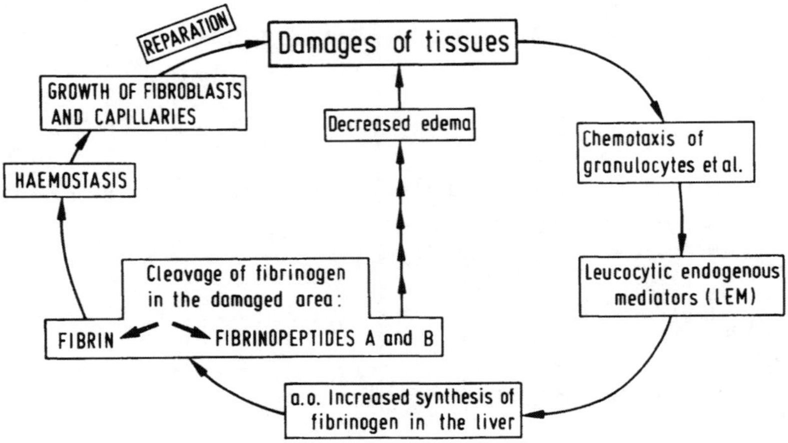

Fig. Scheme of the systemic feed back mechanism of the reparation of damaged tissues (6).

Results

The EAE was induced by sensitization of guinea pigs and rats with encephalitogenic antigens and adjuvant. Human fibrinopeptides A and B were injected daily interperitoneally over a 2-3 week period. In comparison to the saline controls the treated animals showed a significant diminution of the neurological signs of disease, e.g., in the number, the severity and the duration of the exacerbations. The inflammatory response and the extravasation of plasma proteins into the tissues was also less pronounced in the treated animals. The specific immune response, however, to the antigenic challenge was not altered and the characteristic cellular infiltrates of EAE were also similar in both groups. However, a significantly higher titer of circularing immune complexes was observed in the serum of the treated animals. The injection of

FITC-labeled dextran (M.W. 150,000) 5 min before sacrificing the animals, resulted in a remarkable retention of these substances in the vascular bed of the treated animals in comparison with the controls.

These results provide evidence for the hypothesis, that the fibrinopeptides A and B act primarily by reducing the permeability of capillaries in damaged tissues areas and not via an immunological mechanism.

References

1. Scherer, R., Moraraescu, A., Ruhenstroth-Bauer, G.: Klin. Wschr. 53, 265-273 (1975).
2. Scherer, R., Ruhenstroth-Bauer, G.: Clin. Chim. Acta 66, 417-433 (1976).
3. Abd-El-Fattah, M., Scherer, R., Ruhenstroth-Bauer, G.: J. Molec. Med. 1, 211-221 (1976).
4. Rapaport, S.I., and Zivelin, A.: Thrombosis and Haemostasis. 35, 692 (1976).
5. Abd-El-Fattah, Scherer, R., Fouad, M.F., and Ruhenstroth-Bauer, G.: Cancer Research (in press).
6. Scherer, R., and Ruhenstroth-Bauer, G.: Blut 36, 327-330 (1978).
7. Deutsches Patent Nr. 27 22 769 (21.5.1977).
8. Scherer, R., Simon, J., Abd-El-Fattah, M., and Ruhenstroth-Bauer, G.: Clin. exp. Immunol. 40, 49-59 (1980).

INTERACTION OF FIBRINOGEN WITH ITS PLATELET RECEPTOR REQUIRES THE NATIVE STRUCTURE OF THE MOLECULE AND DIVALENT IONS

Gerard Marguerie

Inserm U 143, Institut de Pathologie Cellulaire, Hopital Bicêtre,
94270 Le Kremlin-Bicêtre

Edward F. Plow

The research institute of Scripps Clinic,
La Jolla, California, 92037, USA

One of the primary foci of studies of the structure function relationship of fibrinogen has been the formation of fibrin after the proteolytic cleavage of the fibrinopeptide A and B by thrombin. But numerous investigations have indicated that fibrinogen has the potential for mediating several biological process. This includes the acute phase reaction (1), the immune response (2,3) and the regulation of cellular functions of platelet (4,5,6), macrophage (7,8), leucocyte (9) and lymphocyte (2, 3). Thus, in considering fibrinogen as a multifunctional protein, the implication of the molecule in the responsiveness of a variety of cells becomes a major consideration.

With platelets, the role of fibrinogen as cofactor in the ADP-induced aggregation is well documented. This reaction is generally described from the changes in light transmission of platelet suspensions as being composed of three basic reactions : 1) shape change of platelets from discoid to spherical forms ; 2) aggregation and 3) deaggregation (II). It is known that fibrinogen is not required for shape change but is needed for aggregation. Similarily divalent ions such as calcium or

Fibrinogen - Recent Biochemical and Medical Aspects

magnesium are not required for shape change but are requisite for aggregation. The need for fibrinogen in the aggregation of platelets is indicated by several observations. In individuals with congenital afibrinogenemia platelets do not aggregate or aggregate weakly in response to ADP, but normal aggregation is observed upon addition of fibrinogen in their plasma. Additionally with washed platelets, free of plasma fibrinogen, ADP induced aggregation does not occur unless fibrinogen is added.

Thus the functional implication of fibrinogen in platelet aggregation induced by ADP is evident. To explain this participation on a molecular basis, the existence of a receptor system has been hypothesized. Our previous studies were focused on the capacity of washed platelets to directly interact with fibrinogen and we demonstrated that a single class of specific and saturable receptor for this protein is induced by ADP (11) Several other groups have also independently confirmed the existence of a single class of binding sites (12,13,14) with similar affinity for fibrinogen (Ka ranging from 12.5 to $0.3 \times 10^6\ M^{-1}$), although Niewiarowski et al (15) and Peerschke et al (16) have provided evidences for the existence of two classes of binding sites.

In characterizing the interaction of fibrinogen with this platelet receptor system, the identification of the molecule which actually binds to the cell and the role of divalent ions in this reaction were two major considerations.

Interaction between fibrin and platelet has indeed been reported (17), and the distinction between the binding of fibrinogen as opposed to fibrin monomer had to be established. It is well documented that fibrinogen preparations are often contaminated by small amounts of fibrin monomers which forms calcium dependent soluble polymeric isomers (18). Since the total number of molecules which specifically interacted with stimulated platelets did not exeed 0.5 to 1% of the total number of molecules applied, a preferential and selective binding of any contaminating fibrin monomer by the cell had to be excluded. Thus the radiolabeled material associated with platelet was analyzed for the presence of fibrinopeptide A with a specific antiserum to this fibrinopeptide. These experiments are

Table 1

Analysis for the presence of fibrinopeptide A before and after treatment by thrombin in ^{125}I-fibrinogen (^{125}I-fg), platelet bound ^{125}I-fibrinogen extracted with 0,5 % Triton X 100 (P.^{125}I-fg) and ^{125}I-fibrinogen added to a Triton X 100 extract of platelet (P + ^{125}I fg). The results are expressed in percent of radioactivity bound by a 1 : 1000 dilution of an antifibrinopeptide A

Sample	Percent of ^{125} I bound	
	Before thrombin	After thrombin
^{125}I-fg	66	0
P.^{125}I-fg	66	0
P + ^{125}I-fg	62	0

summarized in Table 1. Of the ^{125}I-labeled fibrinogen, 66% was bound to a 1:1,000 dilution of the antiserum in a control experiment. A similar percent was bound from the radioactive material eluted from the platelets with 0.5 % Triton X 100 as well as a control in which ^{125}I-fibrinogen was added to a Triton X 100 extract of platelets. These experiments have established that the bound radioactivity was primarily fibrinogen, and indicated that the native structure and conformation of the molecule are important in the supporting platelet aggregation via the receptor system. Observation that fibrinogen rapidly looses its capacity to interact with the cell and to support aggregation following plasmic degradation is in support of this conclusion (19).

Other experiments were designed to study several characteristics of the fibrinogen receptor system, including the role of divalent ions in this reaction. In the absence of cations (< 1μM by atomic absorption spectrophotometry) platelets exhibited a well defined shape change in response to ADP but did not aggregate unless fibrinogen and calcium or magnesium were present at mMolar levels. Both cations supported the binding of fibrinogen to ADP stimulated platelets (20). On the other hand dissociation experiments run in parallel indicated that the interaction of fibrinogen with the cell is initially reversible and becomes progressively irreversible through a time dependent reaction, as platelet bound fibrinogen was not displaced by a large excess of unlabeled fibrinogen after 20 min incubation. Observation made with apyrase verified this evolution from a reversible to an irreversible reaction and correlated with the aggregation and diaggregation properties of platelets with the enzyme (20). To integrate these observations into a comprehensive mechanism, the following multistep reaction sequence has been postulated

$$P_1 \overset{ADP}{\rightleftharpoons} P_2 \overset{\substack{ADP \\ Ca^{++}/Mg^{++}}}{\rightleftharpoons} P_3 + fg \overset{Ca^{++}/Mg^{++}}{\rightleftharpoons} P_3\text{-}fg \longrightarrow P_4\ fg$$

In these reactions fg represents fibrinogen and P_1 to P_4 represent different stages of platelet activation. Reaction 1 involves stimulation of the resting platelet P_1 as a result of the first encounter between ADP and the cell culminating in shape change and a conversion of the platelet to state P_2. This reaction is independent of both fibrinogen and divalent ions. Reaction 2 involves the induction of the platelet receptor for fg on P_3. This step may entail a processing of ADP by the platelet by a specific pathway which might be differentially affected by calcium or magnesium. Reaction 3 involves the direct binding of fibrinogen to its platelet receptor and is the event which directly induces platelet aggregation. Reaction 4 describes the evolution of the system to a state in which fibrinogen is non dissociable. This series of reactions provides

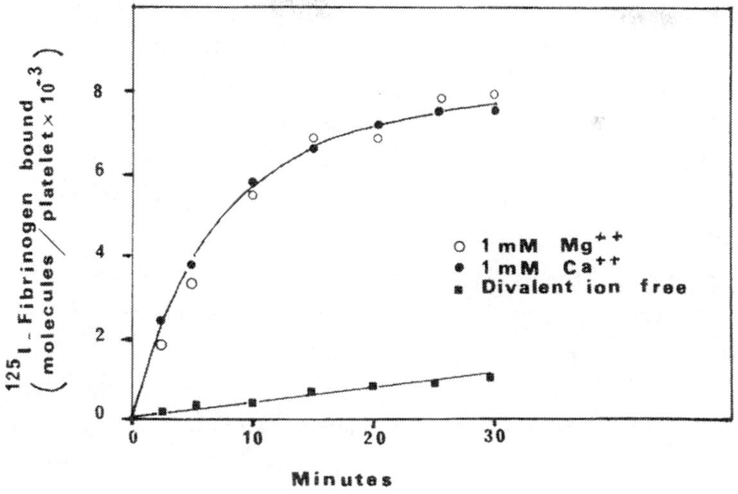

Figure 1 : Time course of the association of ^{125}I-fibrinogen with ADP-stimulated fixed platelets, in the presence or absence of divalent ions. Washed platelets were stimulated with 10 μM ADP, prior to the fixation with paraformaldehyde and the binding experiments were carried out in the presence of 0.2 μM ^{125}I-fibrinogen.

the first testable hypothesis to explain the molecular events associated with ADP-induced platelet aggregation.

To assess the role of divalentions in these reactions platelets were fixed in their various states of activation using paraformaldehyde fixation and the direct interaction of ^{125}I-fibrinogen to the fixed platelets was examined. Non stimulated fixed platelets failed to bind ^{125}I-fibrinogen whereas ADP-stimulated fixed platelets (state P_3) interacted with the labeled protein. As shown in figure 1 the binding was a time dependent process and occured only in the presence of mMolar levels of either calcium or magnesium. This emphasized the need for divalent ions in reaction 3. Fibrinogen can interact with divalent ions and possess high affinity binding sites (21) which are specific for calcium and not for magnesium. The binding of calcium entails conformational rearrangements in the molecule (22) and an effect of the cation on the protein could be postulated. But no difference between the bindings supported by calcium or magnesium was observed (figure 1) suggesting that the effect of divalent ions is on the platelet receptor rather than on the fibrinogen molecule.

The stabilization of the fibrinogen receptor complex described by the reaction 4 is likely divalent ions independent. Evidences for this are observations that EDTA fails to remove platelet bound fibrinogen from non fixed platelets but displaces ^{125}I-fibrinogen from fixed platelets (not shown).

In summary platelet possess an inducible receptor which is specific for the native structure of fibrinogen. This platelet receptor system confers to fibrinogen a new and important role in the hemostatic response, distinct from the fibrin formation. The interaction of the protein with its receptor is part of a complex multistep reaction which involves an initial reversible binding and stabilization of the fibrinogen-receptor complex.Divalent ions are required for the initial reversible interaction and not for the stabilization reaction but the effect of the cations is likely to be on the platelet receptor itself rather than on the structure of fibrinogen.

References

1. Barnhart, M.I., Sulisz, L. and Bluhm, G.B.: in Chemical Biology of inflammation. Forscher, B. ed. Pergamon Press, Oxford, (1968).
2. Girmann, G., Pees, H., Schwarze, G., and Scheurlen, P.G.: Nature 259, 399-401, (1976).

3. Edgington, T.S., Curtiss, L.K., and Plow, E.F.: Thrombos. Haemostas. 38, 170, (1977).

4. Solum, N.O., and Stormorken, H.: Scand. J. Clin. Lab. Invest. 17, 170-182, (1965).

5. Deykin, D., Pritzker, G.R. and Scolnik, E.M.: Nature 208, 296-298, (1965).

6. Mustard, J.F., Perry, D.W., Kinlough-Rathboue, R.L. and Packham, M.A.: Am. J. Physiol. 228, 1757-1765, (1975).

7. Calvin, R.B. and Dyorak, H.F.: J. Exp. Med. 142, 1377-1390, (1975).

8. Sherman, L.A. and Lee, J.: J. Ep. Med. 145, 76-85, (1977).

9. Kay, A.B., Pepper, D.S., Ewart, M.R.: Nature New Biol. 243, 56-57 (1973).

10. Holmsen, H., Salganikoff, L. and Fukami, M.H. in Haemostasis : Biochemistry, Physiology and Pathology - Ogston, D. and Bennett, B. eds, J. Wiley and Sons, N.Y. (1977).

11. Marguerie, G., Plow, E.F. and Edgington, T.S.: J. Biol. Chem. 254, 5357-5363, (1979).

12. Bennett, J.S., and Villaire, G.J. J. Clin. Invest. 64, 1393-1401, (1979).

13. Hawiger, J., Parkinson, S. and Timmons, S.: Nature 283, 195-197, (1980).

14. Marfenist, E.J., Packham, M.A. and Mustard, J.H.: Blood 56, 189-198, (1980).

15. Niewiarowski, S., Budzinski, A.Z., Morinelli, T.A., Brudzynski, T.M., and Stewart, G.J.: J. Biol. Chem. 256, 917-925, (1981).

16. Peerschke, E.I., Zucker, M.B., Grant, R.A., Egan, J.J. and Johnson, M.M.: Blood 55, 841-847, (1980).

17. Niewiarowski, S., Regoeczi, E., Stewart, G.J., Senyi, A.F., and Mustard, F.J.: J. Clin. Invest. 51, 685-700, (1972).

18. Marguerie, G., Benabid, Y. and Suscillon, M.: Biochim. Biophys. Acta 579, 134-141, (1979).

19. Niewiarowski, S., Budzinski, A.Z., and Lipinski, B.: Blood, 49, 635-644, (1977).

20. Marguerie, G., Edgington, T.S., and Plow, E.F.: J. Biol. Chem 255, 154-161, (1980).

21. Marguerie, G., Chagniel, G. and Suscillon, M.: Biochim. Biophys. Acta 490, 94-103, (1977).

22. Marguerie, G.: Biochim. Biophys. Acta 494, 172-181, (1977).

STUDIES ON THE BINDING OF TISSUE PLASMINOGEN ACTIVATOR TO FIBRINOGEN AND FIBRIN

Egbert K.O. Kruithof, Fedor Bachmann

Laboratoire Central d'Hématologie, CHUV
CH-1011 Lausanne, Switzerland

Introduction

Venous stasis, exercise and certain vasoactive substances lead to the release of vascular plasminogen activator (VA) from the vascular endothelium. This activator has a high affinity for fibrin (1). Tissue plasminogen activator (TA), which is immunologically and functionally indistinguishable from VA (2), also shows strong binding to fibrin (3). It has furthermore been shown that the activity of TA and of VA is enhanced in the presence of fibrin but not in the presence of fibrinogen (4, 5, 6). The release of VA during blockage of a vessel, its binding to fibrin and the enhancement of its activity in the presence of fibrin, would greatly increase the specificity of the activator (7, 8). Previous studies on the binding of TA and VA to fibrin did not consider the effect fibrinogen might have on the binding of these activators to fibrin. In a previous report, we have shown that TA binds equally well to fibrin-Sepharose and to fibrinogen-Sepharose (9). Furthermore, we have shown that TA coprecipitates with fibrinogen at a PEG 6000 concentration of 5% and that the amount of TA precipitated correlated with the concentration of the fibrinogen. In both systems, however, the occurrence of artefacts could not be excluded. The fixation of fibrinogen to Sepharose might have modified its binding characteristics. Furthermore, PEG

Fibrinogen - Recent Biochemical and Medical Aspects

6000 is known to increase interactions between proteins. We therefore set up a system in which we could study the binding of TA to fibrin and measure the effect of fibrinogen on this binding.

Materials and Methods

Tissue plasminogen activator was prepared from pig heart. The tissue was delipidated with -20 °C acetone and the resulting dry powder was extracted with 1,4 M potassium acetate, pH 4,2 (10). The extract was fractionated with ammonium sulfate. The fraction between 0,51 M (13%) and 2,54 M (65%) was collected, dissolved in 0,05 M Tris and the pH adjusted to 7,35. This TA preparation was soluble at neutral pH and physiological ionic strength and was stored at -70 °C until use.

The activity of TA was determined using the fibrin plate assay. For the preparation of fibrinogen-Sepharose, human fibrinogen (Kabi, Grade L, 95% clottable) was used. For the other binding studies, fibrinogen was prepared from human plasma by fractional precipitation with β-alanine (11) and subsequent precipitation with 5% PEG 6000 at 0 °C. This fibrinogen was 98% clottable. Both fibrinogen preparations had normal α, β and γ chains as judged by SDS gel electrophoresis. Fibrin monomer was prepared in acetic acid as described by Müller-Berghaus (12). ^{125}I-fibrinogen was obtained from Sorin-Biomedika and was 85-90% clottable. Thrombin (Topostasin, Roche) was further purified by heparin-Sepharose affinity chromatography (13).

Fibrinogen-Sepharose was prepared by reacting fibrinogen with CNBr activated Sepharose (Pharmacia) for 20 hours at 4 °C. After washing, the remaining active groups were blocked with 1 M ethanolamine for 2 hours at room temperature. Fibrin-

Sepharose was made by treating fibrinogen-Sepharose with thrombin (10 U/ml) for 30 min at 25 °C.

Results

Binding of TA to human fibrinogen - and fibrin-Sepharose. TA was passed over 1 ml columns of fibrin-Sepharose or fibrinogen-Sepharose. The fibrin(ogen) concentration in these columns was 33 mg/ml (107 μM). Effluent fractions of 1 ml were collected and their OD and activator content determined. The elution patterns were similar (Fig. 1). The bound TA could be eluted from the columns with 1 M KSCN. TA did not bind to albumin-Sepharose or ethanolamine-Sepharose. This rules out nonspecific interaction of the activator with Sepharose. The binding of TA to fibrinogen- and fibrin-Sepharose was analyzed

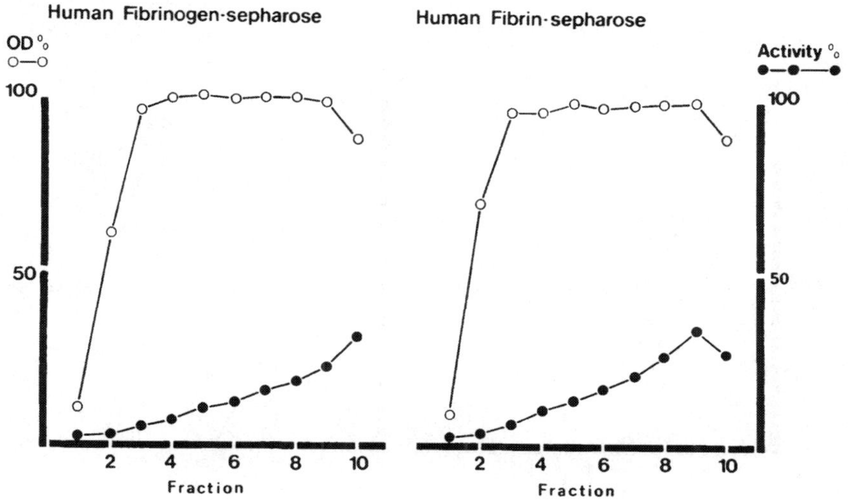

Fig. 1. Binding of TA to human fibrinogen- and fibrin-Sepharose.

by means of a Scatchard plot. The concentration of TA was estimated by assuming a molecular weight of 60 000 and a specific activity of pure TA of 200 000 U/mg. An activator concentration of 1 U/ml then corresponds to 0,08 nM. The ligand concentration in the gel was taken to be the fibrin(ogen) concentration in the gel. The bound activator concentration was determined by subtracting the total amount of activator eluted from the total amount of activator applied, after correction for the column volume. Both Scatchard plots showed a linear relationship between the concentration of bound TA/fibrinogen concentration (r) and r/free activator concentration (Af) (Fig. 2), suggesting one type of binding sites. The data do not exclude, however, the existence of additional low affinity binding sites. The number of binding sites (n) per

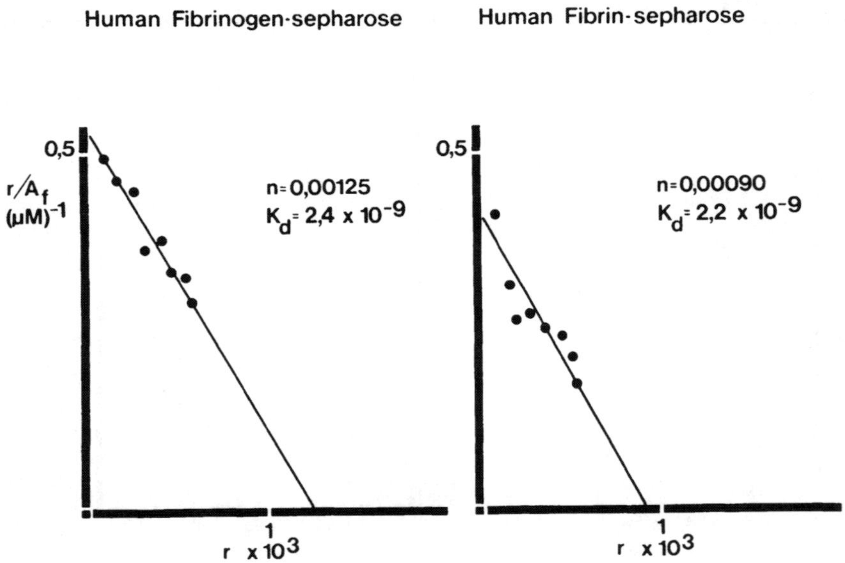

Fig. 2. Scatchard plot of the binding of TA to fibrinogen- and fibrin-Sepharose.

molecule of fibrinogen or fibrin thus obtained was very low (one binding site per thousand molecules of fibrinogen or fibrin). In both instances, the calculated dissociation constants were approximately $2 \cdot 10^{-9}$ M.

Binding of TA to human fibrin. The binding of TA to human fibrin was studied by adding varying amounts of fibrin monomer to a solution of TA diluted in 0,05 M imidazol buffer pH 7,35, containing 0,14 M NaCl. The TA concentration in this mixture was 11 U/ml (=0,9 nM). Prior to the addition of the monomer, a calculated amount of NaOH was added to the TA solution in order to compensate for the acetic acid in the monomer solution. Ca^{++} was added to a final concentration of 5 mM. TA was incubated with the fibrin for 30 minutes at 4 °C. The fibrin formed was spun down (20 min, 3500 g, 4 °C) and the supernatant removed. The tubes were carefully dried with tissue paper. The fibrin was dissolved in 0,017 M acetic acid to the original volume and the activity of the solution tested.

The percentage of TA bound was dependent on the fibrin concentration. Figure 3 shows that 50% binding of TA to fibrin occurred at a fibrin concentration of 150 nM (=50 µg/ml).

Influence of fibrinogen on the binding of TA to fibrin. This was studied by adding fibrin monomer to TA and varying concentrations of fibrinogen. Thrombin had been removed from the fibrin monomer by polymerizing the latter in the presence of 2,7 mM diisopropylfluorophosphate and hirudin (1 ATU/ml) at pH 7,4. After one hour incubation at room temperature, the fibrin was collected and dissolved in 0,017 M acetic acid.

In order to monitor the extent of clotting of the added fibrin, ^{125}I-fibrinogen was added to the human fibrinogen prior to preparation of the fibrin monomer.

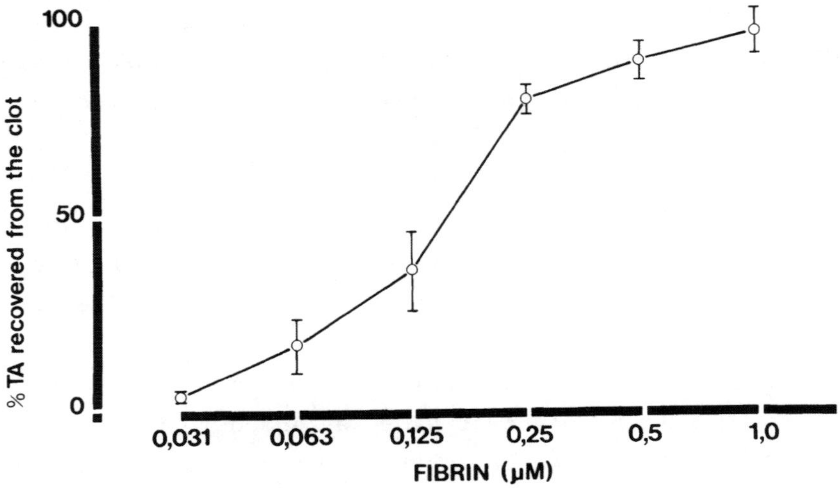

Fig. 3. Binding of TA to human fibrin.

The incubation mixture contained (final concentrations) 2 µM fibrin monomer, 0 to 6 µM fibrinogen, 2,2 U/ml of TA (∿ 1,8 µM), 1 ATU/ml of hirudin and 0 or 10 mM of $CaCl_2$. Incubation was 30 min at 4 °C or 15 min at 23 °C.
The fibrin was spun down and the precipitate dissolved in 0,017 M acetic acid to the original volume. Radioactivity, fibrinolytic activity and optical density at 280 nm were determined for each sample. The results were expressed in percent of the total radioactivity added (17 000 cpm), of the OD of the added fibrin (1,02) and of the activity of the added TA (Fig. 4). Taking into account the apparent dissociation constant of 150 nM found in the previous experiment, more than 90% of the TA was expected to bind to the fibrin. Furthermore, if TA bound equally well to fibrinogen as to fibrin, one would expect that the addition of 6 µM of fibrinogen to 2 µM of fibrin decreased the binding of TA to fibrin by 75%.

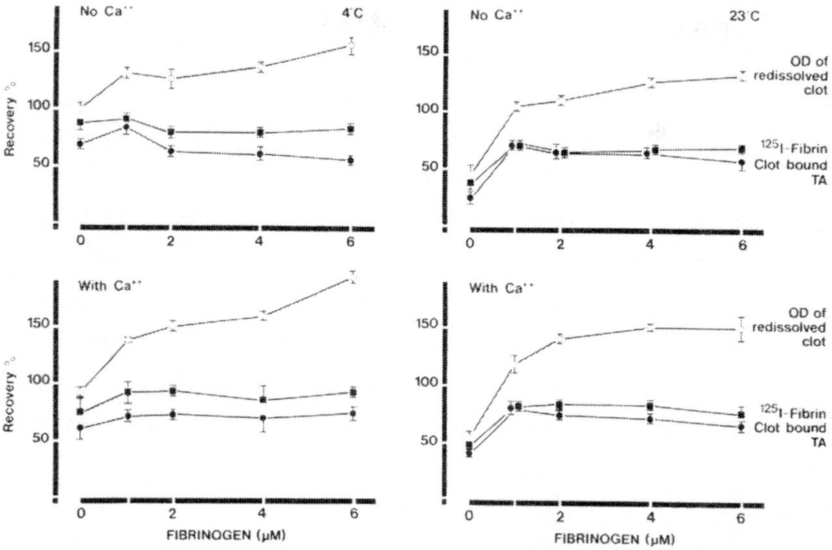

Fig. 4. Influence of fibrinogen on the binding of TA to fibrin at 4 °C and 23 °C in the presence and absence of Ca^{++}.

The OD of the redissolved clot was higher than 100 percent when fibrinogen was added, indicating that part of the fibrinogen had been incorporated into the clot. The amount of fibrinogen incorporated was higher at higher fibrinogen concentrations, in the presence of calcium and at 4 °C. The latter observation excludes that the incorporation is due to the presence of contaminating thrombin. On the other hand, ^{125}I-fibrin was not quantitatively incorporated into the clot. Recovery of ^{125}I-fibrin was lower in the absence of calcium and at 23 °C. Surprisingly, the addition of 1 μM fibrinogen at 23 °C increased the recovery of fibrin from the clot. No explanation for this phenomenon can be given at present. Recovery curves for TA and for ^{125}I-fibrin were parallel. The addition of fibrinogen did not diminish appreciably the binding of TA to fibrin.

Discussion

For these binding studies, a partially purified TA preparation was used, in which TA is soluble at neutral pH and physiological ionic strength. Most highly purified TA preparations require buffers of high ionic strength and / or detergents to keep TA in solution (14, 15), thus making them less suitable for binding studies under physiological conditions. TA had an equal affinity for fibrin-Sepharose and fibrinogen-Sepharose. Analysis of the results, however, indicated a low binding site concentration with a very high affinity. This unexpected result could be due to one or a combination of the following:

a) Binding of fibrinogen and fibrin to Sepharose could result in an almost complete inactivation of binding sites,
b) The fibrinogen used for coupling may have lost its TA binding site(s),
c) Our estimate of the molar concentration of TA may be erroneous,
d) TA may bind only to a contaminating protein in the fibrinogen preparation,
e) Other proteins in the crude TA preparation may compete with TA for binding to fibrin.

The first two hypotheses are unlikely to be the sole reason(s) for the calculated low amount of binding sites. If a and b were true then the calculated dissociation constant of about 2 nM would be correct, which is in contradiction with the K_d of 150 nM found in the experiment of the binding of TA to fibrin (Fig. 3). Furthermore, the presence of an intact α-chain in the fibrinogen preparations, as determined by

SDS-PAGE, precludes degradation of the fibrinogen as an explanation for the low binding site concentration. The estimates for the molecular weight of pure TA and of its specific activity could not explain the big difference in binding site concentration and fibrin concentration. Possibility d and e cannot be excluded by our experimental results. The observation that TA bound effectively to fibrin prepared from monomer with a clottability of more than 99,5% implies that the hypothetical TA binding contaminant also has to be a clottable protein. For hypothesis e to be true, a large excess of contaminating proteins would have to bind to fibrinogen- and fibrin-Sepharose for each TA molecule bound.

In fact, when passing crude TA over fibrin(ogen)-Sepharose, bound TA can be eluted almost quantitatively with 0,017 M acetic acid. However, the obtained purity of the TA is only 300 U/mg. Taking into account a specific activity of 200 000 U/mg for pure TA, it follows that per molecule of TA several hundred molecules of other proteins were binding to fibrinogen-Sepharose. If a substantial part of these proteins were to bind at or near the TA binding site, we would grossly underestimate the actual binding site concentration. As a consequence, the real dissociation constant would be proportionally higher. The apparent dissociation constant for the binding of TA to fibrin was 50-100 times higher than that for the binding of TA to fibrin(ogen)-Sepharose.

If about 50 contaminant protein molecules were to bind at or near the TA binding site for each molecule of TA bound, the real K_d would be about 100 nM. This value is comparable with the K_d derived from the fibrin concentration at which 50% of TA was binding. In the latter experiment where an excess of binding sites is available, the K_d is not influenced by the presence of competing contaminants.

The quantitative assessment of the relative affinities of TA

to fibrin and fibrinogen is complicated by the fact that part of the fibrin remained in solution and that part of the fibrinogen coprecipitated with the fibrin. All curves of percentage TA precipitated and percentage fibrin clotted were parallel and the addition of fibrinogen did not appreciably diminish the precipitation of TA. This proves that the affinity of TA to fibrin is much higher than that to fibrinogen, and suggests that TA binds equally well to fibrin monomer as to polymerized fibrin.
The equal binding of TA to fibrinogen-Sepharose and fibrin-Sepharose is an apparent contradiction with the preferential binding of TA to fibrin. Does the fixation of fibrinogen modify its conformation in such a way that it takes on the fibrin conformation in the TA binding domain ?
The binding of TA to fibrin monomer or oligomer, if confirmed, would be of considerable physiological importance, because it could greatly accelerate the breakdown of circulating fibrin.

References

1. Blix, S.: Acta Med. Scand. suppl. 386, 1-24 (1962).
2. Rijken, D.C., Wijngaards, G., Welbergen, J.: Thromb. Res. 18, 815-830 (1980).
3. Thorsen, S., Glas-Greenwalt, P., Astrup, T.: Thromb. Diathes. Haemorrh. 28, 65-74 (1972).
4. Camiolo, S.M., Thorsen, S., Astrup, T.: Proc. Soc. Exp. Biol. Med. 138, 277-280 (1971).
5. Wallén, P.: in: Thrombosis and Urokinase, Ed.: Paoletti, R., Sherry, S., p. 91-102, Acad. Press, London (1977).
6. Binder, B.R., Spragg, J.: in: Protides of the Biological Fluids, Ed.: Peeters, H., 28, 391-394, Pergamon Press, Oxford (1980).
7. Wiman, B., Collen, D.: Nature 272, 549-550 (1978).

8. Collen, D.: Thromb. Haemost. 43, 77-89 (1980).

9. Kruithof, E.K.O., Bachmann, F.: in: Progress in Fibrinolysis and Thrombolysis, Ed.: Davidson, J. et al, vol. V (in press), Churchill-Livingstone, Edinburg.

10. Bachmann, F., Fletcher, A.P., Alkjaersig, N., Sherry, S.: Biochemistry 3, 1578-1585 (1964).

11. Straughn, W., Wagner, R.H.: Thromb. Diathes. Haemorrh. 16, 198-206 (1966).

12. Müller-Berghaus, G., Róka, L., Lasch, H.G.: Thromb. Diathes. Haemorrh. 29, 375-383 (1973).

13. Iverius, P.H.: Biochem. J. 124, 677-683 (1971).

14. Rijken, D.G., Wijngaards, G., Zaal-de Jong, M., Welbergen, J.: Biochim. Biophys. Acta 580, 140-153 (1979).

15. Cole, E.R., Bachmann, F.W.: J. Biol. Chem. 252, 3729-3737 (1977).

Supported in part by the Swiss National Foundation (grant No. 3.145.077)

FIBRIN BINDING AND TRANSAMIDATION REACTIVE DOMAINS OF FIBRONECTIN

Helmut Hörmann, Martin Seidl, Hartmut Richter
Max Planck-Institut für Biochemie
D-8033 Martinsried bei München, Germany

Introduction

Fibrin has affinity to cold-insoluble globulin (plasma fibronectin) as indicated by binding of this globulin to coagulated (1, 2, 3) or immobilized (4) fibrin and the formation of cold-insoluble complexes of fibrinogen-fibrin monomer and fibronectin (5). These aggregates keep fibrin monomer at least partially in solution (6) and, probably, are important for an elimination of soluble fibrin by cells of the reticulo-endothelial system (7, 8). Fibrinogen binds far less effectively to fibronectin than fibrin does (5).

The binding sites of fibrin which interact with fibronectin are not completely characterized. The higher affinity of fibrin compared to fibrinogen (4, 5) indicates that a thrombin activated region in the N-terminal disulfide knot might be involved. In addition, the flexible C-terminal parts of the α-chains are suggested to interact because fibrin derived from the fibrinogen fractions I-8 and I-9 lacking those sequences was inable to form cryoprecipitates with fibronectin (5).

As shown by Mosher (9) fibrin and fibronectin also can be covalently cross-linked by a transamidase catalyzed reaction. Each of the two molecules contains donator and acceptor sites for transamidation. In fibrin these are located close to the C-terminal ends of the γ-chains (11) as well as in the outer regions of the α-chains (12, 13).

In the present investigation the fibrin binding and cross-linking sites of fibronectin were correlated with special domains in the nearly identical fibronectin subunits of molecular weight 225 000 and 235 000 (for review on fibronectin see l.c. 14, 15, 16).

Results and Discussion

The transamidase acceptor sites of human plasma fibronectin were labelled by coupling with ^{14}C-putrescine in presence of thrombin activated Factor XIIIa. Under optimal conditions this results in an incorporation of 2.6 molecules amine per fibronectin (17) indicating the presence of one or two reactive sites in each subunit. Previous investigations (18) had already shown that most of the activity was present in a terminal peptide of molecular weight ca. 27 000 which was selectively removed by plasmin.

^{14}C-putrescine labelled fibronectin was digested with cathepsin D as outlined by Balian et al. (19) and the peptides obtained (Fig. 1) were fractionated on fibrin-Sepharose. The retained fraction was eluted with 1.0 M potassium bromide at pH 5.3 and analyzed by dodecylsulfate gel electrophoresis. It contained a fragment of Mr 70 000 together with high molecular weight peptides which yielded the same fragment by further cathepsin D digestion. The peptide of Mr 70 000 exhibited radioactivity indicating presence of a transamidase reactive site in addition to its fibrin binding center. The two activities were more closely localized by cleaving the Mr 70 000 fragment with plasmin which resulted in two fragments of Mr 30 000 and Mr 40 000 (Fig. 1). The Mr 30 000 peptide was the only radioactive one and was retained on fibrin-Sepharose although less firmly than the Mr 70 000 fragment. Obviously, the radioactive Mr 30 000 peptide is related to the terminal fragment obtained by plasminolysis of fibronectin directly (17). Several evidences indicate that it originates

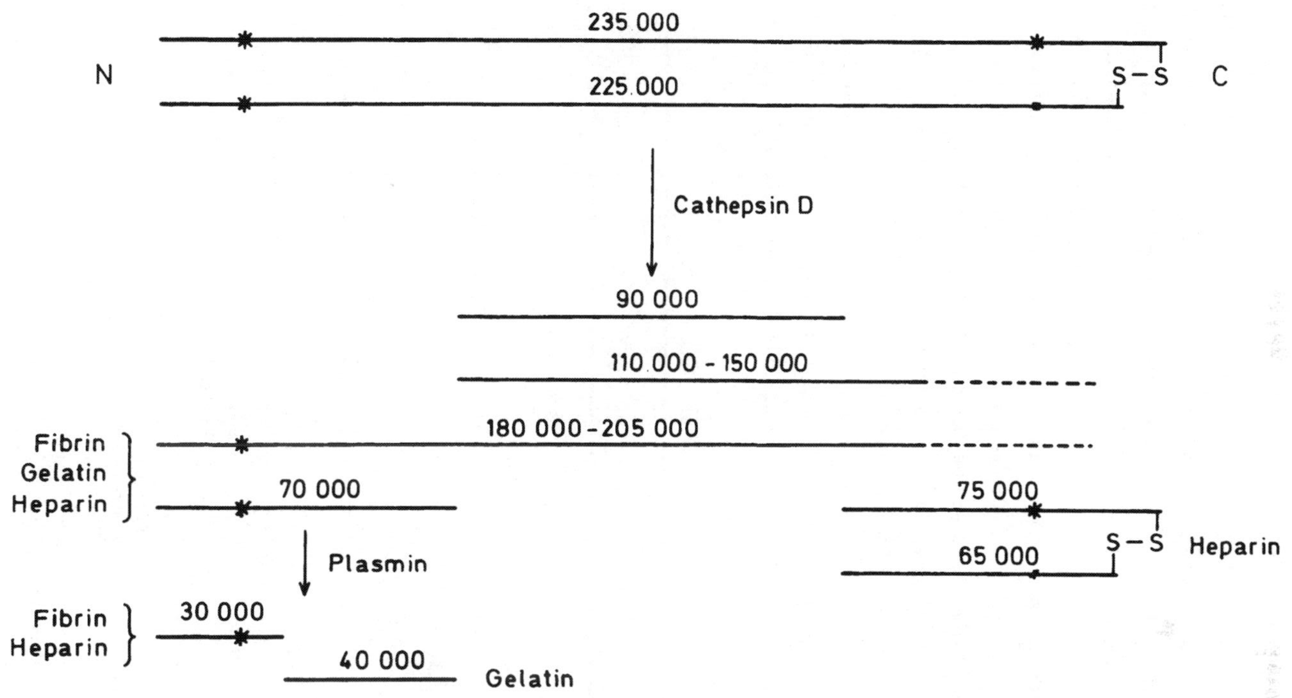

Fig. 1: Cathepsin D digest of ^{14}C-putrescine-labelled plasma fibronectin. Correlation of peptides.
* 14-C-putrescine-labelled sites; fibrin, gelatin and heparin denote immobilized substrates with affinity to the peptides indicated.

from the N-terminus (20, 21, 22).

The other fragment of Mr 40 000 contained a gelatin binding site (23). It was weakly adsorbed by fibrin-Sepharose, too, (19) suggesting that it is a remnant of a second larger fibrin binding domain in accord with previous conclusions of Engvall et al. (3). Indeed, if a chymotrypsin digest of ^{14}C-putrescine-labelled fibronectin was passed over fibrin-Sepharose, a non-radioactive fragment of Mr 60 000 was retained in addition to a radioactive peptide of Mr 30 000 which, evidently, was related to the above mentioned plasminolysis product (Fig. 2). The Mr 60 000 fragment also bound to gelatin-Sepharose and was cleaved by cathepsin D to yield the previously described gelatin binding peptide of Mr 40 000 and a non-binding remnant of Mr 20 000. Obviously, the fragment of Mr

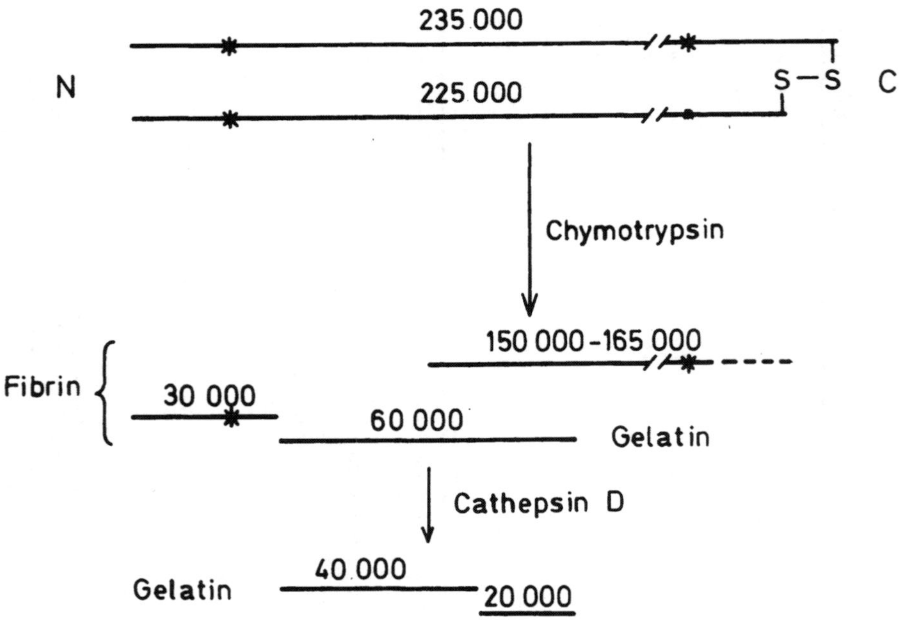

Fig. 2: Chymotrypsin digest of 14-C-putrescine labelled fibronectin. Correlation of several peptides. Symbols and denotions as in Fig. 1. Enzyme:substrate 1:250, 37°C, 15 min.

60 000 represented a second fibrin binding domain extending above the cathepsin D cleavage site which terminates the gelatin binding region on its C-terminal end. This second fibrin binding domain is located adjacent to the N-terminal binding domain of Mr 30 000 which also contained a transamidase reactive site.

In the cathepsin D digest still another ^{14}C-putrescine-labelled peptide of Mr 140 000 was found (Fig. 1) which was not adsorbed by fibrin-Sepharose (24). It was isolated by chromatography on heparin-Sepharose from which it was eluted with 0.5 M NaCl after desorption of other fragments with 0.25 M NaCl. The fragment, if reduced with mercaptoethanol, dissociated into two subpeptides of Mr 75 000 and Mr 65 000. This suggests that it originated from the C-terminal portion of fibronectin where the two subunits are connected by disulfide bonds. The majority of radioactivity was in the longer subpeptide although the shorter one still contained minor amounts. Obviously, the longer chain contains a more effective transamidation reactive site than the shorter. Evidences reported elsewhere (24) indicated that the two subpeptides have a closely related structure and differ only in a region containing an additional stretch of Mr 10 000 in the longer subpeptide.

The given results favor the following distribution of fibrin binding and transamidation reactive sites within fibronectin (Fig. 3): Two domains of Mr 30 000 exhibiting both functions are located at the N-terminus of either peptide chain. They are followed by a domain of Mr 60 000 with a second fibrin affinity site as well as with gelatin binding properties. It lacks a cross-linking site. However, an additional transamidation reactive site is found in the C-terminal region and is expressed predominantly in the longer of the two subunits.

We are indebted to Miss E.M. Wagner for technical assistance. The investigations were supported by Deutsche Forschungsgemeinschaft, Project Ho 740.

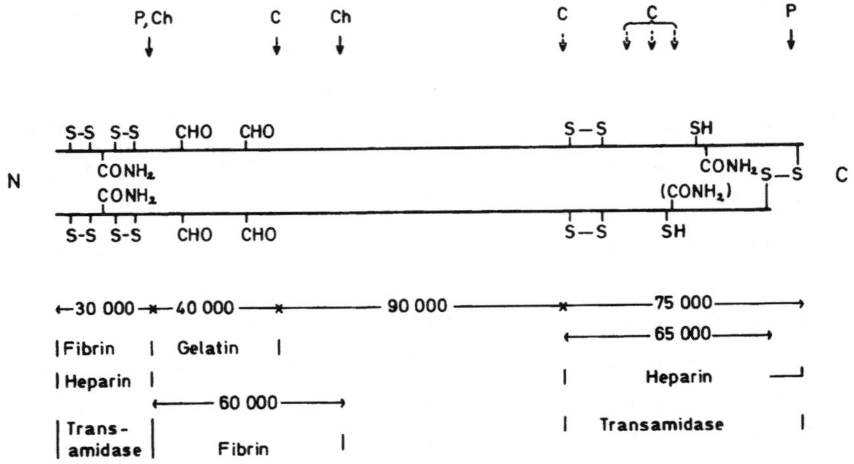

Fig. 3: Domain structure of fibronectin. $CONH_2$ cross-linking acceptor sites, CHO carbohydrates. Cleavage sites: C cathepsin D, Ch chymotrypsin, P plasmin.

References

1. Ruoslahti, E., Vaheri, A.: J. Exp. Med. 141, 497-501 (1975).
2. Dessau, W., Jilek, F., Adelmann, B.C., Hörmann, H.: Biochim. Biophys. Acta 533, 227-237 (1978).
3. Engvall, E., Ruoslahti, E., Miller, E.J.: J. Exp. Med. 147, 1584-1595 (1978).
4. Stemberger, A., Hörmann, H.: Hoppe Seyler's Z. Physiol. Chem. 357, 1003-1005 (1976).
5. Stathakis, N.E., Mosesson, M.W., Chen, A.B., Galanakis, D.K.: Blood 51, 1211-1222 (1978).
6. Kaplan, J.E., Snedeker, P.W.: J. Lab. Clin. Med. 96, 1054-1061 (1980).
7. Saba, T.M., Blumenstock, F.A., Scovill, W.A., Bernard, H.: Science 201, 622-624 (1978).
8. Sherman, L.A., Lee, J.: Thrombosis Haemostasis 42, 121 (1979).
9. Mosher, D.F.: J. Biol. Chem. 250, 6614-6621 (1975).
10. Keski-Oja, J., Mosher, D.F., Vaheri, A.: Cell 9, 29-35 (1976).

11. Chen, R., Doolittle, R.F.: Proc. Natl. Acad. Sci. USA 63, 420-427 (1969).

12. McKee, P.A., Mattock, P., Hill, R.L.: Proc. Natl. Acad. Sci. USA 66, 738-744 (1970).

13. Pisano, J.J., Finlayson, J.S., Peyton, M.P., Nagai, Y.: Proc. Natl. Acad. Sci. USA 68, 770-772 (1971).

14. Mosher, D.F.: Progress in Hemostasis and Thrombosis (Spaet, T.H. ed.) Vol. 5, p. 111-151, Grune & Stratton, New York 1980.

15. Mosesson, M.W.: Blood 56, 145-158 (1980).

16. Ruoslahti, E., Engvall, E., Hayman, E.G.: Collagen Res. 1, 95-128 (1981).

17. Mosher, D.F., Schad, P.E., Vann, J.M.: J. Biol. Chem. 255, 1181-1188 (1980).

18. Jilek, F., Hörmann, H.: Hoppe Seyler's Z. Physiol. Chem. 358, 1165-1168 (1977).

19. Hörmann, H., Seidl, M.: Hoppe Seyler's Z. Physiol. Chem. 361, 1449-1452 (1980).

20. Furie, M.B., Rifkin, D.B.: J. Biol. Chem. 255, 3134-3140 (1980).

21. Balian, G., Click, E.M., Bornstein, P.: J. Biol. Chem. 255, 3234-3236 (1980).

22. Wagner, D.D., Hynes, R.O.: J. Biol. Chem. 255, 4304-4312 (1980).

23. Balian, G., Click, E.M., Crouch, E., Davidson, J.M., Bornstein, P.: J. Biol. Chem. 254, 1429-1432 (1979).

24. Richter, H., Seidl, M., Hörmann, H.: Hoppe Seyler's Z. Physiol. Chem. in press.

AUTHOR INDEX

SUBJECT INDEX

FSC
www.fsc.org
MIX
Papier | Fördert
gute Waldnutzung
FSC® C083411

Kolibri 360 GmbH
Ferdinand-Jühlke-Straße 7
99095 Erfurt, Deutschland
produktsicherheit@kolibri360.de